Contents

Contents

Get the most from this book

Welcome to the AQA GCSE Chemistry Student Book.

This book covers the Foundation and Higher-tier content for the 2016 AQA GCSE Chemistry specification.

The following features have been included to help you get the most from this book.

Prior knowledge

This is a short list of topics you should be familiar with before starting a chapter. The questions will help to test your understanding. Extra help and practice questions can be found online in our AQA GCSE Science Teaching & Learning Resources

KEY TERMS

Important words and concepts are highlighted in the text and clearly explained for you in the margin.

Practical

These practical-based activities will help consolidate your learning and test your practical skills.

Required practical

AQA's required practicals are clearly highlighted.

TIPS

These highlight important facts, common misconceptions and signpost you towards other relevant chapters. They also offer useful ideas for remembering difficult topics.

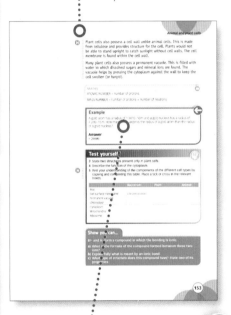

Higher-tier only

Some material in this book is only required for students taking the Higher-tier examination. This content is clearly marked with the blue symbol seen here.

Examples

Examples of questions and calculations that feature full workings and sample answers.

Show you can...

Complete the Show you can tasks to prove that you are confident in your understanding of each topic.

Test yourself questions

These short questions, found throughout each chapter, allow you to check your understanding as you progress through a topic.

Chapter review questions

These questions will test your understanding of the whole chapter. They are colour coded to show the level of difficulty and also include questions to test your maths and practical skills.

Simple questions that everyone should be able to answer without difficulty.

These are questions that all competent students should be able to handle.

More demanding questions for the most able students.

Answers

Answers for all questions and activities in this book can be found online at:

www.hoddereducation.co.uk/aqagcsechemistry

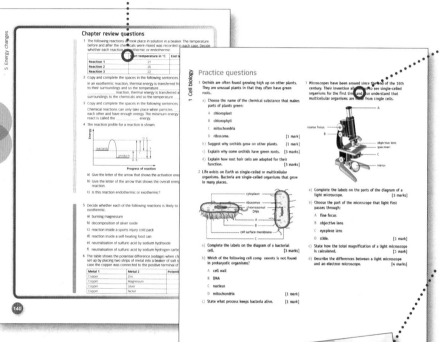

Practice questions

You will find Practice questions at the end of every chapter. These follow the style of the different types of questions you might see in your examination and have marks allocated to each question part.

Working scientifically

In this book, Working scientifically skills are explored in detail in the activity at the end of each chapter. Work through these activities on your own or in groups. You will develop skills such as Dealing with data, Scientific thinking and Experimental skills.

* AQA only approve the Student Book and Student eTextbook. The other resources referenced here have not been entered into the AQA approval process.

1 Atomic structure and the periodic table

Until you reached GCSE, Chemistry was studied at the particle level. In order to take chemistry further, you now need to understand what is inside atoms. The elements in the periodic table are ordered by what is inside their atoms. An understanding of the periodic table allows you to explain and/or work out a lot of chemistry even if you have never studied it.

Specification coverage

This chapter covers specification points 4.1.1 to 4.1.3 and is called Atomic structure and the periodic table.

It covers the structure of atoms, reactions of elements, the periodic table and mixtures.

Writing formulae and equations is covered separately in the Appendix.

Prior knowledge

Previously you could have learned:

> Elements are made of particles called atoms.
> Elements are substances containing only one type of atom – this means they cannot be broken down into simpler substances.
> Each element has its own symbol and is listed in the periodic table.
> Elements are either metals or non-metals.
> Compounds are substances made from atoms of different elements bonded together. (chemically bonded together)
> Compounds have different properties from the elements from which they are made.
> Compounds are difficult to break back down into their elements.
> Substances in mixtures are not chemically joined to each other.
> Substances in mixtures can be separated easily by a range of techniques.

Test yourself on prior knowledge

1 What is an element?
2 What is a compound?
3 Why do compounds have different properties from the elements from which they are made?
4 List some differences between metals and non-metals.
5 Why is it easy to separate the substances in a mixture but not to break apart a compound?
6 Name four methods of separating mixtures.

Structure of atoms

TIP ✓
Remember that:
protons are **p**ositive
neutrons are **neu**tral
leaving electrons as negative

TIP ✓
The charge of a proton can be written as + or +1. The charge of an electron can be written as – or –1.

KEY TERM ⭐
Atom The smallest part of an element that can exist. A particle with no electric charge made up of a nucleus containing protons and neutrons surrounded by electrons in energy levels.

◯ Protons, neutrons and electrons

Atoms are the smallest part of an element that can exist. Atoms are made up of smaller particles called **protons**, **neutrons** and **electrons**. The table below shows the relative mass and electric charge of these particles. The mass is given relative to the mass of a proton. Protons and neutrons have the same mass as each other while electrons are much lighter (Table 1.1).

Table 1.1

	Proton	Neutron	Electron
Relative mass	1	1	very small
Relative charge	+1	0	−1

◯ The structure of atoms

Atoms are very small. Typical atoms have a radius of about 0.1 nm (0.000 000 000 1 m, that is 1×10^{-10} m). Atoms have a central **nucleus** which contains protons and neutrons (Figure 1.1). The nucleus is surrounded by electrons. The electrons move around the nucleus in **energy levels** or **shells**.

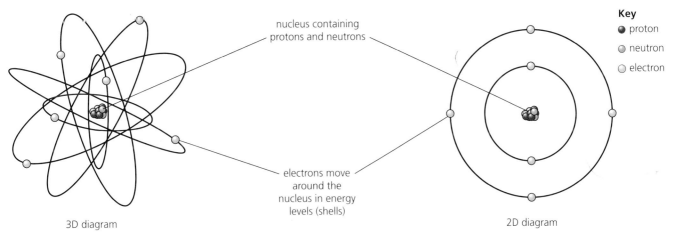

nucleus containing
protons and neutrons

electrons move
around the
nucleus in energy
levels (shells)

3D diagram

2D diagram

▲ Figure 1.1

KEY TERMS

Proton Positively charged particle found inside the nucleus of atoms.

Neutron Neutral particle found inside the nucleus of atoms.

Electron Negatively charged particle found in energy levels (shells) surrounding the nucleus inside atoms.

Nucleus Central part of an atom containing protons and neutrons.

Energy level (shell) The region an electron occupies surrounding the nucleus inside an atom.

TIP ✓

$1\,nm = 1 \times 10^{-9}\,m$ $(0.000\,000\,001\,m)$

TIP ✓

The SI units for length are metres (m).

▲ Figure 1.2 The size of the nucleus compared to the atom is like a pea compared to a football pitch.

KEY TERM ⭐

Atomic number Number of protons in an atom.

The nucleus is tiny compared to the size of the atom as a whole. The radius of the nucleus is less than 1/10 000th of that of the atom $(1 \times 10^{-14}\,m)$. This difference in size between a nucleus and an atom is equivalent to a pea placed in the middle of a football pitch (Figure 1.2).

The nucleus contains protons and neutrons. These are much heavier than electrons. This means that most of the mass of the atom is contained in the tiny nucleus in the middle.

Test yourself

1 Carbon atoms have a radius of 0.070 nm. Write this in standard form in the units of metres.
2 The radius of a hydrogen atom is $2.5 \times 10^{-11}\,m$. Write this in nanometres.
3 The radius of a chlorine atom is $1 \times 10^{-10}\,m$ and the radius of a silicon atom is 0.060 nm. Which atom is bigger?
4 Sodium atoms have a radius of 0.180 nm. The nucleus of an atom is about 10 000 times smaller. Estimate the radius of the nucleus of a sodium atom. Write your answer in both nanometres and metres.
5 A copper atom has a diameter of 0.256 nm. A copper wire has a diameter of 0.0440 cm.
   a Write the diameter of the atom and the wire in metres.
   b How many times wider is the copper wire than a copper atom? Give your answer to 3 significant figures.
6 A gold atom has a diameter of 0.270 nm. The largest gold bar in the world in 45.5 cm long. How many gold atoms fit into 45.5 cm? Give your answer to 3 significant figures.

○ Atomic number and mass number

The number of protons that an atom contains is called its atomic number. Atoms of different elements have different numbers of protons. It is the number of protons that determines which element an atom is. For example, all atoms with 6 protons are carbon atoms, while all atoms with 7 protons are nitrogen atoms.

All atoms are neutral, which means they have no overall electric charge. This is because the number of protons (which are positively charged) is the same as the number of electrons (which are negatively charged).

Most of the mass of an atom is due to the protons and neutrons. Protons and neutrons have the same mass as each other. The **mass number** of an atom is the sum of the number of protons and neutrons in an atom. For example, an atom of sodium has 11 protons and 12 neutrons and so has a mass number of 23.

ATOMIC NUMBER = number of protons

MASS NUMBER = number of protons + number of neutrons

The atomic number and mass number of an atom can be used to work out the number of protons, neutrons and electrons in an atom:

- number of protons = atomic number
- number of neutrons = mass number − atomic number
- number of electrons = atomic number (*only for atoms, not ions*).

The mass number and atomic number of atoms can be shown as in Figure 1.3.

TIP
A carbon atom is neutral because it contains 6 protons (charge + 6) and 6 electrons (charge − 6) and so has no overall charge.

KEY TERM
Mass number Number of protons plus the number of neutrons in an atom.

TIP
Atoms are sometimes shown in the form ^{23}Na. As the atom is a sodium atom, it must have an atomic number of 11 and so it is not necessary to include the atomic number.

mass number = 23

Na

atomic number = 11

mass number: protons + neutrons
atomic number: protons
number of protons = 11
number of neutrons = 23 − 11 = 12
number of electrons = 11

▲ **Figure 1.3**

Example

How many protons, neutrons and electrons are there in an atom of $^{81}_{35}Br$?

Answer

Number of protons: 35 (we could also find this by looking at the atomic number in the periodic table if it was not shown).

Number of neutrons: 81 − 35 = 46 (the mass number minus the number of protons).

Number of electrons: 35 (the same as the number of protons).

Isotopes

For most elements there are atoms with different numbers of neutrons. Atoms with the same number of protons but a different number of neutrons are called **isotopes**. This means that isotopes have the same atomic number but a different mass number.

For example, carbon has three isotopes and so there are three different types of carbon atoms. These are shown in the table below. These three isotopes are all carbon atoms because they all contain 6 protons, but they each have a different number of neutrons (Table 1.2).

Table 1.2

Atom	$^{12}_{6}C$	$^{13}_{6}C$	$^{14}_{6}C$
Protons	6	6	6
Neutrons	6	7	8
Electrons	6	6	6

Isotopes have a different mass, but their chemical properties are the *same* because they contain the same number of electrons.

KEY TERM
Isotopes Atoms with the same number of protons, but a different number of neutrons.

TIP
Relative atomic mass is not the same as mass number. Mass numbers are integers because they are the number of protons plus the number of neutrons. Relative atomic mass is an average mass that takes into account all the isotopes of an element and so is not an integer.

Relative atomic mass

The **relative atomic mass** (A_r) of an element is the average mass of atoms of that element taking into account the mass and amount of each isotope it contains.

KEY TERM
Relative atomic mass The average mass of atoms of an element taking into account the mass and amount of each isotope it contains (on a scale where the mass of a ^{12}C atom is 12).

This can be calculated as shown:

$$\text{Relative atomic mass } (A_r) = \frac{\text{total mass of all atoms of element}}{\text{total number of atoms of that element}}$$

Example

Find the relative atomic mass of chlorine which is found to contain 75% of atoms with mass number 35, and 25% of atoms with mass number 37. Give the answer to one decimal place.

Answer

$$\text{Relative atomic mass } (A_r) = \frac{[(75 \times 35) + (25 \times 37)]}{75 + 25} = \frac{3550}{100} = 35.5$$

TIP

When you calculate relative atomic mass, the answer should have a value somewhere between the mass of the lightest isotope and the heaviest isotope.

Test yourself

7 List the three particles found inside atoms.

8 Identify the particle found inside the nucleus of atoms that has no charge.

9 Atoms contain positive and negative particles. Explain why atoms are neutral.

10 How many protons, neutrons and electrons are there in an atom of $^{31}_{15}P$?

11 What is it about the atom $^{39}_{19}K$ that makes it an atom of potassium?

12 Describe the similarities and differences between atoms of the isotopes $^{35}_{17}Cl$ and $^{37}_{17}Cl$.

13 The element copper contains 69% ^{63}Cu and 31% ^{65}Cu. Find the relative atomic mass of copper to one decimal place. Show your working.

14 The element magnesium contains 79% ^{24}Mg, 10% ^{25}Mg and 11% ^{26}Mg. Find the relative atomic mass of magnesium to one decimal place. Show your working.

15 Explain why mass number is an integer, but relative atomic mass is not.

Show you can...

Copy and complete the table for each of the elements listed.

Element	Atomic number	Mass number	Number of protons	Number of electrons	Number of neutrons
$^{7}_{3}Li$					
$^{24}_{12}Mg$					
$^{27}_{13}Al$					
$^{39}_{19}K$					
$^{107}_{47}Ag$					

◯ Electron arrangement

The electrons in an atom are in energy levels, also known as shells. Electrons occupy the lowest available energy levels. The lowest energy level (the first shell) is the one closest to the nucleus and can hold up to two electrons. Up to eight electrons occupy the second energy level (the second shell) with the next eight electrons occupying the third energy level (third shell). The next two electrons occupy the fourth energy level (fourth shell).

The arrangement of electrons in some atoms are shown in Table 1.3. The electron structure can be drawn on a diagram or written using numbers. For example, the electron structure of aluminium is 2,8,3 which means that it has two electrons in the first energy level, eight electrons in the second energy level and three electrons in the third energy level (Table 1.3).

TIP ✓

It is usual to write ion charges with the number before the +/– sign, such as 2+, but it is not wrong to write it as +2.

KEY TERM ★

Ion An electrically charged particle containing different numbers of protons and electrons.

TIP ✓

Positive ions have more protons than electrons. Negative ions have more electrons than protons. This is because electrons are negatively charged.

Table 1.3

Atom	He	F	Al	K
Atomic number	2	9	13	19
Number of electrons	2	9	13	19
Electron structure (written)	2	2,7	2,8,3	2,8,8,1
Electron structure (drawn)				

Ions

Ions are particles with an electric charge because they do not contain the same number of protons and electrons. Remember that protons are positive and electrons are negative. Positive ions have more protons than electrons. Negative ions have more electrons than protons.

For example:

- An ion with 11 protons (total charge 11+) and 10 electrons (total charge 10–) will have an overall charge of 1+.
- An ion with 16 protons (total charge 16+) and 18 electrons (total charge 18–) will have an overall charge of 2–.

Table 1.4 shows some common ions.

Table 1.4

Ion	Li^+	Al^{3+}	Cl^-	O^{2-}
Atomic number	3	13	17	8
Number of protons	3 (charge 3+)	13 (charge 13+)	17 (charge 17+)	8 (charge 8+)
Number of electrons	2 (charge 2–)	10 (charge 10–)	18 (charge 18–)	10 (charge 10–)
Overall charge	1+	3+	1–	2–
Electron structure (written)	2	2,8	2,8,8	2,8
Electron structure (drawn)				

Simple ions (those made from single atoms) have the same electron structure as the elements in Group 0 of the periodic table (Table 1.5). The elements in Group 0 are called the noble gases. The noble gases have very stable electron structures.

Table 1.5

Group 0 element	He	Ne	Ar
Electron structure	2	2,8	2,8,8
Common ions with the same electron structure	Li^+, Be^{2+}	O^{2-}, F^-, Na^+, Mg^{2+}, Al^{3+}	S^{2-}, Cl^-, K^+, Ca^{2+}

The hydrogen ion (H⁺) is the only simple ion that does not have the electron structure of a noble gas. It does not have any electrons at all. This makes it a very special ion with special properties, and it is the H⁺ ion that is responsible for the behaviour of acids.

Test yourself

18 What is the charge of a particle with 19 protons and 18 electrons?
19 What is the charge of a particle with 7 protons and 10 electrons?
20 What is the electron structure of the P^{3-} ion?
21 How many protons, neutrons and electrons are there in the $^{19}_{9}F^{-}$ ion?
22 What is the link between the electron structure of ions and the Group 0 elements (the noble gases)?

Show you can...

Table 1.6 gives some information about six different particles, A, B, C, D, E and F. Some particles are **atoms** and some are **ions**. (The letters are not chemical symbols).

Table 1.6

Particle	Atomic number	Mass number	Number of protons	Number of neutrons	Number of electrons	Electron structure
A	18	40				2,8,8
B		27	13			2,8
C			20	20	20	
D		35	17			2,8,7
E	16	32			18	
F	17			20	17	

a) Copy and complete the table.
b) Particle C is an atom. Explain, using the information in the table, why particle C is an atom.
c) Particle E is a negative ion. What is the charge on this ion?
d) Which two atoms are isotopes of the same element?

negatively charged electrons

positively charged ball

▲ **Figure 1.4** Plum-pudding model of the atom (1897).

Development of ideas about the structure of atoms

The idea that everything was made of particles called atoms was accepted in the early 1800s after work by John Dalton. At that time, however, people thought that atoms were the smallest possible particles and the word *atom* comes from the Greek word *atomos* which means something that cannot be divided.

However, in 1897 the electron was discovered by J.J. Thompson while carrying out experiments on the conduction of electricity through gases. He discovered that electrons were tiny, negatively charged particles that were much smaller and lighter than atoms. He came up with what was called the 'plum-pudding' model of the atom. In this model, the atom was a ball of positive charge with the negative electrons spread through the atom (Figure 1.4).

A few years later in 1911, this model was replaced following some remarkable work from Hans Geiger and Ernest Marsden working with Ernest Rutherford. They fired alpha particles (He^{2+} ions) at a very thin piece of gold foil. They expected the particles to pass straight through the foil but a tiny fraction were deflected or even bounced back. This did not fit in with the plum-pudding model. Rutherford worked out that the scattering of some of the alpha particles meant that there must be a tiny, positive nucleus at the centre of each atom. This new model was known as the nuclear model (Figure 1.5).

plum-pudding model (1897)

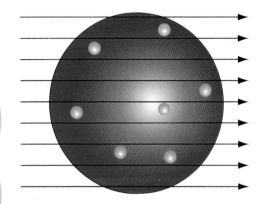

The alpha particles would all be expected to travel straight through the gold foil according to the plum-pudding model

nuclear model (1911, but revised since)

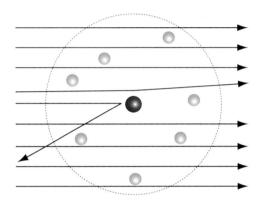

A tiny fraction of alpha particles were deflected or bounced back. Rutherford worked out that there must be a tiny, positive nucleus to explain this

▲ Figure 1.5

Key
● proton
● neutron
● electron

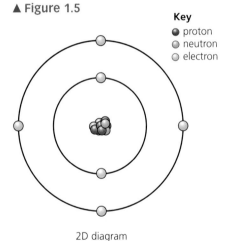

2D diagram

Today's model of an atom

▲ **Figure 1.6** Today's model of an atom.

In 1913, Neils Bohr adapted the nuclear model to suggest that the electrons moved in stable orbits at specific distances from the nucleus called shells. Bohr's theoretical calculations agreed with observations from experiments.

Further experiments led to the idea that the positive charge of the nucleus was made up from particles which were given the name protons.

Scientists realised that there was some mass in atoms that could not be explained by this model, and in 1932 James Chadwick discovered a new particle inside the nucleus that had the same mass as a proton but had no electric charge. This particle was given the name neutron.

The model has been developed further since then, but the basic idea of atoms being made up of a tiny central nucleus containing protons and neutrons surrounded by electrons in shells remains (Figure 1.6).

The development of ideas about atomic structure shows very well how scientific models and theories develop over time. When new discoveries are made, models and theories may have to be altered or sometimes completely replaced if they do not fit in with the new discoveries.

Show you can...

Use a table to compare and contrast the plum-pudding model, the nuclear model and today's model of an atom.

Test yourself

23 What was discovered that led to scientists realising that atoms were made up of smaller particles?
24 Why was the plum-pudding model replaced?
25 Why would a nucleus deflect an alpha particle?

Reactions of elements

KEY TERM

Element A substance containing only one type of atom; a substance that cannot be broken down into simpler substances by chemical methods.

TIP

When writing symbols you must be very careful to ensure your capital letters cannot be mistaken for small letters and vice versa.

Elements in the periodic table

An **element** is a substance containing only one type of atom. For example, in the element carbon all the atoms are carbon atoms meaning that all the atoms have 6 protons and so have the atomic number 6. Elements cannot be broken down into simpler substances.

Atoms are known with atomic numbers up to just over 100. This means that there are just over 100 elements. All the elements are listed in the periodic table. The elements are listed in order of atomic number (Figure 1.7).

Atoms of each element are given their own symbol, each with one, two or three letters. The first letter is always a capital letter with any further letters being small letters. For example, carbon has the symbol C while copper has the symbol Cu.

Group 1	Group 2											Group 3	Group 4	Group 5	Group 6	Group 7	Group 0
							1 **H** hydrogen 1										4 **He** helium 2
7 **Li** lithium 3	9 **Be** beryllium 4											11 **B** boron 5	12 **C** carbon 6	14 **N** nitrogen 7	16 **O** oxygen 8	19 **F** fluorine 9	20 **Ne** neon 10
23 **Na** sodium 11	24 **Mg** magnesium 12											27 **Al** aluminium 13	28 **Si** silicon 14	31 **P** phosphorus 15	32 **S** sulfur 16	35.5 **Cl** chlorine 17	40 **Ar** argon 18
39 **K** potassium 19	40 **Ca** calcium 20	45 **Sc** scandium 21	48 **Ti** titanium 22	51 **V** vanadium 23	52 **Cr** chromium 24	55 **Mn** manganese 25	56 **Fe** iron 26	59 **Co** cobalt 27	59 **Ni** nickel 28	63.5 **Cu** copper 29	65 **Zn** zinc 30	70 **Ga** gallium 31	73 **Ge** germanium 32	75 **As** arsenic 33	79 **Se** selenium 34	80 **Br** bromine 35	84 **Kr** krypton 36
85 **Rb** rubidium 37	88 **Sr** strontium 38	89 **Y** yttrium 39	91 **Zr** zirconium 40	93 **Nb** niobium 41	96 **Mo** molybdenum 42	[98] **Tc** technetium 43	101 **Ru** ruthenium 44	103 **Rh** rhodium 45	106 **Pd** palladium 46	108 **Ag** silver 47	112 **Cd** cadmium 48	115 **In** indium 49	119 **Sn** tin 50	122 **Sb** antimony 51	128 **Te** tellurium 52	127 **I** Iodine 53	131 **Xe** xenon 54
133 **Cs** caesium 55	137 **Ba** barium 56	139 **La*** lanthanum 57	178 **Hf** hafnium 72	181 **Ta** tantalum 73	184 **W** tungsten 74	186 **Re** rhenium 75	190 **Os** osmium 76	192 **Ir** iridium 77	195 **Pt** platinum 78	197 **Au** gold 79	201 **Hg** mercury 80	204 **Ti** thallium 81	207 **Pb** lead 82	209 **Bi** bismuth 83	[209] **Po** polonium 84	[210] **At** astatine 85	[222] **Rn** randon 86
[223] **Fr** francium 87	[226] **Ra** radium 88	[227] **Ac**** actinium 89	[261] **Rf** rutherfordium 104	[262] **Db** dubnium 105	[266] **Sg** seaborgium 106	[264] **Bh** bohrium 107	[277] **Hs** hassium 108	[268] **Mt** meitnerium 109	[271] **Ds** darmstadtium 110	[272] **Rg** roentgenium 111	[285] **Cn** copernicium 112	[284] **Uut** ununtrium 113	[289] **Fl** flerorium 114	[288] **Uup** ununpentium 115	[293] **Lv** livermorium 116	[294] **Uus** ununseptium 117	[294] **Uuo** ununoctium 118

*	140 **Ce** cerium 58	141 **Pr** praseodymium 59	144 **Nd** neodymium 60	(145) **Pm** promethium 61	150 **Sm** samarium 62	152 **Eu** europium 63	157 **Gd** gadolinium 64	159 **Tb** terbium 65	162 **Dy** dysprosium 66	165 **Ho** holmium 67	167 **Er** erbium 68	169 **Tm** thulium 69	173 **Yb** ytterbium 70	175 **Lu** lutetium 71
**	232 **Th** thorium 90	231 **Pa** protactinium 91	238 **U** uranium 92	237 **Np** neptunium 93	(244) **Pu** plutonium 94	(243) **Am** americium 95	(247) **Cm** curium 96	(247) **Bk** berkelium 97	(251) **Cf** californium 98	(252) **Es** einsteinium 99	(257) **Fm** fermium 100	(258) **Md** mendelevium 101	(259) **No** nobelium 102	(260) **Lr** lawrencium 103

Key
relative atomic mass
atomic symbol
name
atomic (proton) number

Metal   Non-metal   Difficult to classify

▲ **Figure 1.7**

TIP

The periodic table shows relative atomic mass and not mass number. This is the average mass of the isotopes of each element.

Metals and non-metals

Over three-quarters of the elements are metals, with most of the rest being non-metals. Typical properties of metals and non-metals are shown in Table 1.7, although there are some exceptions.

Table 1.7

	Metals	Non-metals
Melting and boiling points	High	Low
Conductivity	Thermal and electrical conductor	Thermal and electrical insulator (except graphite)
Density	High density	Low density
Appearance	Shiny when polished	Dull
Malleability	Can be hammered into shape	Brittle as solids
Reaction with non-metals	React to form positive ions in ionic compounds	React to form molecules
Reaction with metals	No reaction	React to form negative ions in ionic compounds
Acid-base properties of oxides	Metal oxides are basic	Non-metal oxides are acidic

There are a few elements around the dividing line between metals and non-metals, such as silicon and germanium, that are hard to classify as they have some properties of metals and some of non-metals.

Test yourself

26 Is each of the following elements a metal or non-metal?
  a) Element **1** is a dull solid at room temperature that easily melts when warmed.
  b) Element **2** is a dense solid that is a thermal conductor.
  c) Element **3** reacts with oxygen to form an oxide which dissolves in rain water to form acid rain.
  d) Element **4** reacts with chlorine to form a compound made of molecules.
  e) Element **5** reacts with sodium to form a compound made of ions.

Show you can...

Figure 1.8 shows magnesium and oxygen reacting to form a single product.

a) State two differences in the physical properties of magnesium and oxygen.
b) Suggest the name of the product of this reaction.
c) Is the product acidic or basic?
d) Does the product consist of ions or molecules?

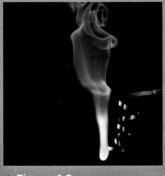

▲ Figure 1.8

○ Reactions between elements

When elements react with each other they form compounds. Compounds are substances made from different elements bonded together. A chemical reaction takes place when elements combine to form compounds. Chemical reactions always involve the formation of one or more new substances and there is usually a detectable energy change.

When elements react with each other, electrons are either shared with other elements or transferred from one element to another. This is done so that atoms obtain the stable electron structure of the noble gases (Group 0 elements).

Table 1.8 shows what happens in general when elements react with each other.

Table 1.8

Elements reacting	What happens to the electrons to obtain noble gas electron structures	Type of particles formed	Type of compound formed	Example
Non-metal + non-metal	Electrons shared	Molecules (where atoms are joined to each other by covalent bonds)	Molecular compound *covalent bond*	Hydrogen reacts with oxygen by sharing electrons and forming molecules of water
Metal + non-metal	Electrons transferred from metal to non-metal	Positive and negative ions	Ionic compound *Ionic bond*	Sodium reacts with chlorine by transferring electrons from sodium to chlorine to form sodium chloride which is made of ions
Metal + metal	No reaction as both metals cannot lose electrons		*Metallic bond*	

Show you can...

The electron structures of the atoms of 5 different elements, A, B, C, D and E, are shown below.

A 2,8,8   B 2,8,8,1   C 2,6
D 2,1   E 2,8,7

Using the letters A, B, C, D or E choose:

a) An unreactive element.
b) Two elements found in the same Group of the periodic table.
c) An element whose atoms will form ions with a charge of 2–.
d) Two elements that react to form an ionic compound.
e) Two elements that react to form a covalent compound.

Test yourself

27 Do the following elements react with each other, and if they do, what type of compound is formed?
   a) potassium + oxygen
   b) bromine + iodine
   c) oxygen + sulfur
   d) sulfur + magnesium
   e) calcium + potassium
   f) nitrogen + hydrogen

The periodic table

○ Electron structure and the periodic table

The elements are placed in the periodic table in order of increasing atomic number (the number of protons). Figure 1.9 shows the first 36 elements in the periodic table.

Group 1	Group 2										Group 3	Group 4	Group 5	Group 6	Group 7	Group 0
								H 1								He 2
Li 3	Be 4										B 5	C 6	N 7	O 8	F 9	Ne 10
Na 11	Mg 12										Al 13	Si 14	P 15	S 16	Cl 17	Ar 18
K 19	Ca 20	Sc 21	Ti 22	V 23	Cr 24	Mn 25	Fe 26	Co 27	Ni 28	Cu 29	Zn 30	Ga 31	Ge 32	As 33	Se 34	Br 35	Kr 36

▲ Figure 1.9

The table can be seen as arranging the elements by electron structure. At the end of each period, a noble gas stable electron structure is reached and then a new energy level (shell) starts to be filled at the start of the next period. The electron structure of the first 20 elements is shown.

Group 1	Group 2										Group 3	Group 4	Group 5	Group 6	Group 7	Group 0
								H 1								He 2
Li 2,1	Be 2,2										B 2,3	C 2,4	N 2,5	O 2,6	F 2,7	Ne 2,8
Na 2,8,1	Mg 2,8,2										Al 2,8,3	Si 2,8,4	P 2,8,5	S 2,8,6	Cl 2,8,7	Ar 2,8,8
K 2,8,8,1	Ca 2,8,8,2															

▲ Figure 1.10

TIP ✓

The chemical properties of the elements in the periodic table repeat at regular (periodic) intervals. This is why it is called the periodic table.

Elements in the same group have the same number of electrons in their outer shell. The number of electrons in the outer shell equals the Group number. For example, all the elements in Group 1 have 1 electron their outer shell (Li = 2,1; Na = 2,8,1; K = 2,8,8,1) (Figure 1.10). All the elements in Group 7 have 7 electrons in their outer shell (F = 2,7; Cl = 2,8,7). The only exception to this is Group 0 where all the elements have 8 electrons in their outer shell except helium which has 2 electrons (but the first shell can only hold 2 electrons).

All the elements in the same group have similar chemical properties because they have the same number of electrons in their outer shell.

Show you can...

Element A has electron structure 2,8,1.

a) Explain why element A is not found in Group 5.

b) Determine the atomic number of A.

Test yourself

28 In what order are the elements in the periodic table?

29 In which group of the periodic table do the elements with these electron structures belong?

    a) 2,8,4   **b)** 2,8,8,1   **c)** 2,8,18,3

30 Explain why the periodic table has the word periodic in its name.

○ Group 0 – the noble gases

The main elements of Group 0 are helium, neon, argon, krypton, xenon and radon (Figure 1.11 and Table 1.9). They are known as the noble gases (Table 1.10). These atoms all have stable electron structures. Helium's outer shell is full with 2 electrons while the others have 8 electrons in their outer shells.

KEY TERM

Noble gases The elements in Group 0 of the periodic table (including helium, neon and argon).

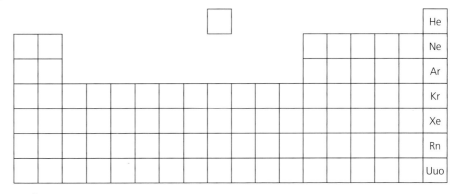

▲ Figure 1.11

Table 1.9

Element	Formula	Appearance at room temperature	Number of electrons in outer shell	Relative mass of atoms	Boiling point in °C
Helium	He	Colourless gas	2	4	−269
Neon	Ne	Colourless gas	8	20	−246
Argon	Ar	Colourless gas	8	40	−190
Krypton	Kr	Colourless gas	8	84	−157
Xenon	Xe	Colourless gas	8	131	−111
Radon	Rn	Colourless gas	8	222	−62

Table 1.10 Properties of the noble gases

Metals or non-metals?	All the elements are non-metals.
Boiling points	The noble gases are all colourless gases with low boiling points. The boiling points increase as the atoms get heavier going down the group.
Reactivity	The Group 0 elements are very unreactive and do not easily react to form molecules or ions because their atoms have stable electron arrangements.

Show you can...

Copy and complete the table:

Element	He	Ar
Reactive or unreactive?		
Metal or non-metal?		
Solid, liquid or gas at room temperature?		
Electron structure		

Test yourself

31 Why are the noble gases unreactive?

32 Suggest a reason why the noble gases are refered to as being in Group 0 rather than Group 8.

33 Some atoms of element 118 (Uuo) have been produced. Element 118 is in Group 0. Predict the chemical and physical properties of this element.

○ Group 1 – the alkali metals

The main elements of Group 1 are lithium, sodium, potassium, rubidium and caesium (Table 1.11). They are known as the alkali metals (Table 1.12). The Group 1 elements have similar chemical and physical properties because they all have one electron in their outer shell. They are all soft metals that can be cut with a knife (Figure 1.12). They are very reactive and so are stored in bottles of oil to stop them reacting with water and oxygen.

KEY TERM

Alkali metals The elements in group 1 of the periodic table (including lithium, sodium and potassium).

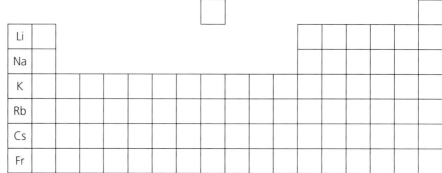

▲ **Figure 1.12** The alkali metals are all soft and can be cut with a knife.

Table 1.11

Element	Formula	Appearance at room temperature	Number of electrons in outer shell	Relative mass of atoms	Melting point in °C	Density in g/cm³
Lithium	Li	Silvery-grey metal	1	7	180	0.53
Sodium	Na	Silvery-grey metal	1	23	98	0.97
Potassium	K	Silvery-grey metal	1	39	63	0.89
Rubidium	Rb	Silvery-grey metal	1	85	39	1.53
Caesium	Cs	Silvery-grey metal	1	133	28	1.93

Table 1.12 Properties of the alkali metals

Metals or non-metals?	All the elements are metals.
Melting points	The alkali metals are all solids with relatively low melting points at room temperature. The melting points decrease as the atoms get bigger going down the group.
Density	The alkali metals have low densities for metals. Lithium, sodium and potassium all float on water as they are less dense than water.
Reaction with non-metals	The metals all react easily with non-metals by the transfer of electrons from the metal to the non-metal forming compounds made of ions. Alkali metals always form 1+ ions (e.g. Li^+, Na^+, K^+, Rb^+, Cs^+) as they have one electron in their outer shell which they lose when they react to obtain a noble gas electron structure.
Reaction with oxygen	The alkali metals all burn in oxygen to form metal oxides which are white powders: metal + oxygen → metal oxide e.g. $4Na + O_2 \rightarrow 2Na_2O$ The metals burn with different colour flames. For example lithium burns with a crimson-red flame, sodium with a yellow-orange flame and potassium with a lilac flame.
Reaction with chlorine	The alkali metals all burn in chlorine to form metal chlorides which are white powders: metal + chlorine → metal chloride e.g. $2Na + Cl_2 \rightarrow 2NaCl$
Reaction with water	The alkali metals all react with water, releasing hydrogen gas and forming a solution containing a metal hydroxide: metal + water → metal hydroxide + hydrogen: e.g. $2Na + 2H_2O \rightarrow 2NaOH + H_2$ The solution of the metal hydroxide that is formed is alkaline.
Compounds made from Group 1 metals	Compounds made from alkali metals: are ionic are white solids dissolve in water to form colourless solutions

Reactivity trend of the alkali metals

The alkali metals get more reactive the further down the group (Figure 1.13). This can be seen when the alkali metals react with water.

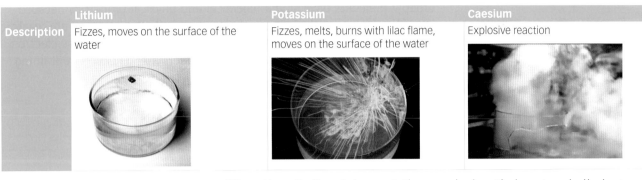

	Lithium	Potassium	Caesium
Description	Fizzes, moves on the surface of the water	Fizzes, melts, burns with lilac flame, moves on the surface of the water	Explosive reaction

lithium (2, 1)

sodium (2, 8, 1)

potassium (2, 8, 8, 1)

▲ **Figure 1.13** The further down the group, the further the outer electron is from the nucleus.

When the alkali metals react they are losing their outer shell electron in order to get a noble gas electron structure. The further down the group, the further away the outer electron is from the nucleus as the atoms get bigger. This means that the outer electron is less strongly attracted to the nucleus and so easier to lose. The easier the electron is to lose, the more reactive the alkali metal.

Test yourself

34 Why are the alkali metals reactive?

35 Write word and balanced symbol equations for the reaction of potassium with water.

36 Explain why the solution formed when potassium reacts with water has a high pH.

37 Potassium reacts with chlorine to form an ionic compound. Explain why this reaction happens.

38 Explain why potassium is more reactive than sodium.

39 Francium is the last element in Group 1. Predict the chemical and physical properties of francium.

Show you can...

This question gives information about the reaction of Group 2 elements with water (these reactions are not on the specification) and tests your ability to interpret data.

Element	Reactivity with water	Name of product
Be	No reaction	No products
Mg	Reacts very slowly with cold water	Magnesium hydroxide and hydrogen
Ca	Reacts moderately with cold water	Calcium hydroxide and hydrogen
Sr	Reacts rapidly with cold water	Strontium hydroxide and hydrogen
Ba	Reacts very rapidly with cold water	Barium hydroxide and hydrogen

Use the information in the table and your own knowledge of Group 1 elements to compare and contrast the reactions of Group 1 and Group 2 elements with water.

In your answer compare:
a) The products formed.
b) The reactivity of the Group 1 elements compared to the Group 2 elements.
c) The trend in reactivity down both groups.

▲ **Figure 1.14** Bromine is a dark brown liquid that easily vaporises to give an orange gas.

Group 7 – the halogens

The main elements of Group 7 are fluorine, chlorine, bromine (Figure 1.14) and iodine (Table 1.13). They are known as the halogens (Table 1.14). The Group 7 elements have similar chemical and physical properties because they all have seven electrons in their outer shell. The halogens are reactive because they only need to gain one electron to gain a noble gas electron structure. The particles in each of the elements are molecules containing two atoms (diatomic molecules), such as F_2, Cl_2, Br_2 and I_2.

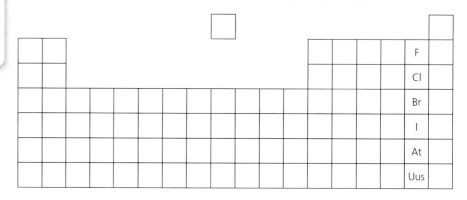

Table 1.13

Element	Formula	Appearance at room temperature	Number of electrons in outer shell	Relative mass of molecules	Melting point in °C	Boiling point in °C
Fluorine	F_2	Pale yellow gas	7	38	−220	−188
Chlorine	Cl_2	Pale green gas	7	71	−102	−34
Bromine	Br_2	Dark brown liquid	7	160	−7	59
Iodine	I_2	Grey solid	7	254	114	184

Table 1.14 Properties of the halogens

Metals or non-metals?	Fluorine, chlorine, bromine and iodine are all non-metals.
Toxicity	Each of the halogens is toxic.
Melting and boiling points	The halogens have low melting and boiling points. The melting and boiling points increase as the molecules get heavier going down the group.
Reaction with non-metals	The halogens react with other non-metals by sharing electrons to form compounds made of molecules.
Reaction with metals	The halogens all react easily with metals by the transfer of electrons from the metal to the halogen forming compounds made of ions. Halogens always form 1− ions (e.g. F^-, Cl^-, Br^-, I^-, all known as halide ions) as they have seven electrons in their outer shell and gain one more electron when the react to get a noble gas electron structure.

Reactivity trend of the halogens

The halogens get less reactive the further down the group. This can be seen by looking at which halogens can displace each other from compounds. Compounds containing halogens, such as sodium chloride and potassium bromide are often called halides or halide compounds.

A more reactive element can displace a less reactive element from a compound. You have seen this (before GCSE) with metals when a more reactive metal can displace a less reactive metal from a compound.

For example, aluminium can displace iron from iron oxide because aluminium is more reactive than iron.

aluminium + iron oxide → aluminium oxide + iron

In a similar way, a more reactive non-metal can displace a less reactive non-metal from a compound. This means that a more reactive halogen can displace a less reactive halogen from a halide compound.

This can be seen when aqueous solutions of the halogens react with aqueous solutions of halide compounds (aqueous means dissolved in water) (Table 1.15 and Figure 1.15).

▲ **Figure 1.15** Colourless chlorine water reacts with colourless potassium bromide solution to form a yellow solution of bromine.

Table 1.15

	Chlorine (aq)	Bromine (aq)	Iodine (aq)
Potassium chloride (aq)		**No reaction** Bromine cannot displace chlorine	**No reaction** Iodine cannot displace chlorine
Potassium bromide (aq)	chlorine + potassium bromide → potassium chloride + bromine $Cl_2 + 2KBr \rightarrow 2KCl + Br_2$ ($Cl_2 + 2Br^- \rightarrow 2Cl^- + Br_2$) Yellow solution formed (due to production of bromine) Chlorine displaces bromine		**No reaction** Iodine cannot displace bromine
Potassium iodide (aq)	chlorine + potassium iodide → potassium chloride + iodine $Cl_2 + 2KI \rightarrow 2KCl + I_2$ ($Cl_2 + 2I^- \rightarrow 2Cl^- + I_2$) Brown solution formed (due to production of iodine) Chlorine displaces iodine	bromine + potassium iodide → potassium bromide + iodine $Br_2 + 2KI \rightarrow 2KBr + I_2$ ($Br_2 + 2I^- \rightarrow 2Br^- + I_2$) Brown solution formed (due to production of iodine) Bromine displaces iodine	

It can be seen from these reactions that the trend in reactivity for these three halogens is:

Most reactive   Chlorine

Bromine

Least reactive   Iodine

In general, the further down the group the less reactive the halogen (Figure 1.16). The higher up the group, the more reactive the halogen. This means that fluorine is the most reactive halogen. You will not do experiments with fluorine because it is very reactive and toxic.

When the halogens react they gain one electron in order to get a noble gas electron structure. The further down the group, the electron gained is further away from the nucleus as the atoms get bigger. This means that the electron gained is less strongly attracted to the nucleus and so harder to gain. The harder the electron is to gain, the less reactive the halogen.

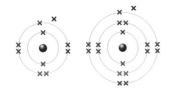

fluorine (2,7)   chlorine (2,8,7)

▲ **Figure 1.16** The further down the group, the further the electron gained is from the nucleus.

TIP

The explanation for the reactivity trend in Group 7 is about the distance between the nucleus and the electron gained which is from **outside** the atom – it is not about the outer shell electrons.

Test yourself

40 Why are the halogens reactive?
41 All the halogens are made of diatomic molecules. What are diatomic molecules?
42 Predict what would happen and why if fluorine and sodium chloride were mixed.
43 Bromine reacts with chlorine to form a molecular compound. Explain why this reaction happens.
44 Explain why chlorine is more reactive than bromine.

Practical

Reactions of the halogens

The diagram shows chlorine gas being passed through a dilute solution of potassium iodide. The upper layer is a hydrocarbon solvent. A colour change occurs in the potassium iodide solution due to the displacement reaction that occurs.

1 a) What is the most important safety precaution, apart from wearing safety glasses, which must be taken when carrying out this experiment?

   b) Explain why a displacement reaction occurs between chlorine and potassium iodide.

   c) Name the products of the displacement reaction which occurs.

   d) What is the colour change that occurs in the potassium iodide solution?

   e) Write a balanced symbol equation for the reaction between chlorine and potassium iodide.

   f) If this experiment was repeated using potassium bromide, instead of potassium iodide, explain if the observations would be different.

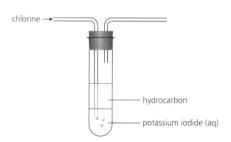

2 The halogens are more soluble in hydrocarbon solvents than in water and produce coloured solutions. After the reaction of chlorine with aqueous potassium iodide, the aqueous layer is shaken with the hydrocarbon solvent and most of the displaced halogen dissolves in the upper layer.

   a) Explain the meaning of the word solvent.

   b) Use the information in the table to suggest what happens to the hydrocarbon solvent, after shaking.

Halogen	Colour of hydrocarbon when halogen dissolves
Chlorine	Pale green
Bromine	Orange
Iodine	Purple

⦿ The transition metals

The transition metals are in the block in the middle of the periodic table between Groups 2 and 3 (Figure 1.17). They are all metals including many common metals such as chromium (Cr), manganese (Mn), iron (Fe), cobalt (Co), nickel (Ni) and copper (Cu).

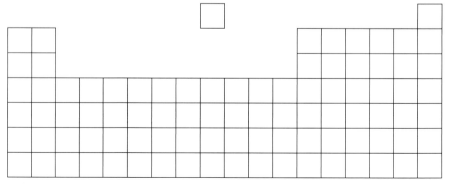

▲ Figure 1.17

Transition metals have some properties in common with the alkali metals, but many differences (Table 1.16).

Table 1.16

	Group 1 – alkali metals	Transition metals
Similarities	Thermal conductor	
	Electrical conductor	
	React with non-metals to form ionic compounds	
	Shiny when polished	
Differences	Low melting points	High melting points (except mercury)
	Low density	High density
	Very soft	Stronger and harder
	Very reactive (e.g. with water and oxygen)	Low reactivity (e.g. with water and oxygen)
	React to form 1+ ions (e.g. Na^+)	React to form ions with different charges (e.g. iron forms Fe^{2+} and Fe^{3+})
	Compounds are white	Compounds are coloured (Figure 1.18)
	Do not act as catalysts	Metals and their compounds are often catalysts (catalysts speed up reactions but are not used up themselves) (Figure 1.19)

▲ **Figure 1.18** The Statue of Liberty in New York is coated with a copper compound which is green.

KEY TERM

Catalyst Substance that speeds up a chemical reaction but is not used up.

▲ **Figure 1.19** Margarine is made by reacting plant oils (e.g. sunflower oil) with hydrogen using a nickel catalyst.

Test yourself

45 What are the transition metals?
46 List some ways in which the transition metals are similar to the alkali metals.
47 List some ways in which the transition metals are different from the alkali metals.
48 What is a catalyst?

Show you can...

The diagram shows part of the periodic table divided into five sections A, B, C, D and E.

Select from A to E the section in which you would find:

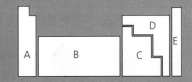

a) A gas that forms no compounds.
b) A metal that is used as a catalyst in industrial reactions.
c) A metal that reacts rapidly with cold water.
d) A coloured gas.
e) A metal that forms coloured compounds.

○ History of the periodic table

As more elements were discovered, scientists tried to classify the elements into some sort of order and pattern. This was originally done before the discovery of protons, neutrons and electrons. Scientists' first attempts were based on the use of the atomic weights of elements which we now know as relative atomic mass.

John Newlands spotted that the properties of elements seemed to repeat every eighth element when placed in order of atomic weight. He called this the 'law of octaves' as it was similar to notes in musical scales. One of the successes of his table was that he had lithium, sodium and potassium in the same group, each of which has very similar properties (Figure 1.20).

No.		No.		No.		No.		No.		No.		No.		No.
H	1	F	8	Cl	15	Co & Ni	22	Br	29	Pd	36	I	42	Pt & Ir	50
Li	2	Na	9	K	16	Cu	23	Rb	30	Ag	37	Cs	44	Os	51
G	3	Mg	10	Ca	17	Zn	24	Sr	31	Cd	38	Ba & V	45	Hg	52
Bo	4	Al	11	Cr	19	Y	25	Ce & La	33	U	40	Ta	46	Tl	53
C	5	Si	12	Ti	18	In	26	Zr	32	Sn	39	W	47	Pb	54
N	6	P	13	Mn	20	As	27	Di & Mo	34	Sb	41	Nb	48	Bi	55
O	7	S	14	Fe	21	Se	28	Ro & Ru	35	Te	43	Au	49	Th	56

▲ **Figure 1.20** Newlands' table (1865).

At the time though, only about 50 elements were known and his table was not accepted because it only worked for the first 20 or so of the known elements. After that there were problems, such as copper being in the same group as lithium, sodium and potassium. Copper has very different properties to those metals. For example, copper does not react with water but the other three react vigorously with water.

A few years later in 1869, a Russian chemist called Dimitri Mendeleev devised a table which has become the basis for the periodic table today (Figure 1.21). He also placed elements in order of atomic weight, but crucially he did two things differently to Newlands.

1 Mendeleev left gaps for elements he predicted had yet to be discovered. He also predicted the properties of these elements.

2 Mendeleev was prepared to alter slightly the order of the elements if he thought it fitted the properties better. For example, he swapped around the order of iodine (atomic weight 127) and tellurium (atomic weight 128). He placed iodine after tellurium so that it was in the same group as fluorine, chlorine and bromine which have very similar properties. Mendeleev actually believed that the atomic weights must have been measured incorrectly.

Tabelle II.

Reihen	Gruppe I. — R^2O	Gruppe II. — RO	Gruppe III. — R^2O^3	Gruppe IV. RH^4 RO^2	Gruppe V. RH^3 R^2O^5	Gruppe VI. RH^2 RO^3	Gruppe VII. RH R^2O^7	Gruppe VIII. — RO^4
1	H=1							
2	Li=7	Be=9,4	B=11	C=12	N=14	O=16	F=19	
3	Na=23	Mg=24	Al=27,3	Si=28	P=31	S=32	Cl=35,5	
4	K=39	Ca=40	—=44	Ti=48	V=51	Cr=52	Mn=55	Fe=56, Co=59, Ni=59, Cu=63.
5	(Cu=63)	Zn=65	—=68	—=72	As=75	Se=78	Br=80	
6	Rb=85	Sr=87	?Yt=88	Zr=90	Nb=94	Mo=96	—=100	Ru=104, Rh=104, Pd=106, Ag=108.
7	(Ag=108)	Cd=112	In=113	Sn=118	Sb=122	Te=128	J=127	
8	Cs=133	Ba=137	?Di=138	?Ce=140	—	—	—	— — —
9	(—)	—	—	—	—	—	—	
10	—	—	?Er=178	?La=180	Ta=182	W=184	—	Os=195, Ir=197, Pt=198, Au=199.
11	(Au=199)	Hg=200	Tl=204	Pb=207	Bi=208	—	—	
12	—	—	—	Th=231	—	U=240	—	— — —

der chemischen Elemente.

▲ **Figure 1.21** Mendeleev's table (1869).

Over the next few years, elements were discovered that Mendeleev had predicted would exist (Table 1.17). These included gallium (1875), scandium (1879) and germanium (1886). In each case the properties of the element closely matched Mendeleev's predictions. Table 1.17 shows some of the properties that Mendeleev predicted for the element he called eka-silicon and that we call germanium.

Table 1.17

	Element	Appearance	Atomic weight	Melting point in °C	Density in g/cm³	Formula of oxide	Formula of chloride
Mendeleev's predictions	Eka-silicon (Es)	Grey metal	72	high	5.5	EsO_2	$EsCl_4$
Actual properties	Germanium (Ge)	Grey-white metal	73	947	5.4	GeO_2	$GeCl_4$

As these elements were discovered, Mendeleev's ideas and table became well accepted and formed the basis of the periodic table as we know it today. We now know that Mendeleev placed the elements in order of atomic number (the number of protons in an atom) even though he did not know about the existence of protons. It is the atomic number rather than atomic weight that matters, because elements are made of a mixture of isotopes.

The story of Mendeleev illustrates how strong support can come for a scientific idea if predictions made using that theory are later found to be correct.

Show you can...

State some features of the periodic table developed by Mendeleev which are different from today's modern periodic table

Test yourself

49 In what order did Mendeleev put the elements?
50 Why did Mendeleev not stick to this order for some elements?
51 Why did Mendeleev leave some gaps in his table?
52 Why did Mendeleev's ideas become accepted?

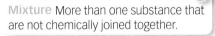

Mixtures

○ Mixtures compared to compounds

KEY TERM

Mixture More than one substance that are not chemically joined together.

A mixture consists of two or more substances that are mixed together and not chemically combined. In a mixture, each substance has its own properties. Mixtures are very different from compounds (Table 1.18).

Table 1.18

	Compound	Mixture
Description	A substance made from two or more elements chemically bonded together. A compound is a single substance with its own unique properties	Two or more substances each with their own properties (the different substances are not chemically joined to each other)
Proportions	Each compound has a fixed proportion of elements (so each compound has a fixed formula)	There can be any amount of each substance in the mixture
Separation	Compounds can only be separated back into elements by chemical reaction because the elements are chemically joined.	No chemical reaction is needed as the substances in the mixture are not chemically joined. They can be separated by physical methods (e.g. filtration, distillation).

Sodium is a very reactive, dangerous, grey metal that reacts vigorously with water. Chlorine is a pale green, toxic gas that is very reactive. In a mixture of sodium and chorine each substance keeps its own properties as a grey metal and green gas, respectively. It is easy to separate the sodium and chlorine because they are not chemically joined together.

However, if heated together sodium reacts with chlorine to make the compound sodium chloride. Sodium chloride is very different from both sodium and chlorine. Sodium chloride is a white solid that is not very reactive and is safe to eat. It is very difficult to break sodium chloride back down into the elements because the sodium and chlorine are chemically joined together.

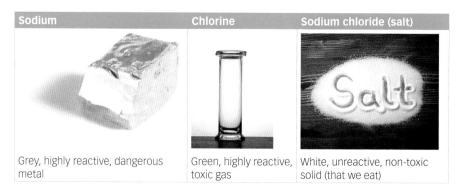

Sodium	Chlorine	Sodium chloride (salt)
Grey, highly reactive, dangerous metal	Green, highly reactive, toxic gas	White, unreactive, non-toxic solid (that we eat)

Show you can...

For each of the substances A, B, C, D decide if it is an element, compound or mixture.

If any substance is a mixture decide if it is a mixture of elements, a mixture of elements and compounds, or a mixture of compounds.

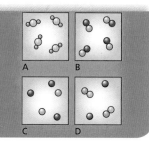

○ Separating mixtures

The substances in a mixture are quite easy to separate because the substances are not chemically joined to each other. Different methods are used depending on what type of mixture there is (Table 1.19).

Table 1.19

Type of mixture	Insoluble solid and liquid	Soluble solid dissolved in a solvent	Soluble solids dissolved in a solvent	Two miscible liquids (liquids that mix)	Two immiscible liquids (liquids that do not mix)
Method of separation	Filtration	Evaporation (to obtain solid) Crystallisation (to obtain sold) Simple distillation (to obtain solvent)	Chromatography	Fractional distillation	Separating funnel

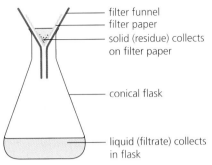

▲ Figure 1.22 Filtration.

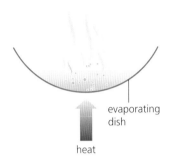

▲ Figure 1.23 Evaporation.

▲ Figure 1.24 Crystallisation.

Filtration

This method is used to separate an insoluble solid from a liquid. For example, it could be used to separate sand from water.

The mixture is poured through a funnel containing a piece of filter paper. The liquid (called the filtrate) passes through the paper and the solid (called the residue) remains on the filter paper (Figure 1.22).

Evaporation

This method is used to separate a dissolved solid from the solvent it is dissolved in. For example, it could be used to separate salt from water.

The mixture is placed in an evaporating dish and heated until all the solvent has evaporated or boiled, leaving the solid in the evaporating basin (Figure 1.23).

Crystallisation

This method is also used to separate a dissolved solid from the solvent it is dissolved in. For example, it could be used to separate copper sulfate crystals from a solution of copper sulfate (Figure 1.24).

The mixture is heated to boil off some of the solvent to create a hot, saturated solution. A saturated solution is one in which no more solute can dissolve at that temperature. As it cools down, the solute becomes less soluble and so cannot remain dissolved, so some of the solute crystallises out of the solution as crystals. The crystals can then be separated from the rest of the solution by filtration.

Simple distillation

This method is used to separate the solvent from a solution. For example, it could be used to separate pure water from sea water.

The mixture is heated and the solvent boils. The vaporised solvent passes through a water-cooled condenser where it cools and condenses. The condenser directs the condensed solvent into a container away from the original solution (Figure 1.25).

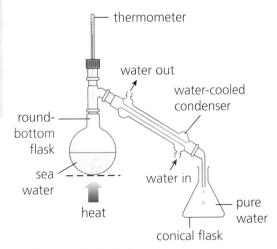

▲ Figure 1.25 Distillation of sea water.

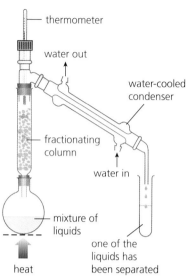

▲ **Figure 1.26** Fractional distillation.

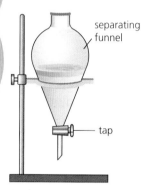

▲ **Figure 1.27** Separating funnel.

▲ **Figure 1.28** Substances separate as they move up the paper with the solvent at different speeds.

KEY TERMS

Miscible Liquids that mix together.

Immiscible Liquids that do not mix together and separate into layers.

Separating funnel Glass container with a tap used to separate immiscible liquids.

Fractional distillation

Liquids that mix together are called **miscible** liquids. Water and alcohol are examples of miscible liquids. Fractional distillation is used to separate mixtures of miscible liquids. It works because the liquids have different boiling points.

The apparatus used is similar to that for simple distillation, but a long column (called a fractionating column) is used to help separate different liquids as they boil. The fractionating column often contains glass beads.

In industry, such as in the fractional distillation of crude oil (see Chapter 7), the whole mixture is vaporised and then condensed in a fractionating column which is hot at the bottom and cold at the top. The liquids will condense at different heights in the fractionating column (Figure 1.26).

Separating funnel

Liquids that do not mix together are called **immiscible** liquids. A hydrocarbon and water is an example of liquids that are immiscible with each other. They can be separated in a **separating funnel**. The liquids form two layers and the bottom layer can be removed using the tap at the bottom of the funnel. The liquid with the greater density is the lower layer (Figure 1.27).

Chromatography

There are many forms of chromatography. Paper chromatography is used to separate mixtures of substances dissolved in a solvent.

A piece of chromatography paper, with the mixture on, is placed upright in a beaker so that the bottom of the paper is in the solvent. Over time, the solvent soaks up the paper. The substances move up the paper at different speeds and so are separated (Figure 1.28). Chromatography is studied further in Chapter 8.

Test yourself

53 How would you separate the following mixtures?
   a) alcohol from a mixture of alcohol and water
   b) magnesium hydroxide from a suspension of insoluble magnesium hydroxide in water
   c) pure dry cleaning solvent from waste dry cleaning solvent containing dirt that dissolved in the solvent from clothes
   d) sunflower oil and water
   e) food colourings in a sweet.

Show you can...

Three common methods of separation include filtration, distillation and fractional distillation.

For each of these separation methods pick **two** words or phrases from the list and insert them into a copy of the table with an explanation of their meaning. Also include the type of mixture separated by each method:

condenser, distillate, fractionating column, filtrate, miscible liquids, residue.

	Filtration	Distillation	Fractional distillation
Type of mixture separated			
Important word and definition			
Important word and definition			

Rock salt

Common salt is sodium chloride and is found naturally in large amounts in seawater or in underground deposits. Sodium chloride can be extracted from underground by the process of solution mining.

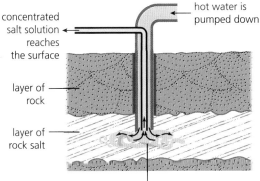

1 a) On what physical property of sodium chloride does this process depend?

   b) Suggest one reason why solution mining uses a lot of energy.

   c) Suggest one negative effect which solution mining has on the environment.

   d) Suggest how sodium chloride is obtained from the concentrated salt solution.

2 Rock salt is a mixture of salt, sand and clay. To separate pure salt from rock salt, the method listed below can be used in the laboratory.

Method:

   i   Place 8 spatulas of rock salt into a mortar and grind using a pestle.

   ii Place the rock salt into a beaker and quarter fill with water.

   iii Place on a gauze and tripod and heat, stirring with a glass rod. Stop heating when the salt has dissolved – the sand and clay will be left undissolved.

   iv Allow to cool and then filter.

   v   Heat until half the volume of liquid is left.

   vi Place the evaporating basin on the windowsill to evaporate off the rest of the water slowly. Pure salt crystals should be left.

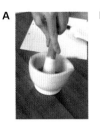  

Choose one step of the method (i to vi) which is best represented in each photograph A–C.

3 a) Why is rock salt considered to be a mixture?

   b) What was the purpose of grinding the rock salt?

   c) Why was the mixture heated and stirred?

   d) State what the filtrate contains.

   e) State what the residue contains.

   f) Explain why the salt obtained may still be contaminated with sand and suggest how you would improve your experiment to obtain a purer sample of salt.

Chapter review questions

1 Choose from the following list of elements to answer the questions below:

   bromine   calcium   krypton   nickel   nitrogen   potassium   silicon

   a) Which element is most like lithium?

   b) Which element is most like iron?

   c) Which element is most like helium?

   d) Which element is most like fluorine?

   e) Which element is most like carbon?

2 In which group or area of the periodic table would you find these elements?

   a) Element **A** has 7 electrons in its outer shell.

   b) Element **B** reacts vigorously with water to give off hydrogen gas and an alkaline solution.

   c) Element **C** is a metal with 4 electrons in its outer shell.

   d) Element **D** is a colourless gas that does not react at all.

   e) Element **E** forms coloured compounds.

   f) Element **F** is toxic and is made of diatomic molecules.

   g) Element **G** forms 1– ions when it reacts with metals to form ionic compounds

   h) Element **H** can form both 1+ and 2+ ions

   i) Element **I** is a metal that floats on water

   j) Element **J** has the electron structure 2,8,18,6

   k) Element **K** has 12 protons

   l) Element **L** has a full outer shell

   m) Element **M** can act as a catalyst

3 Identify a mixture that could be separated by each of the following methods.

   a) simple distillation

   b) filtration

   c) crystallisation

   d) evaporation

   e) chromatography

   f) fractional distillation

4 Look at the following atoms and ions.

   $^{12}_{6}C$ $^{14}_{6}C$ $^{16}_{8}O^{2-}$ $^{19}_{9}F^{-}$ $^{20}_{10}Ne$

   Which of these atoms and ions, if any,

   a) are isotopes?

   b) have 9 protons?

   c) have 10 electrons?

   d) have 10 neutrons?

   e) have more protons than electrons?

5 Caesium atoms are among the largest atoms. A caesium atom has a radius of 0.260 nm. Write this in metres in standard form.

6 Predict whether each of the following pairs of elements will (i) react by sharing electrons; (ii) react by transferring electrons; or (iii) not react:

a) magnesium + oxygen

b) sulfur + hydrogen

c) aluminium + magnesium

d) argon + oxygen

e) bromine + phosphorus

f) fluorine + lithium

7 Sodium (Na) reacts with bromine (Br_2) to form the ionic compound sodium bromide (NaBr). Sodium is in Group 1 of the periodic table. Bromine is in Group 7 of the periodic table.

a) What names are often given to Groups 1 and 7?

b) Describe in detail what happens in terms of electrons when sodium reacts with bromine.

c) Write a balanced equation for the reaction between sodium and bromine.

d) Potassium reacts with bromine more vigorously than sodium. Explain why potassium reacts more vigorously than sodium.

8 This question is about the metals sodium (Na) and nickel (Ni).

a) Identify the group or region of the periodic table that each element belongs to.

b) Which one of the two elements

i)   has the higher melting point?

ii) has the higher density?

iii) is the most reactive?

iv) has one electron in its outer shell?

v)   forms one common ion only?

vi) can act as a catalyst in some reactions?

vii) forms coloured compounds?

viii) can be cut with a knife?

c) Write a balanced equation for the reaction of sodium with water.

9 A yellow solution of bromine water was added dropwise to a colourless solution of sodium iodide. The solution darkened to pale brown.

a) Explain why the solution darkened.

b) Write an ionic equation for the reaction that took place.

c) Explain, in terms of electrons, why this reaction took place.

10 The following four substances are mixed together: salt; water; cyclohexane; diethyl ether. Cyclohexane and diethylether are liquids made of organic molecules. Salt is soluble in water but not in cyclohexane or diethyl ether. Cyclohexane and diethyl ether are miscible, but water is not miscible with cyclohexane or diethyl ether. Describe how the four substances could be separated.

Practice questions

1 How many electrons are there in a potassium ion (K^+)?

A	18	B	19
C	20	D	39	[1 mark]

2 In which of the following atoms is the number of protons greater than the number of neutrons?

A	^2_1H	B	^3_2He
C	$^{10}_5B$	D	$^{16}_8O$	[1 mark]

3 An aluminium atom contains three types of particle.

a) Copy and complete the table below to show the name, relative mass and relative charge of each particle in an aluminium atom. [4 marks]

Particle	Relative charge	Relative mass
Proton		1
		Very small
Neutron	0	

b) Complete the sentences below about an aluminium ion by choosing one of the words or numbers in bold. [4 marks]

i) In an aluminium atom, the protons and neutrons are in the **nucleus/shells**.

ii) The number of protons in an aluminium atom is the **atomic number/group number/mass number**.

iii) The sum of the number of protons and neutrons in an aluminium atom is the **atomic number/group number/mass number**.

iv) The number of electrons in an aluminium atom is **13/14/27**.

4 The structure of the atom has caused debate for thousands of years. In the late 19th century the 'plum-pudding model' of the atom was proposed. This was replaced at the beginning of the 20th century with the nuclear model of the atom which is the basis of the model we use today.

a) Describe the differences between the 'plum-pudding' model of the atom and the model of the atom we use today. [5 marks]

b) The diagram represents an atom of an element. The electrons are missing from the diagram.

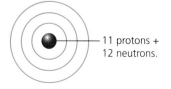

11 protons + 12 neutrons.

i) State the atomic number of this element. [1 mark]

ii) State the mass number of this element. [1 mark]

iii) Name the part of the atom in which the protons and neutrons are found. [1 mark]

iv) Copy and complete the diagram to show the electron configuration of the atom, using x to represent an electron. [1 mark]

c) The table shows some information for several atoms and simple ions. Copy and complete the table. [6 marks]

Atom/Ion	Number of protons	Electron structure
	7	2,5
S^{2-}		
Ca^{2+}		
	12	2,8

5 Mixtures may be separated in the laboratory in many different ways. Three different methods of separating mixtures are shown below.

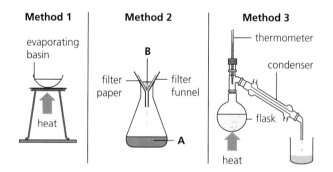

a) Name each method of separation. [3 marks]

b) Which method (**1**, **2** or **3**) would be most suitable for obtaining water from potassium chloride solution? [1 mark]

c) Which method would be most suitable for removing sand from a mixture of sand and water? [1 mark]

d) What general term is used for liquid **A** and solid **B** in method 2? [2 marks]

e) State why method 2 would **not** be suitable to separate copper(II) chloride from copper(II) chloride solution. [1 mark]

6 To determine if two different orange drinks **X** and **Y** contained the food colourings E102, E101 or E160 a student put a drop of each orange drink and a drop of each food colouring along a pencil line on filter paper.

The filter paper was placed in a tank containing 1 cm depth of solvent. The solvent soaked up the paper and carried different components with it. After 5 minutes, the filter paper was removed and allowed to dry. The results are shown.

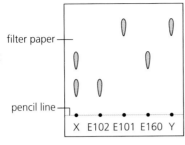

filter paper

pencil line

X E102 E101 E160 Y

a) What is the name of the process used by the student to analyse the two orange drinks? [1 mark]

b) i) Orange drink X contains the food colouring E102. How do the results show this? [1 mark]

ii) What other food colouring does orange drink X contain? [1 mark]

iii) Re-draw the diagram and add a spot to show that orange drink Y also contained food colouring E160. [1 mark]

iv) The line across the bottom of the filter paper was drawn with a pencil not with ink. Why should the line not be drawn with ink? [1 mark]

7 When Group 1 elements react, the atom forms an ion. For example when potassium reacts with water, potassium ions are formed from potassium atoms.

a) Why is potassium stored under oil in the laboratory? [1 mark]

b) Before reacting Group 1 elements with water a risk assessment is carried out. Give two safety precautions, apart from wearing safety glasses, which must be included in the risk assessment for reacting potassium with water. [2 marks]

c) Equal sized pieces of three Group 1 metals are added to separate troughs of water which contain universal indicator. The observations made are recorded in the table. Use the information in the table to answer the questions that follow.

Group 1 metal	Observation on reacting with water	Colour of universal indicator solution
Potassium	Melts Burns with a lilac flame Moves on the surface of the water Disappears quickly	Changes colour from green to purple
Lithium	Floats Moves on the surface of the water Eventually disappears	Changes colour from green to purple
Sodium	Melts Moves on the surface of the water Disappears	Changes colour from green to purple

i) What happens to the reactivity of the Group 1 elements as the Group is descended? [1 mark]

ii) Explain fully why the universal indicator changed colour from green to purple. [2 marks]

iii) Give one more observation which could be added to the table for all three reactions. [1 mark]

iv) Write a word equation to describe the reaction between sodium and water. [1 mark]

v) Write a balanced symbol equation to describe the reaction between sodium and water. [2 marks]

8 The modern periodic table has been in use for over 100 years. Its development included the work of several chemists including that of Dmitri Mendeleev.

a) Fill in the blanks in the following passage

The modern periodic table arranges the elements in order of increasing atomic _____ whereas early versions of the periodic table arranged them in order of increasing atomic _____. [2 marks]

b) State one other difference between the modern periodic table and Mendeleev's table. [1 mark]

c) Elements in the periodic table are arranged in groups. The table gives details of some of the groups of the periodic table. Copy and complete the table. [6 marks]

Group number	Name of group	Number of electrons in the outer shell of an atom of this group	Reactive or non-reactive group?
1			
		7	

d) Many trends in reactivity and physical properties are apparent as a group is descended.

i) State the trend in reactivity as Group 7 is descended. [1 mark]

ii) Name the least reactive element in Group 1. [1 mark]

iii) Astatine is found at the bottom of Group 7. Predict its state at room temperature and pressure. [1 mark]

9 Dmitri Mendeleev produced a table that is the basis of the modern periodic table. Describe the key features of Mendeleev's table and explain why his table came to be accepted over time by scientists. [6 marks]

10 The alkali metals and the transition metals are all metals. Describe some ways in which these metals are similar and some ways in which they are different. [6 marks]

Working scientifically: How theories change over time

A version of the periodic table hangs on the wall of almost every chemistry laboratory across the world – it is a powerful icon and a single document that summarises much of our knowledge of chemistry. The history of the periodic table can be traced back over centuries and illustrates how scientific theories change over time.

After the discovery of the new element phosphorus in 1649, scientists began to think about the definition of an element. In 1789 Antoine-Laurent de Lavoisier produced a table similar to that below of simple substances, or elements, which could not be broken down further by chemical reactions.

Acid-making elements	Gas making elements	Metallic elements	Earth elements
Sulfur	Light	Cobalt, mercury, tin	Lime (calcium oxide)
Phosphorus	Caloric (heat)	Copper, nickel, iron	Magnesia (magnesium oxide)
Charcoal (carbon)	Oxygen	Gold, lead, silver, zinc	Barytes (barium sulfate)
	Azote (nitrogen)	Manganese, tungsten	Argila (aluminium oxide)
	Hydrogen	Platina (platinum)	Silex (silicon dioxide)

In addition to many elements which form the basis of our modern periodic table, Lavoisier's list also included 'light' and 'caloric' (heat) which at the time were believed to be material substances. Lavoisier incorrectly classified some compounds as elements because high temperature smelting equipment or electricity was not available to break down these compounds. The incorrect classification of these compounds as elements was due to a lack of technology as much as a lack of knowledge.

1 What is an element?

2 Which elements in Lavoisier's table also appear in today's periodic table?

3 Which group of elements did Lavoisier classify correctly?

4 Why do you think sulfur, phosphorus and charcoal are described as 'acid-making' elements?

5 Which substances in Lavoisier's list, from your own modern knowledge, are compounds? Why do you think Lavoisier thought these were elements?

Following on from the work of Lavoisier, in the early 19th century Johann Döbereiner noted that certain elements could be arranged in groups of three because they have similar properties. For example

▶ lithium, sodium and potassium – very reactive metals which produce alkalis with water

▶ calcium, strontium and barium – reactive metals but with higher melting points and different formulae of their oxides

▶ chlorine, bromine and iodine – low melting point, coloured, reactive non-metals.

He also noted that the 'atomic weight' of the middle element was close to the average of the other two:

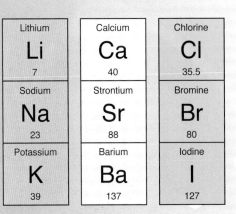

Lithium	Calcium	Chlorine
Li	Ca	Cl
7	40	35.5
Sodium	Strontium	Bromine
Na	Sr	Br
23	88	80
Potassium	Barium	Iodine
K	Ba	I
39	137	127

$Li = 7; \ Na = 23; \ K = 39; \dfrac{7 + 39}{2} = 23$

$Ca = 40; \ Sr = 88; \ Ba = 137; \dfrac{40 + 137}{2} = 88.5$

These groups were called triads and were the first partial representation of a group of elements with similar properties.

6 State the group represented by each of the three Döbereiner triads.

7 Does the final triad listed above follow Döbereiner's atomic weight rule? Show your working.

You have already learned of the work of Newlands and Mendeleev in the development of the periodic table. It is important to realise that their work was built on the theory and tables suggested by Lavoisier, Döbereiner and others. Use the following questions to think about the ways in which scientific theories and methods develop over time.

8 In what ways is Newlands' periodic table superior to Lavoisier's classification of the elements?

9 Why is Newlands' classification superior, yet building on Johann Döbereiner's work?

10 State as many features as you can think of in which the modern periodic table is superior to Mendeleev's periodic table.

In the 1940s Glenn Seaborg was part of a research team working on 'nuclear synthesised' elements with atomic masses beyond the naturally occurring limit of uranium. When isolating the elements americium and curium, he wondered if these elements belonged to a different series which would explain why their chemical properties were different from what was expected. In 1945, against the advice of colleagues, he proposed a significant change to Mendeleev's table – the actinide series. Today this series is well accepted and included in the periodic table.

Through the history of the periodic table we can easily see

▶ the ways in which scientific methods and theories develop over time

▶ how a variety of concepts and models are used to develop scientific explanations and understanding.

2 Bonding, structure and the properties of matter

Atoms are so small that we cannot see them. We cannot see them even using the most powerful light microscope because atoms are much smaller than the wavelength of light. However, being able to picture what the particles are like in a substance and how they are bonded to each other is vital to understand chemistry. In this chapter we will examine what the particles are and how they bond together in different substances to help us understand the properties of these different substances.

Specification coverage

This chapter covers specification points 4.2.1 to 4.2.4 and is called Bonding, structure and the properties of matter.

It covers ionic, molecular, giant covalent and metallic substances, as well an overview of types of bonding and structures, nanoscience and the different forms of carbon.

Related work on writing formulae and equations can be found in the Appendix.

Ionic substances

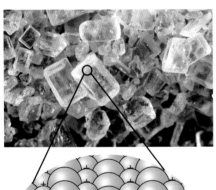

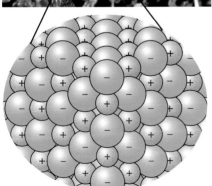

▲ **Figure 2.1** Ionic substances are made up of a giant lattice of positive and negative ions.

⚪ What are ionic substances?

Many substances are made of ions. Ions are electrically charged particles which have a different number of protons (which are positively charged) and electrons (which are negatively charged).

Most compounds made from a combination of metals and non-metals have an ionic structure. For example, sodium chloride is made from sodium (metal) and chlorine (non-metal) and is ionic. Copper sulfate is made from copper (metal), sulfur (non-metal) and oxygen (non-metal) and is also ionic.

⚪ The structure of ionic substances

In substances made of ions, there are lots of positive and negative ions in a giant lattice. A giant lattice contains a massive number of particles in a regular structure that continues in all directions throughout the substance (Figure 2.1).

The positive and negative ions are attracted to each other by electrostatic attraction because opposite charges attract. **Ionic bonding** is the attraction between positive and negative ions. Each ion is attracted to all the ions of opposite charge around it. This attraction is strong and so all ionic substances are solids at room temperature.

Four alternative ways of representing the ions in a lattice are shown in Table 2.1 (using sodium chloride as an example).

Table 2.1

	Dot and cross diagram	2D space-filling structure	3D space-filling structure	Ball and stick structure
Diagram ⊖ Cl⁻ ion ⊕ Na⁺ ion	$[Na]^+$ (2,8) $[\overset{\times\times}{\underset{\times\times}{\times}Cl\overset{\times}{\times}}]^-$ (2,8,8)			The ions are shown with gaps between them Lines are drawn between the ions to show how they are arranged
Advantages of this representation	Shows the electron structure of the ions	Very easy to draw	Gives very good representation of how the ions are packed together	Helps to show how the ions are arranged relative to each other
Disadvantages of this representation	Can give the impression that the structure is made of pairs of ions rather than being a continuous structure containing a massive number of ions	Can give the impression that the structure is limited to a few ions rather than being a continuous structure with a massive number of ions		May make you think there are covalent bonds between the ions (there are **NO** covalent bonds in an ionic lattice) May make you think the ions are a long way apart (but they are packed close together)
		Only shows the structure in 2D		

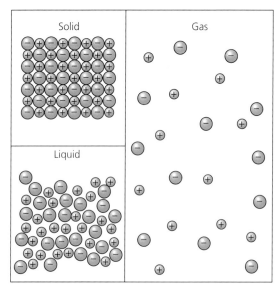

▲ **Figure 2.2** The structure of an ionic substance as a solid, liquid and gas.

◯ **The properties of ionic substances**

Melting and boiling points

In order to melt and boil ionic substances, the strong attraction between the positive and negative ions has to be overcome (Figure 2.2). This is difficult and requires a lot of energy and so ionic substances have high melting and boiling points. For example, sodium chloride melts at 801°C and aluminium oxide melts at 2072°C.

Electrical conductivity

An electric current is the flow of electrically charged particles such as ions or electrons. Ionic substances are made of ions, but as a solid the ions cannot move so they cannot conduct electricity. However, when melted, the ions can move and carry charge, so ionic substances will conduct electricity when molten. Many ionic substances dissolve in water and will conduct electricity if they dissolve because the ions can move (Figure 2.3).

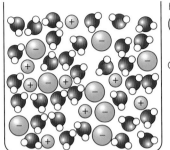

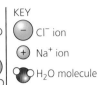

KEY

⊖ Cl⁻ ion

⊕ Na⁺ ion

H₂O molecule

▶ **Figure 2.3** When an ionic substance dissolves in water the ions separate, mix in with the water molecules and move around.

Test yourself

1 What are ions?
2 Explain why ionic substances have high melting points.
3 Explain why ionic substances conduct electricity when molten or dissolved.
4 Explain why ionic substances do not conduct electricity as solids.
5 Ionic substances are made of a giant lattice of ions. What is a giant lattice?
6 Which of the following substances are likely to be ionic: CO_2, PH_3, Fe_2O_3, CH_4O, SiO_2, $MgBr_2$?

The formula of ionic substances

The charge on ions

You can work out the charge on some ions easily. For example, all the elements in

- Group 1 have one electron in their outer shell and so lose one electron when they react forming 1+ ions (e.g. Na^+, K^+).
- Group 2 have two electrons in their outer shell and so lose two electrons when they react forming 2+ ions (e.g. Ca^{2+}, Mg^{2+}).
- Group 6 have six electrons in their outer shell and so gain two electrons when they react forming 2– ions (e.g. O^{2-}, S^{2-}).
- Group 7 have seven electrons in their outer shell and so gain one electron when they react forming 1– ions (e.g. Cl^-, Br^-).

These charges and those of other common ions are shown in the Tables 2.2 and 2.3.

Table 2.2 Positive ions

Group 1 ions (form 1+ ions)		Group 2 ions (form 2+ ions)		Group 3 ions (form 3+ ions)		Others	
Li^+	lithium	Mg^{2+}	magnesium	Al^{3+}	aluminium	NH_4^+	ammonium
Na^+	sodium	Ca^{2+}	calcium			H^+	hydrogen
K^+	potassium	Ba^{2+}	barium			Cu^{2+}	copper(ɪɪ)
						Fe^{2+}	iron(ɪɪ)
						Fe^{3+}	iron(ɪɪɪ)
						Ag^+	silver
						Pb^{2+}	lead
						Zn^{2+}	zinc

Table 2.3 Negative ions

Group 6 ions (form 2– ions)		Group 7 ions (form 1– ions)		Others	
O^{2-}	oxide	F^-	fluoride	CO_3^{2-}	carbonate
S^{2-}	sulfide	Cl^-	chloride	OH^-	hydroxide
		Br^-	bromide	NO_3^-	nitrate
		I^-	iodide	SO_4^{2-}	sulfate

TIP ✓

You need to be able to work out the charge on ions of elements in Groups 1, 2, 6 and 7. You will be provided with the charges on other ions.

What the formula of an ionic substance means

The formula of an ionic substance represents the ratio of the ions in the lattice. For example, in sodium chloride the formula NaCl means that the ratio of sodium (Na^+) ions to (Cl^-) chloride ions in the lattice is 1:1. In aluminium oxide, the formula Al_2O_3 means that the ratio of aluminium (Al^{3+}) ions to oxide (O^{2-}) ions in the lattice is 2:3.

Working out the formula of an ionic substance

In an ionic substance the total number of positive charges must equal the total number of negative charges. This allows us to work out the formula of ionic substances.

TIP ✓

The charge on the nitrate (NO_3^-) ion is 1– not 3–. The charge on the ammonium (NH_4^+) ion is 1+ not 4+.

Some ions contain atoms of different elements. Examples include sulfate (SO_4^{2-}), hydroxide (OH^-) and nitrate (NO_3^-). If you need to write more than one of these in a formula, then those ions should be placed in a bracket (Table 2.4).

Table 2.4

Name	Positive ions		Negative ions		Formula
Sodium chloride	Na^+	(1+ charge)	Cl^-	(1– charge)	NaCl
Magnesium chloride	Mg^{2+}	(2+ charges)	Cl^- Cl^-	(2– charges)	$MgCl_2$
Magnesium sulfide	Mg^{2+}	(2+ charges)	S^{2-}	(2– charges)	MgS
Copper(II) sulfate	Cu^{2+}	(2+ charges)	SO_4^{2-}	(2– charges)	$CuSO_4$
Sodium carbonate	Na^+ Na^+	(2+ charges)	CO_3^{2-}	(2– charges)	Na_2CO_3
Ammonium sulfate	NH_4^+ NH_4^+	(2+ charges)	SO_4^{2-}	(2– charges)	$(NH_4)_2SO_4$
Calcium nitrate	Ca^{2+}	(2+ charges)	NO_3^- NO_3^-	(2– charges)	$Ca(NO_3)_2$
Aluminium oxide	Al^{3+} Al^{3+}	(6+ charges)	O^{2-} O^{2-} O^{2-}	(6– charges)	Al_2O_3
Iron(III) hydroxide	Fe^{3+}	(3+ charges)	OH^- OH^- OH^-	(3– charges)	$Fe(OH)_3$

Test yourself

7 What will be the charge on ions of strontium, astatine, selenium and rubidium?

8 What does the formula K_2O mean?

9 Write the formula of the following substances: sodium sulfide, calcium fluoride, magnesium hydroxide, potassium carbonate, barium nitrate, caesium oxide.

Copy and complete the table to give information about some copper compounds.

Name of compound	Formula of positive ion in compound	Formula of negative ion in compound	Formula of compound	What the numbers in the formula represent
Copper(II) carbonate				
Copper(II) hydroxide		OH⁻		
Copper(II) nitrate	Cu²⁺			
Copper(II) oxide				
Copper(II) sulfate			CuSO₄	1 copper, 1 sulfur, 4 oxygen
Copper(II) sulfide				

TIP ✓

When metal atoms lose electrons they form positive ions. When non-metal atoms gain electrons they form negative ions.

TIP ✓

When atoms gain electrons to form ions, the name changes to end in -ide. For example, oxygen atoms gain electrons to form oxide ions, chlorine atoms gain electrons to form chloride ions, sulfur gains electrons to form sulfide ions.

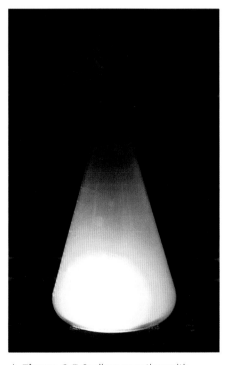

▲ **Figure 2.5** Sodium reacting with chlorine.

The reaction between metals and non-metals

Ionic compounds can be formed when metals react with non-metals. In these reactions, electrons are transferred from the outer shell of the metal atom to the outer shell of the non-metal atom, to produce ions. These ions have the electron structure of the noble gases (Group 0 elements).

For example, when sodium reacts with chlorine, each sodium atoms loses one electron and each chlorine atom gains one electron (Figures 2.4–2.6). This produces sodium ions and chloride ions which have noble gas electron structures.

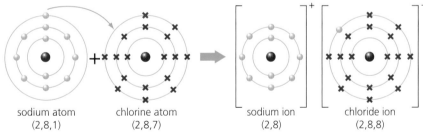

sodium atom (2,8,1)   chlorine atom (2,8,7)   sodium ion (2,8)   chloride ion (2,8,8)

▲ **Figure 2.4** Electron transfer during the formation of sodium chloride.

This can be drawn as a 'dot and cross' diagram which only shows the outer shell electrons. Electrons from one atom are shown as dots (●) and electrons from the other atom as crosses (✗) (Figure 2.6).

(2,8,1)   (2,8,7)   (2,8)   (2,8,8)

▲ **Figure 2.6** Sodium atoms react with chlorine atoms in the ratio 1 : 1 to form NaCl.

Some other examples of these reactions are shown in Figure 2.7.

$$Mg_{(2,8,2)} + F_{(2,7)} + F_{(2,7)} \rightarrow [Mg]^{2+}_{(2,8)} + [F]^{-}_{(2,8)} + [F]^{-}_{(2,8)}$$

(a) Magnesium atoms react with fluorine atoms in the ratio 1:2 to form MgF$_2$.

$$Ca_{(2,8,8,2)} + O_{(2,6)} \rightarrow [Ca]^{2+}_{(2,8,8)} + [O]^{2-}_{(2,8)}$$

(b) Calcium atoms react with oxygen atoms in the ratio 1:1 to form CaO.

$$K_{(2,8,8,1)} + K_{(2,8,8,1)} + S_{(2,8,6)} \rightarrow [K]^{+}_{(2,8,8)} + [K]^{+}_{(2,8,8)} + [S]^{2-}_{(2,8,8)}$$

(c) Potassium atoms react with sulfur atoms in the ratio 2:1 to form K$_2$S.

▲ **Figure 2.7**

Test yourself

10 Explain why ions are formed when metals react with non-metals.
11 Draw a diagram to show what happens in terms of electrons when lithium reacts with oxygen.

Show you can...

a) Explain fully what happens when magnesium atoms react with oxygen atoms to produce magnesium oxide.
b) Explain fully what happens when magnesium atoms react with fluorine atoms to produce magnesium fluoride.
c) Highlight any similarities and differences between the reaction of magnesium with oxygen and the reaction with fluorine.

Molecular substances

What are molecular substances?

Many substances are made of molecules. Molecules are neutral particles made from atoms joined together by covalent bonds. A covalent bond is two shared electrons that join atoms together.

Many substances are made of molecules. Some non-metal elements are made of molecules. The most common ones are listed in the periodic table as shown in Figure 2.8.

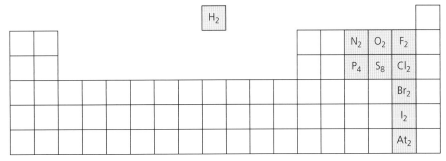

▲ Figure 2.8

Most compounds that are made from a combination of non-metals are made of molecules. For example, water (H_2O) is made from hydrogen (non-metal) and oxygen (non-metal); glucose ($C_6H_{12}O_6$) is made from carbon (non-metal), hydrogen (non-metal) and oxygen (non-metal).

The structure of molecular substances

A molecular substance is made of many identical molecules that are not joined to each other (Figure 2.9). Within each molecule, the atoms are joined together by the very strong covalent bonds. However, the molecules are not bonded to each other. There are only some weak forces between the molecules – these weak forces are intermolecular forces.

Molecules are often quite small, containing just a few atoms, but some substances are made of big molecules (e.g. wax and many polymers).

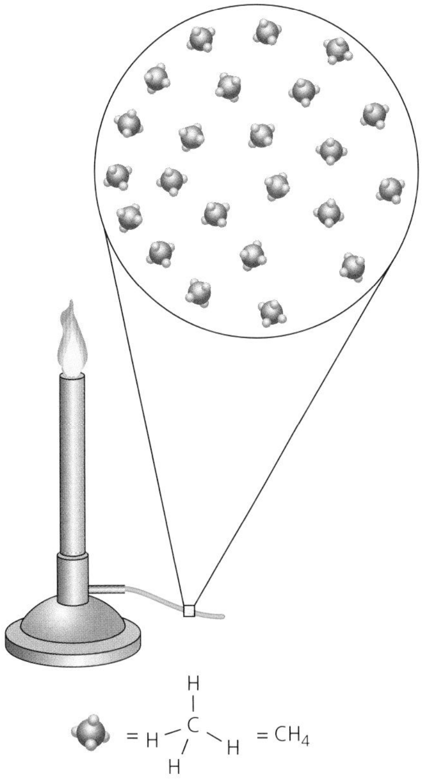

▲ Figure 2.9 In molecular substances such as methane (natural gas, CH_4) there are many separate molecules. The molecules are not bonded to each other.

$$= H - \overset{\overset{\textstyle H}{|}}{\underset{\underset{\textstyle H}{|}}{C}} - H \quad = CH_4$$

The properties of molecular substances

Melting and boiling points

Molecules are not bonded to each other. The intermolecular forces (forces between molecules) are only weak and so are easy to overcome. This means that molecular substances have low melting and boiling points. Many molecular substances with small molecules are gases and liquids at room temperature. For example, methane boils at −162°C and water boils at 100°C.

Generally, the bigger the molecules, the stronger the forces between the molecules and so the higher the melting and boiling points. Molecules of glucose are quite large and it melts at 146°C.

When molecular substances change state, the covalent bonds do **NOT** break. For example, water molecules are identical as H_2O whether it is steam, water or ice (Figure 2.10). No covalent bonds are broken when water changes state.

TIP ✓

The covalent bonds in molecules do NOT break when molecules change state.

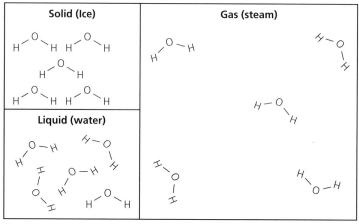

▲ **Figure 2.10** The structure of water, a molecular substance as a solid, liquid and gas. No covalent bonds are broken when it changes state.

Electrical conductivity

Molecules are electrically neutral which means that molecular substances do not conduct electricity at all.

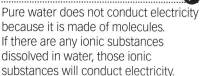

TIP ✓

Pure water does not conduct electricity because it is made of molecules. If there are any ionic substances dissolved in water, those ionic substances will conduct electricity.

Show you can...

Metal oxides and non-metal oxides have different properties. Sulfur dioxide, a non-metal oxide has a melting point of –72°C and calcium oxide, a metal oxide, has a melting point of 2613°C. Explain why the melting point of sulfur dioxide is low but that of calcium oxide is high.

Test yourself

12 What are molecules?

13 What is a covalent bond?

14 Explain why molecular substances have low melting and boiling points.

15 What happens to the covalent bonds in a molecular substance when it melts and boils?

16 Explain why molecular substances do not conduct electricity.

17 Which of the following substances are likely to be molecular: H_2S, Na_2O, KNO_3, $ZnBr_2$, CO, N_2H_4, C_2H_6O?

◯ **The formula of molecular substances**

Molecular substances have two formulae, the empirical formula and the molecular formula. The molecular formula is the one that is normally used.

Table 2.5

	Diagram of molecule	Molecular formula (gives the number of atoms of each element in each molecule)	Empirical formula (gives the simplest ratio of the atoms of each element in the substance)
Butane	H—C—C—C—C—H (with H atoms above and below each C)	**C_4H_{10}** There are 4 C atoms and 10 H atoms in each molecule	**C_2H_5** Simplest ratio of C : H = 2 : 5
Water	H—O—H	**H_2O** There are 2 H atoms and 1 O atom in each molecule	**H_2O** Simplest ratio of H : O = 2 : 1
Glucose	(ring structure of glucose)	**$C_6H_{12}O_6$** There are 6 C atoms, 12 H atoms and 6 O atoms in each molecule	**CH_2O** Simplest ratio of C : H : O = 1 : 2 : 1

Test yourself

18 Benzene has the molecular formula C_6H_6. What does this tell us about benzene?

19 What is the empirical formula of benzene? What does this tell us about benzene?

Show you can...

a) Copy and complete the table:

Molecular formula	Empirical formula
C_2H_6	
$C_{21}H_{22}N_2O_2$	
$C_2H_4O_2$	

b) Are the empirical formula and the molecular formula of a substance always different? Using an example, explain your answer.

○ Drawing molecules

KEY TERM

Covalent bond Two shared electrons joining atoms together.

When atoms join together to form molecules they share electrons in order to obtain noble gas electron structures. A covalent bond is two shared electrons joining atoms together. Table 2.6 shows how many covalent bonds atoms typically form.

Table 2.6

Atoms	H	Group 4 atoms (e.g. C, Si)	Group 5 atoms (e.g. N, P)	Group 6 atoms (e.g. O, S)	Group 7 atoms (e.g. F, Cl, Br, I)
Number of electrons in their outer shell	1	4	5	6	7
Number of electrons needed to obtain a noble gas electron structure	1	4	3	2	1
Number of covalent bonds formed	1	4	3	2	1

There are several ways to show how atoms join together by sharing electrons in covalent bonds to form molecules (Table 2.7).

Table 2.7

Dot and cross diagram showing all electrons and shell circles	Dot and cross diagram showing only outer shell electrons and shell circles	Dot and cross diagram showing only outer shell electrons	Stick diagram	Ball and stick diagram	Space-filling diagram

Note that the ○ and ✕ represent electrons that came from different atoms			Each stick (or line) represents one covalent bond (i.e. 2 shared electrons)	A good representation of a molecule showing how atoms merge into each other, but the covalent bonds are not visible	

When drawing stick diagrams, each atom makes the number of covalent bonds shown in Table 2.6. When drawing dot-cross diagrams, each single covalent bond is made up of two electrons. Atoms can make double and triple covalent bonds (Figure 2.11). A double covalent bond contains four electrons (two from each atom), while a triple bond contains six electrons (three from each atom).

Any outer shell electrons that are not used up in making covalent bonds are found in non-bonding electron pairs, often called lone pairs.

TIP ✓

In molecules, the number of electrons around the outer shell of each atom is usually 8, apart from hydrogen atoms where it is 2 electrons.

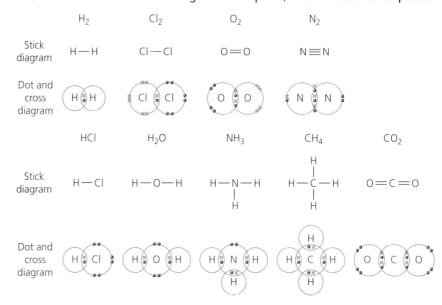

▶ **Figure 2.11**

Test yourself ⚙

20 What does each stick represent in a stick diagram?

21 What do the dots and crosses represent in dot-cross diagrams?

22 Draw a stick diagram and a dot-cross diagram for H_2S.

23 Draw a stick diagram and a dot-cross diagram for CS_2.

Show you can...

Phosphorus bonds with hydrogen to form phosphine. PH_3 is a colourless gas which has an unpleasant, rotting fish odour. Phosphorus also bonds with chlorine to form phosphorus trichloride which is a toxic colourless liquid.

a) Draw a dot and cross diagram to show the bonding in PH_3.
b) Suggest the formula of phosphorus trichloride.
c) Draw a dot and cross diagram to show the bonding in phosphorus trichloride.
d) Using your diagram from c), explain what is meant by a covalent bond and a non-bonding electron pair.

Polymers

There are many different types of polymer (plastics), including polythene, PVC, Perspex, Teflon and polystyrene. Polymers contain very large molecules, often with hundreds or thousands or atoms. Within each molecule, the atoms are joined to each other by covalent bonds (Figure 2.12).

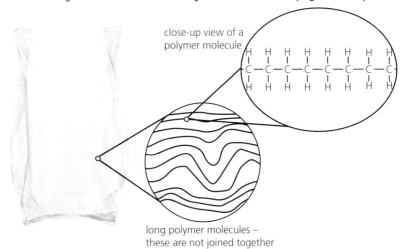

close-up view of a polymer molecule

long polymer molecules – these are not joined together

▲ **Figure 2.12** Polymers contain very large molecules, within which atoms are joined together by covalent bonds.

Thermosoftening polymers soften or melt when heated. In thermosoftening polymers, the molecules are not joined together. However, because the molecules are very large the forces between the polymer molecules are relatively strong. This means that they are solids at room temperature, although they will melt quite easily if heated.

Practical

Testing the electrical conductivity of ionic and molecular covalent substances

To investigate the conduction of electricity by a number of compounds in aqueous solution.

The apparatus was set up as shown in the diagram.

Questions

1 Describe the experimental method which you would use to test the solutions using the apparatus shown.

2 Copy and complete the results table.

3 Using the results from column three and four of the table write a conclusion for this experiment stating and explaining any trends shown in the results.

4 Would the results be different if solid copper(ɪɪ) sulfate was used in place of copper(ɪɪ) sulfate solution? Explain your answer.

5 Predict and explain the results you would obtain for calcium nitrate solution.

6 Predict and explain the results you would obtain for bromine solution.

Test solution	Does the bulb light?	Does the substance conduct electricity?	Does the substance contain ionic or covalent bonding?
Copper(ɪɪ) sulfate	yes		
Ethanol (C_2H_5OH)	no		
Magnesium sulfate	yes		
Potassium iodide	yes		
Glucose ($C_6H_{12}O_6$)	no		
Sodium chloride	yes		

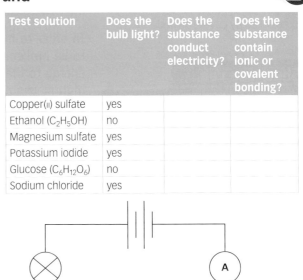

electrodes

test solution

Giant covalent substances

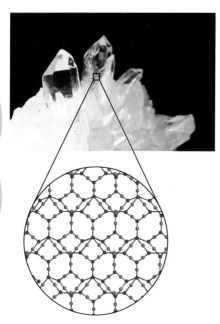

▲ **Figure 2.13** Giant covalent substances are made up of a continuous network of atoms linked by covalent bonds.

Show you can...

Carbon and silicon both form dioxides; carbon dioxide and silicon dioxide. Carbon dioxide is a gas at room temperature and silicon dioxide is a solid with melting point of 1610°C.

a) Copy and complete the table:

Substance	Type of bonding	Structure
Carbon dioxide		
Silicon dioxide		

b) Explain why silicon dioxide is a solid with a high melting point and carbon dioxide is a gas with a low melting point.

○ What are giant covalent substances?

There a few substances that have atoms joined by covalent bonds in a continuous network. Common examples are

- diamond, C (a form of carbon) – studied in detail later in the chapter (Different forms of carbon section)
- graphite, C (a form of carbon) – studied in detail later in the chapter (Different forms of carbon section)
- silicon, Si
- silicon dioxide, SiO_2 (also known as silica).

○ The structure of giant covalent substances

In a giant covalent substance all the atoms are in a giant lattice. They are all joined together by covalent bonds in a continuous network throughout the structure (Figure 2.13).

These substances are **not** molecules. In molecular substances, there are lots of separate molecules with the atoms in each molecule joined by covalent bonds but the molecules are not joined together. In a giant covalent substance, there is one continuous network.

○ The properties of giant covalent substances

Melting and boiling points

In order to melt a giant covalent substance, many covalent bonds have to be broken. Covalent bonds are very strong and so it takes a lot of energy to break them. Therefore, giant covalent substances have very high melting and boiling points. For example, diamond melts at over 3500°C.

Electrical conductivity

Most giant covalent substances do not conduct electricity because they do not contain any delocalised electrons. However, graphite does as it does have some delocalised electrons. Delocalised electrons are able to move throughout the substance.

Test yourself

24 Describe the structure of a giant covalent substance.
25 Why do giant covalent substances have very high melting points?
26 Why do giant covalent substances, except graphite, not conduct electricity?

Metallic substances

⃝ What are metallic substances?

Metals are metallic substances. Over three-quarters of all the elements are metals and have a metallic structure.

⃝ The structure of metallic substances

Metals consist of a giant lattice of atoms arranged in a regular pattern. The outer shell electrons from each atom are delocalised which means they are free to move throughout the whole structure (Figures 2.14 and 2.15).

KEY TERMS

Delocalised Free to move around.

Metallic bonding The attraction between the nucleus of metal atoms and delocalised electrons.

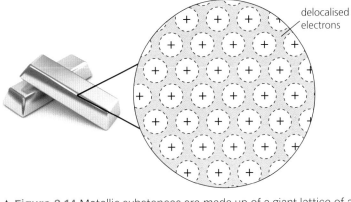

delocalised electrons

▲ **Figure 2.14** Metallic substances are made up of a giant lattice of atoms in a cloud of delocalised electrons.

There is a strong attraction between the positive nucleus of these atoms and the delocalised electrons. This attraction between the nucleus and the delocalised electrons is called metallic bonding.

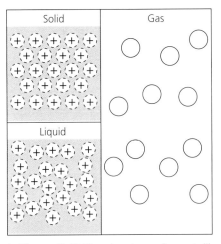

▲ **Figure 2.15** The structure of a metallic substance as a solid, liquid and gas.

⃝ The properties of metallic substances

Melting and boiling points

In metals, the metallic bonding is strong. This means that most metals have high melting and boiling points. For example, aluminium melts at 660°C and iron melts at 1538°C.

Electrical conductivity

Metals are good conductors of electricity because the delocalised electrons are able to move through the structure and carry electrical charge through the metal (Figure 2.16).

▲ **Figure 2.16** Copper is used in electrical cables because it is an excellent electrical conductor, has a high melting point and can be shaped into wires easily.

Thermal conductivity

Metals are also good thermal conductors of heat. The thermal energy is transferred by the delocalised electrons.

Malleability

Metals are **malleable**, which means they can be bent and hammered into shape. This is because the layers of atoms can slide over each other while maintaining the metallic bonding (Figure 2.17). This makes metals soft.

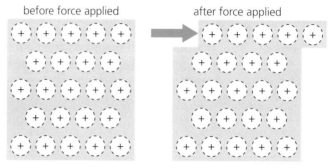

before force applied   after force applied

▶ **Figure 2.17** Metals are soft because atoms can slide over each other.

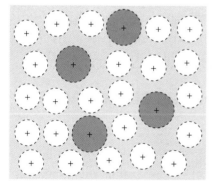

▲ **Figure 2.18** Alloys are harder than pure metals because atoms cannot slide over each other as easily due to the presence of some different sized atoms.

KEY TERMS

Malleable Can be hammered into shape.

Alloy A mixture of a metal with small amounts of other elements, usually other metals.

◯ Alloys

Pure metals are very malleable. This can make them too soft for most uses as they lose their shape easily. Metals can be made more useful by making them into alloys.

An **alloy** is a mixture of a metal with small amounts of other elements, usually other metals. Pure metals such as aluminium, iron, copper and gold are rarely used, and alloys of these metals are used instead. For example, steels are alloys made from iron. Alloys of gold are used for making jewellery as pure gold would lose its shape too easily.

Alloys also have metallic structures. However, some of the atoms in the alloy are a different size to those of the metal. This distorts the layers in the structure and makes it much more difficult for the layers of atoms to slide over each other (Figure 2.18).

Test yourself

27 What is metallic bonding?
28 Why do metals have high melting points?
29 Why do metals conduct electricity?
30 Why are metals malleable?
31 What are alloys?
32 Why are alloys harder than pure metals?

Show you can...

Aluminium is used in overhead electricity cables and to make saucepans.

a) Explain why aluminium is a good conductor of electricity.
b) State two other reasons why aluminium is used in overhead electricity cables.
c) Explain in terms of structure why aluminium is malleable.
d) Aluminium oxide is a compound of aluminium. Compare and contrast how aluminium and aluminium oxide conduct electricity.

Practical

The physical properties of Group 1 elements
The table shows some physical properties of Group 1 elements.

Element	Melting point in °C	Boiling point in °C	Density in g/cm³	Electrical conductivity
Li	180	1340	0.53	Good
Na	98	880	0.97	Good
K	63	766	0.89	Good
Rb	39	686	1.53	Good
Cs	28	669	1.93	Good

Questions
1 Use the table to state a property of Group 1 metals which is common to all metals.
2 Describe the type of bonding found in Group 1 metals.
3 What is meant by the term melting?
4 State the trend in melting point as the group is descended.
5 Write the electron structure of Li, Na and K and state what happens to the distance of the outer electron from the nucleus as the group is descended. Now try and explain the trend in melting point down the group.
6 State the trend in density as the group is descended.
7 Use the data in the table to plot a bar chart to show the density of the different Group 1 metals.
8 State if the Group 1 metals will all float on water (the density of water is 1 g/cm³).

Overview of types of bonding and structures

◯ Types of bonding
There are three types of bonding that are summarised in the Table 2.8.

Table 2.8

	Ionic bonding	Covalent bonding	Metallic bonding
Description	The electrostatic attraction between positive and negative ions	Atoms that are joined together by sharing pairs of electrons	The attraction between positive nucleus of metal atoms and delocalised outer shell electrons
Which substances have this bonding	Ionic compounds	Molecular substances Giant covalent substances	Metallic substances

◯ Types of structure
There are five types of structure that are summarised in the Table 2.9. The elements in Group 0 of the periodic table (the noble gases) have a unique structure called monatomic which is very similar to molecular except that the particles are individual atoms and not molecules.

Table 2.9

	Monatomic	Molecular	Giant covalent	Ionic	Metallic
Which substances have this structure	Group 0 elements	Many non-metal elements Most compounds made from a combination of non-metals	Diamond Graphite Silicon Silicon dioxide	Most compounds made from a combination of metals with non-metals	Metals and alloys
Description of structure	Made up of many separate atoms. The atoms are not bonded to each other. There are very weak forces of attraction between the atoms	Made up of many separate molecules. The atoms within each molecule are joined by covalent bonds. There are no bonds between molecules. There are only weak forces of attraction between the molecules	Made of a giant lattice of atoms joined to each other by covalent bonds	Made of a giant lattice of positive and negative ions. There are strong electrostatic forces of attraction between the positive and negative ions	Made of a giant lattice of metal atoms with a cloud of delocalised outer shell electrons. There are strong forces of attraction between the positive nucleus of the metal atoms and the delocalised electrons
Melting and boiling points	VERY LOW as it is requires little energy to overcome the very weak forces between atoms	LOW as it requires little energy to overcome the weak forces between molecules	VERY HIGH as it requires a lot of energy to break lots of strong covalent bonds	HIGH as it is requires a lot of energy to overcome the strong attraction between the positive and negative ions	HIGH as it requires a lot of energy to overcome the strong attraction between the positive nucleus of the metal atoms and delocalised electrons
Electrical conductivity	NON-CONDUCTOR as atoms are neutral and there are no delocalised electrons	NON-CONDUCTOR as molecules are neutral and there are no delocalised electrons	NON-CONDUCTOR (except graphite) as there are no delocalised electrons (graphite does have delocalised electrons)	Solid = NON-CONDUCTOR as ions cannot move Liquid/solution = CONDUCTOR as ions can move and carry the charge	CONDUCTOR as outer shell electrons are delocalised and can carry charge through the metal

Test yourself

33 For each of the following substances, state the type of bonding and the type of structure it is likely to have:

a) copper(ɪɪ) oxide (CuO)

b) diamond (C)

c) lead carbonate ($PbCO_3$)

d) phosphorus oxide (P_4O_{10})

e) argon (Ar)

f) copper (Cu)

Show you can...

Substances may be classified in terms of their physical properties. Use the table to answer the following questions:

Substance	Melting point in °C	Boiling point in °C	Electrical conductivity as solid	Electrical conductivity as liquid
A	3720	4827	Good	Poor
B	−95	69	Poor	Poor
C	327	1760	Good	Good
D	3550	4827	Poor	Poor
E	801	1413	Poor	Good

a) **Which substance could be sodium chloride? Explain your answer.**

b) **Which substance consists of small covalent molecules? Explain your answer.**

c) **Explain why substance A could not be diamond.**

d) **Which substance is a metal?**

○ States of matter

The three states of matter are solid, liquid and gas (Figure 2.19). Substances change state at their melting and boiling points. A substance is a:

● solid at temperatures below its melting point
● liquid at temperatures between its melting and boiling point
● gas at temperatures above its boiling point.

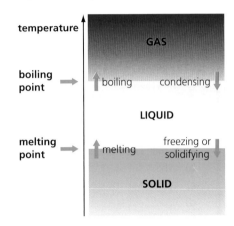

▶ **Figure 2.19** Changes of state.

The amount of energy needed for substances to melt and boil depends on the strength of the forces or bonds between their particles. The stronger the forces or bonds between the particles, the higher their melting and boiling points. For example, giant covalent substances have very high melting and boiling points as strong covalent bonds have to be broken. Molecular substances have low melting and boiling points as there are only weak forces between the molecules that are easy to overcome.

You may have used a very simple model to represent the particles in solids, liquids and gases like the top row in Figure 2.20. It can help to understand how particles are arranged when substances are in each state.

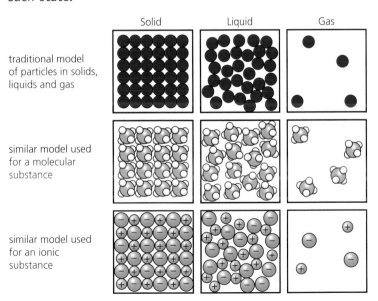

▶ **Figure 2.20**

As you have seen in this chapter, the top row of the table in Figure 2.20 is particularly over-simplistic. The particles in most substances are actually molecules or ions and they are not solid spheres. These diagrams also do not show that there are forces or bonds between the particles. They also cannot show whether the particles are moving around or not.

Test yourself

34 Use the data in the table to answer the questions that follow.

Substance	Melting point in °C	Boiling point in °C
A	45	137
B	595	984
C	−30	56
D	−189	−186
E	186	302

a) Which substance(s) is/are gases at room temperature?

b) Which substance(s) is/are liquids at 100°C?

c) Which substance(s) is/are solids at 100°C?

d) Which substance is a liquid over the widest temperature range?

Show you can...

The melting points and boiling points of six substances are shown in the table.

Substance	Melting point in °C	Boiling point in °C	Type of bonding present
N_2	−210	−196	
CS_2	−112	46	
NH_3	−78	−34	
Br_2	−7	59	covalent
LiCl	605	1137	
Cu	1084	2562	

a) Copy and complete the table by deciding if the bonding in each substance is ionic, covalent or metallic.

b) i) Which element is a solid at room temperature?

   ii) Which compound is a liquid at room temperature?

   iii) Which compound is a gas at room temperature?

   iv) Which element will condense when cooled to room temperature from 100°C?

   v) Which compound will freeze first on cooling from room temperature to a very low temperature?

Nanoscience

What is nanoscience?

Nanoscience is the study of nanoparticles. Nanoparticles are structures that are between 1 nm and 100 nm in size. Nanoparticles typically contain a few hundred atoms. Table 2.10 compares the size of nanoparticles to other particles.

Table 2.10

Type	Atoms and molecules	Nanoparticle	Fine particles (PM$_{2.5}$)	Coarse particles (PM$_{10}$)
Relative size (but not to actual scale)	·	●	●	●
Diameter of structures	0.05–1 nm (5×10^{-11} to 1×10^{-9} m)	1–100 nm (1×10^{-9} to 1×10^{-7} m)	100–2500 nm (1×10^{-7} to 2.5×10^{-6} m)	2500–10000 nm (2.5×10^{-6} to 1×10^{-5} m)
Examples	Carbon atoms, gold atoms, water molecules, carbon dioxide molecules	Carbon nanotubes, gold nanoparticles	Soot	Dust

Nanoparticles

In a normal piece of gold metal there is a huge number of gold atoms in a giant lattice. Only a tiny fraction of the gold atoms are on the surface of the gold.

one gold nanoparticle

Nanoparticles of gold can be made and contain a few hundred gold atoms (Figure 2.21). In the nanoparticles, a much higher fraction of the atoms is on the surface.

Bulk gold metal is very unreactive. However, gold nanoparticles have different properties and have many uses. For example, gold nanoparticles can be used to catalyse some chemical reactions.

Why do nanoparticles behave differently to the bulk material?

Gold nanoparticles are formed as a suspension spread throughout a liquid so that they do not join together

▲ Figure 2.21 Nanoparticles of gold.

In a bulk material, such as a piece of gold, there is one large structure where only a tiny fraction of the atoms are on the surface. However in a nanoparticle, which is much smaller, a much higher fraction of the atoms are on the surface. This explains why nanoparticles have different properties to the bulk material.

Another way to think about this is to look at the surface area to volume ratio of the structure. In Table 2.11 we are comparing a large cube with one that has sides that are ten times shorter in length. As the sides of a cube get shorter by a factor of 10, the surface area to volume ratio increases by a factor of 10. Again this shows that in a smaller structure such as a nanoparticle, a higher fraction of the material is at the surface than in the larger bulk material.

Table 2.11 The surface area to volume ratio of the smaller cube is ten times greater than the bigger cube.

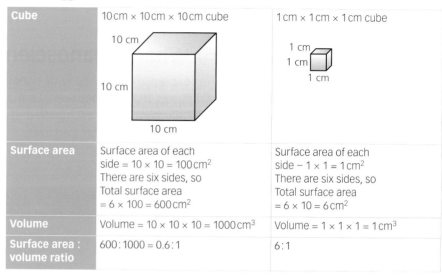

Cube	10 cm × 10 cm × 10 cm cube	1 cm × 1 cm × 1 cm cube
Surface area	Surface area of each side = 10 × 10 = 100 cm² There are six sides, so Total surface area = 6 × 100 = 600 cm²	Surface area of each side − 1 × 1 = 1 cm² There are six sides, so Total surface area = 6 × 10 = 6 cm²
Volume	Volume = 10 × 10 × 10 = 1000 cm³	Volume = 1 × 1 × 1 = 1 cm³
Surface area : volume ratio	600 : 1000 = 0.6 : 1	6 : 1

A much smaller amount of a substance is needed in nanoparticle form than bulk form because a much higher fraction of the atoms are at the surface.

◯ Uses of nanoparticles

There are many uses for nanoparticles and finding more new applications of nanoparticles is a very active and important area of scientific research. Some applications of nanoparticles at present are discussed here.

In fuel cells

Fuel cells are very important and believed to be vital in the years to come as an energy source (Figure 2.22). They use the electrochemical reaction between hydrogen and oxygen to release electrical energy. Most fuels cells use platinum metal as a catalyst for the reaction, but platinum is very expensive. Nanoparticle materials are being used in two ways to lower the cost of the fuel cells. Firstly, platinum nanoparticles can be used as the catalyst, meaning that less platinum is needed. Secondly, nanoparticles of alternative materials are being developed to replace the platinum.

Delivery of drugs

Nanoparticles can be used to deliver drugs to specific cells in the body. This will reduce the amount of the drug needed and reduce any side effects of the drug. Gold nanoparticles are one example of nanoparticles that are being used in this way.

In sun creams

Nanoparticles of titanium dioxide (TiO_2) or zinc oxide (ZnO) are used to absorb harmful UV radiation in sun creams. They give better skin coverage and better protection from UV radiation than normal sun creams, but they are also clear and colourless unlike traditional sun cream (Figure 2.23).

Synthetic skin

Synthetic skin is used to treat patients with severe burns. Nanoparticles (e.g. carbon nanotubes) are being used to create better synthetic skin

▲ **Figure 2.22** Some cars are powered by fuel cells that run on hydrogen as a fuel.

▲ **Figure 2.23** Sun cream made from nanoparticles absorbs harmful UV light but is clear and colourless.

that is stronger but more flexible. Research is even taking place to use nanoparticles to create synthetic skin that senses touch and heat.

Cosmetics

Some examples of the use of nanoparticles in cosmetics include:

- in face creams in emulsions that contain vitamins
- in moisturisers to kill bacteria
- in foundations to diffuse light to partially disguise wrinkles.

Clothing

Some clothes contain silver nanoparticles. Examples include socks, sports clothing, underwear and pyjamas. The silver nanoparticles kill bacteria preventing the build-up of unpleasant odours.

Deodorants

Some deodorants also contain silver nanoparticles which kill bacteria preventing the build-up of unpleasant odours.

Electronics

Nanoparticles are being used to improve electronic components (Figure 2.24). For example, some circuit boards are now printed using nanoparticles and some chips now contain nanoparticles. The use of nanoparticles is allowing smaller components to be made.

◯ The safety of nanoparticles

There are some concerns about the use of nanoparticles. It might have been assumed that if a bulk material is safe to use then nanoparticles of that material would be also. However, if the nanoparticles behave differently to the bulk material, then it is reasonable to assume nanoparticles may well have harmful effects on humans or the environment that the bulk material does not. For example, bulk gold is safe but gold nanoparticles, which have many uses, may well not be safe.

One way in which nanoparticles could be harmful is that some may be able to penetrate cell membranes to enter cells whereas the bulk material may not. There is now some evidence that some nanoparticles may cause problems even though the bulk material does not.

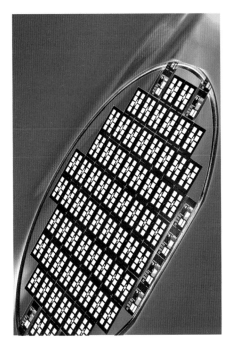

▲ **Figure 2.24** Nanotechnology is being used to make smaller, more powerful silicon chips.

Show you can...

Sun creams should show on the label that they contain nanoparticles. Why should this information be included? To answer this question, pick two statements from the following list:

a) Nanoparticles do not occur in nature.
b) Nanoparticles have a smaller surface area than larger particles.
c) Nanotechnology increases the cost of the sun creams.
d) Not all the effects of nanoparticles are fully understood.
e) Creams containing nanoparticles are easy to apply.
f) Nanoparticles can occur naturally.
g) Nanoparticles may be harmful.

Test yourself

35 Soot from incomplete combustion of fuels is found in the air. Soot is an example of fine particles also referred to as $PM_{2.5}$. Describe what fine particles are.

36 What are nanoparticles?

37 Explain why nanoparticles have different properties to the bulk materials.

38 Show that the surface area to volume ratio of a 2 cm × 2 cm × 2 cm cube is ten times greater than a 20 cm × 20 cm × 20 cm cube.

39 Explain why there are fears about the use of nanoparticles.

The different forms of carbon

There are several different forms of the element carbon which usefully illustrate several parts of this chapter.

○ Diamond

Diamond is probably the best known form of carbon. It has a giant covalent structure with all the carbons joined by covalent bonds in a giant lattice. This can be thought of as a continuous network of atoms linked by covalent bonds (Figure 2.25).

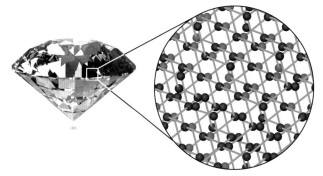

▲ Figure 2.25 Diamond has a giant covalent structure with carbon atoms joined by covalent bonds in a giant lattice.

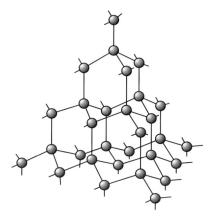

▲ Figure 2.26 A tiny part of the diamond lattice. Each C atom is bonded to four others.

Close examination of the structure of diamond shows that each carbon atom is covalently bonded to four other carbon atoms. Figure 2.26 shows a very small part of the diamond lattice to help see how the carbon atoms are bonded together.

○ Graphite

Graphite is another form of carbon. It is the grey substance that runs through the inside of a pencil and rubs off onto the paper. Like diamond, graphite has a giant covalent structure with all the carbons joined by covalent bonds in a giant lattice (Figure 2.27). However, the carbon atoms are bonded together in flat layers. The layers of atoms are not bonded to each other and there are only very weak attractive forces between these layers of atoms.

Close examination of the structure of graphite shows that each carbon atom is covalently bonded to three other carbon atoms (Figure 2.28).

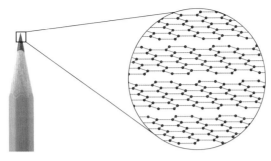

▲ **Figure 2.27** Graphite has a giant covalent structure with carbon atoms joined by covalent bonds within layers in a giant lattice.

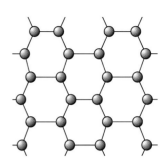

▲ **Figure 2.28** A tiny part of the graphite lattice. Each C atom is bonded to three others.

The diagram shows a very small part of the graphite lattice to help see how the carbon atoms are bonded together.

This bonding leaves one outer shell electron on each carbon atom that is not used in bonding. These electrons become delocalised and are free to move along the layers.

Table 2.12 Comparison of the physical properties of diamond and graphite

	Diamond	Graphite
Melting point	Very high melting point (over 3500°C) because lots of strong covalent bonds need to broken	Very high melting point (over 3500°C) because lots of strong covalent bonds need to be broken
Hardness	Very hard (the hardest natural substance) because the atoms are arranged in a very rigid continuous network held together by strong covalent bonds	Soft because the layers of atoms are not bonded together and so can easily slide over each other
Conductivity	Does not conduct because it contains no delocalised electrons	Conducts because it contains delocalised electrons (one from each atom) that move along the layers and carry charge through the graphite
Uses	Drill and saw tips – due to its hardness	Electrodes – as it conducts electricity In pencils – it rubs off easily onto the paper

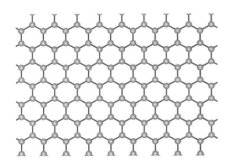

▲ **Figure 2.29** Graphene is a single layer of graphite.

▲ **Figure 2.30** C_{60} has the shape that resembles a football.

Graphene

Graphene is a new substance. It is a single layer of graphite (Figure 2.29). Scientists at the University of Manchester won a Nobel Prize in 2010 for their work on graphene.

Graphene has some remarkable properties. It is extremely thin being just one atom thick, but is extremely strong due to its giant covalent structure. It is also semi see-through as it is so thin. In a similar way to graphite, it is a thermal and electrical conductor due to having some delocalised electrons.

Graphene is a very exciting new material and a lot of research is being done to make use of it. Its properties make it very useful in electronics (e.g. touchscreens) and composite materials (e.g. carbon fibres).

Fullerenes

The molecule C_{60} was identified in the 1980s as another form of carbon. The molecule has a shape that resembles a football (Figure 2.30). It was named buckminsterfullerene after the American architect Richard Buckminster Fuller who built domes that had similar structures. C_{60} is often referred to as a buckyball and was the first fullerene produced. Scientists at the University of Sussex won a Nobel Prize in 1996 for their work on fullerenes.

A whole family of similar molecules, such as C_{70} and C_{84}, have been produced. These molecules are all called **fullerenes**. The structure of fullerenes is based on carbon atoms in hexagonal rings, but some rings have five or seven atoms. They all have a hollow part in the centre of the molecule.

Fullerenes are being used:

- For delivery of drugs into specific parts of the body and/or cells – the drugs are often carried inside the hollow centre of the fullerene molecule.
- In lubricants to reduce friction when metal parts of machines move past each other – the spherical shape of the molecule allows molecules to roll past each other.
- As catalysts – a lot of research is taking place into the use of fullerenes as catalysts and a wide range of potential applications are being found.

○ **Carbon nanotubes**

Carbon nanotubes are cylindrical fullerenes, sometimes called buckytubes (Figure 2.31). They have very high length to diameter ratios, significantly higher than for any other material. They can also be thought of as being tubes of graphene sheets.

These carbon nanotubes have some excellent properties making them very useful. They have:

- High tensile strength – in other words it is very strong when it is pulled – this is due to the many strong covalent bonds throughout its structure.
- High thermal and electrical conductance – this is due to some of the electrons being delocalised.

Carbon nanotubes have many uses, for example to reinforce the materials used to make sports equipment like tennis racquets (Figure 2.32) and golf clubs.

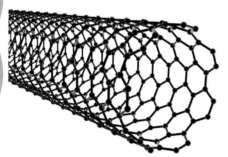

▲ **Figure 2.31** A carbon nanotube.

▲ **Figure 2.32** Carbon nanotubes are very strong and are used to reinforce tennis racquets.

Test yourself

40 Explain why diamond and graphite both have very high melting points.
41 Explain why diamond is hard but graphite is soft.
42 Explain why graphite conducts electricity but diamond does not.
43 What are fullerenes?
44 Carbon nanotubes are used to reinforce and strengthen tennis racquets. Explain why carbon nanotubes strengthen materials.
45 A typical carbon nanotube is 12 cm long and has a diameter of 1 nm. Calculate the length to diameter ratio of this carbon nanotube.

Show you can...

Copy and complete the table to give information about the different structures and uses of carbon.

	Graphite	Diamond	Graphene	Fullerenes	Carbon nanotubes
Description of structure					
Example of use					

Chapter review questions

1   Four structure types are: ionic, metallic, molecular and giant covalent.

   a) Which of these structure types usually have low melting and boiling points?

   b) Which of these structure types usually conduct electricity as solids?

   c) Which of these structure types usually conduct electricity when melted?

2   Carbon dioxide is a molecular compound with the formula CO_2.

   a) Explain why carbon dioxide has a low boiling point.

   b) Explain why carbon dioxide does not conduct electricity.

3   Iron is a metal with the formula Fe.

   a) Explain why iron has a high melting point.

   b) Explain why iron conducts electricity.

   c) Explain why pure iron is soft.

   d) Steels are alloys of iron. Explain why steels are harder than pure iron.

4   Potassium fluoride is an ionic compound containing potassium (K^+) ions and fluoride (F^-) ions.

   a) Give the electron structure of potassium (K^+) ions.

   b) Give the electron structure of fluoride (F^-) ions.

   c) Give the formula of potassium fluoride.

   d) Explain why potassium fluoride has a high melting point.

   e) Explain why potassium fluoride does not conduct electricity as a solid.

   f) Explain why potassium fluoride conducts electricity when molten or dissolved.

5   Decide whether each of the following substances has an ionic, molecular, giant covalent or metallic structure.

   a) zinc (Zn)

   b) ethane (C_2H_6)

   c) diamond (C)

   d) magnesium oxide (MgO)

   e) iodine trifluoride (IF_3)

   f) potassium carbonate (K_2CO_3)

6   Decide whether each of the following substances has an ionic, molecular, giant covalent or metallic structure.

Substance	Melting point in °C	Boiling point in °C	Electrical conductivity as solid	Electrical conductivity as liquid
A	838	1239	Does not conduct	Conducts
B	89	236	Does not conduct	Does not conduct
C	678	935	Conducts	Conducts
D	1056	1438	Does not conduct	Conducts
E	2850	3850	Does not conduct	Does not conduct
F	−39	357	Conducts	Conducts

7 Draw a stick and a dot-cross diagram for each of the following molecules.

   a) phosphine (PH_3)

   b) bromine (Br_2)

   c) carbon dioxide (CO_2)

8 Calcium reacts with chlorine to form the ionic compound calcium chloride. Draw diagrams to show the electron structure in calcium atoms, chlorine atoms, calcium ions and chloride ions.

9 Work out the formula of the following ionic compounds. The charge of some ions is given (sulfate = SO_4^{2-}, hydroxide = OH^-)

   a) potassium oxide

   b) magnesium fluoride

   c) lithium sulfide

   d) iron(III) sulfate

   e) copper(II) hydroxide

10 Poly(propene) is a polymer made of molecules which typically melts at 130°C.

   a) Describe the bonding within the polymer molecules.

   b) Explain why poly(propene) has a relatively high melting point for a molecular substance.

11 Gold nanoparticles have different properties to bulk gold.

   a) What are nanoparticles?

   b) Explain why nanoparticles often have different properties from the bulk material.

   c) Why are some people concerned about the use of nanoparticles?

   d) Calculate how much smaller the surface area to volume is of a cube that measures 10 cm × 10 cm × 10 cm compared to one that has sides that are each half that length.

12 Ethyne is a molecule with the molecular formula C_2H_2.

   a) Explain what this formula tells us about ethyne.

   b) Draw a stick diagram to show the covalent bonds in a molecule of ethyne.

   c) Draw a dot-cross diagram to show the outer shell electrons in a molecule of ethyne.

13 The diagram shows part of a carbon nanotube. They are used to reinforce the materials used to make tennis racquets as they have high tensile strength.

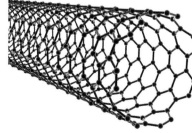

   a) How many covalent bonds does each carbon atom make in a carbon nanotube?

   b) Explain clearly by considering your answer to (a) why nanotubes can conduct electricity.

   c) Explain clearly why carbon nanotubes have high tensile strength.

Practice questions

1 The elements P and Q are in found in Groups 2 and 7 respectively, of the periodic table. Which one of the following shows the formula and the bonding type of the compound that they form? [1 mark]

   A PQ_2 covalent   B PQ_2 ionic

   C P_2Q covalent   D P_2Q ionic

2 Which one of the following does not have a giant covalent structure? [1 mark]

   A diamond   B graphene

   C graphite   D sulfur dioxide

3 A dot and cross diagram is given here:

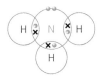

  a) Name and write the formula for this compound. [2 marks]

  b) On a copy of the diagram above, use an arrow to label:

     i) A covalent bond.

     ii) A non-bonded pair of electrons. [2 marks]

  c) Using a line to represent a single covalent bond, redraw the diagram shown above. [1 mark]

  d) What is meant by the term single covalent bond? [2 marks]

4 Nanoscience is the study of nanoparticles.

  a) What is the size of particles studied in nanoscience? [1 mark]

  b) Nanoparticles can be added to other materials. Adding nanoparticles changes the properties of these materials. Describe two examples of products, other than skincare creams, that have nanoparticles added to them. Explain how adding nanoparticles changes the properties of these products, and suggest why this is useful. [6 marks]

5 The following diagram shows some changes between the states of matter.

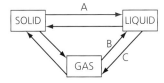

  a) What is the name for each of the changes labelled A, B and C? [3 marks]

  b) Which of the changes A, B, or C is achieved by a decrease in temperature? [1 mark]

  c) Draw a diagram to represent the arrangement of atoms in a solid using small solid spheres to represent atoms. [1 mark]

  d) Explain the limitations of this simple particle theory in relation to change in state. [2 marks]

  e) Explain why the melting points of some solids are greater than others. [2 marks]

  f) Describe the limitations of the simple particle theory of states of matter. [2 marks]

6 The diagrams show two different structures of the Group 4 element carbon and two compounds of Group 4 elements.

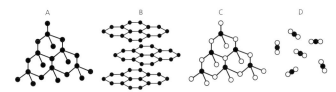

  a) Which two of the diagrams represent different structures of the element carbon? [1 mark]

  b) The substances are carbon dioxide, diamond, graphite and silicon dioxide. Name each substance, A to D. [4 marks]

  c) Name the type of bonding which occurs between the atoms in all of the substances A to D. [4 marks]

  d) Name the type of structure for each substance A to D. [4 marks]

7 The photographs show two uses of graphite. It is used in pencils and for electrodes in the electrolysis of sodium chloride solution.

  a) With reference to the structure of graphite, explain why it is used in pencils. [3 marks]

  b) Graphite electrodes conduct electricity. Explain why graphite is a good conductor of electricity. [2 marks]

c) Describe, as fully as you can, what happens when sodium atoms react with chlorine atoms to produce sodium chloride. You may use a diagram in your answer. [3 marks]

d) Explain why sodium chloride solution will conduct electricity, but sodium chloride solid will not. [2 marks]

8 The table gives some of the properties of the Period 3 element magnesium and one of its compounds, magnesium chloride.

Property	Magnesium	Magnesium chloride
Melting point in °C	649	714
Electrical conductivity when solid	Conducts	Does not conduct
Electrical conductivity when molten	Conducts	Conducts

Use ideas about structure and bonding to explain the similarities and differences between the properties of magnesium and magnesium chloride. [6 marks]

9 a) Chlorine is a green gas which exists as diatomic molecules.

i) Suggest what is meant by the term diatomic. [1 mark]

ii) Use a dot and cross diagram to clearly show how atoms of chlorine combine to form chlorine molecules. [2 marks]

b) Chlorine can form a range of compounds with both metals and non-metals.

i) Describe, as fully as you can, what happens when calcium atoms react with chlorine atoms to produce calcium chloride. You may use a diagram in your answer. [3 marks]

ii) Name the type of bonding found in calcium chloride. [1 mark]

iii) Name the type of structure for calcium chloride. [1 mark]

iv) Use a dot and cross diagram to show how atoms of chlorine combine with atoms of carbon to form tetrachloromethane CCl_4. [2 marks]

v) Name the type of bonding found in CCl_4. [1 mark]

c) The properties of compounds depend very closely on their bonding. Redraw the following table with only the correct words to show some of the properties of calcium chloride and tetrachloromethane. [2 marks]

Compound	Solubility in water	Relative melting point
Calcium chloride	Soluble/insoluble	Low/high
Tetrachloromethane	Soluble/insoluble	Low/high

d) The bonding in the elements calcium and carbon is very different. Describe the bonding in calcium and in carbon (in the form of graphite). [6 marks]

e) Both calcium and graphite can conduct electricity. State two properties of calcium which are different from those of graphite. [2 marks]

f) Why might fullerenes be used in new drug delivery systems? Choose the correct statement A, B or C. [1 mark]

A They are made from carbon atoms.

B They are hollow.

C They are very strong.

g) How does the structure of nanotubes make them suitable as catalysts? Choose the correct statement A, B or C. [1 mark]

A They have a large surface area to volume ratio.

B They are made from reactive carbon atoms.

C They have strong covalent bonds.

Working scientifically:
Units: Using prefixes and powers of ten for orders of magnitude

Standard form

Standard form is used to express very large or very small numbers so that they are more easily understood and managed. It is easier to say that a speck of dust has a mass of 1.2×10^{-6} grams than to say it has a mass of 0.0000012 grams. It uses powers of ten.

Standard form must always look like this:

A must always be between 1 and 10

n is the number of places the decimal point moves (for numbers less than 1, the value of n will be negative)

$$A \times 10^n$$

▲ **Figure 2.33** The mass of the Earth is $5\,972\,200\,000\,000\,000\,000\,000\,000$ kg. This is more conveniently written in scientific form as 5.9722×10^{24} kg.

Example

Write $4\,600\,000$ in standard form.

Answer

- Write the non-zero digits with a decimal place after the first number and then write × 10 after it:

  4.6×10

- Then count how many places the decimal point has moved and write this as the *n* value. The *n* value is positive as $4\,600\,000$ is greater than 1.

  $4\,600\,000 = 4.6 \times 10^6$

Example

Write 0.000345 in standard form.

Answer

- Write the non-zero digits with a decimal place after the first number and then write × 10 after it:

  3.45×10

- Then count how many places the decimal point has moved and write this as the *n* value. The *n* value is negative as 0.000345 is less than 1.

  $0.000345 = 3.45 \times 10^{-4}$

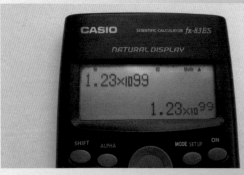

▲ Figure 2.34 Make sure that you are familiar with how standard form is presented on your calculator. This calculator reads 1.23×10^{99} to 3 significant figures.

SI units

The International System of Units (SI) is a system of units of measurements that is widely used all over the world, and the one which you will use in your study of chemistry. It uses several base units of measure, for example metres, grams and seconds. When a numerical unit is very small or large, the units may be modified by using a **prefix**. Some prefixes are shown in the table.

Prefix name	Prefix symbol	Scientific notation 10^n	Decimal
tera	T	10^{12}	1 000 000 000 000
giga	G	10^9	1 000 000 000
mega	M	10^6	1 000 000
kilo	k	10^3	1000
centi	c	10^{-2}	0.01
milli	m	10^{-3}	0.001
micro	μ	10^{-6}	0.000 001
nano	n	10^{-9}	0.000 000 001

A prefix goes in front of a basic unit of measure to indicate a multiple of the unit. For example instead of writing 1000 grams we can add the prefix kilo (10^3) and write 1 kg.

For example using scientific notation:

▶ 1 cm = 1 centimetre = 1×10^{-2} m = 0.01 m

▶ 1 μl = 1 microlitre = 1×10^{-6} l = 0.000 001 l

▶ 1 ns = 1 nanosecond = 1×10^{-9} s = 0.000 000 001 s

▶ 1 μm = 1 micrometre = 1×10^{-6} m = 0.000 001 m

Questions

1 Write the numbers below in standard form.
   a) 0.000 24
   b) 3 230 000 000
   c) 0.02
   d) 0.000 000 007
   e) 24 000
2 Write the numbers below as decimals.
   a) 2.3×10^{-3}
   b) 4.6×10^5
   c) 9.5×10^{-5}
   d) 5.34×10^4
   e) 3.3×10^3

Questions

3 Each prefix has a symbol. Name the units represented by
   a) μs
   b) Mm
   c) mg
   d) nm
   e) Gs

Questions

4 Write the following quantities in units with the appropriate prefixes.
   a) 31 000 000 m
   b) 0.001 g
   c) 9700 m
   d) 0.000 000 002 s
5 a) Atoms are very small with a radius of 0.1 nm. What is this in metres?
   b) The radius of a nucleus is less than 1/10000 of that of an atom. What is this value in nanometres and metres?
6 a) Fine particles have a diameter of between 100 nm and 2500 nm. What are these values in metres?
   b) A coarse particle has a diameter of 2.5×10^{-6} m. Write this using a prefix.

3 Quantitative chemistry

Water companies regularly analyse samples of water supplied to homes to check that it is safe to drink. Their analytical chemists have to be able to carry out very accurate experiments, including titrations, to analyse the water. They also need to be able to carry out calculations using their results to work out how much of each substance is in the water. In this chapter you will learn how to perform a range of calculations and how to carry out titrations.

Specification coverage

This chapter covers specification points 4.3.1 to 4.3.5 and is called Quantitative chemistry. It also covers how to carry out a titration from 4.4.2.

It covers relative mass and moles, conservation of mass, reacting masses, yield, atom economy, gas volumes and the concentration of solutions.

Related work on writing and balancing equations can be found in the Appendix.

Prior knowledge

Previously you could have learned:

> Atoms are tiny.
> Nearly all of the mass of an atom is from the mass of the protons and neutrons.
> The mass number of an atom is the number of protons plus the number of neutrons in an atom.
> Mass is conserved in chemical reactions.
> Balanced equations show the ratios in which substances react.

Test yourself on prior knowledge

1 What does the following equation tell you about the reaction:
$CH_4 + 2O_2 \rightarrow CO_2 + 2H_2O$
2 What is the principle of conservation of mass?
3 In a reaction, hydrogen reacts with oxygen to make water and nothing else. What mass of water will be made if 4 g of hydrogen reacts with 32 g of oxygen?
4 How many protons, neutrons and electrons are there in an atom of $^{39}_{19}K$?

Relative mass and moles

○ Relative atomic mass

Individual atoms have a tiny mass. For example, an atom of ^{12}C has a mass of about 2×10^{-23} g (that is 0.000 000 000 000 000 000 000 02 g). As numbers like this are awkward to use, scientists measure the mass of atoms relative to each other. They use a scale where the mass of a ^{12}C atom is defined as being exactly 12. On this scale, an atom of ^{24}Mg has a relative mass of 24 and is twice as heavy as a ^{12}C atom, whereas an atom of ^{1}H has a relative mass of 1 and is 12 times lighter than a ^{12}C atom. In effect, the relative mass of a single atom equals the mass number of that atom (mass number is the number of protons plus the number of neutrons).

Many elements are made up of a mixture of atoms of different isotopes. For example, 75% of chlorine atoms are ^{35}Cl with relative mass 35 and the remaining 25% are ^{37}Cl atoms with relative mass 37. The average relative mass of chlorine atoms is 35.5 as there are more chlorine atoms with relative mass 35 than 37.

The relative atomic mass (A_r) of an element is the average mass of atoms of that element taking into account the mass and amount of each isotope it contains on a scale where the mass of a ^{12}C atom is 12.

○ Relative formula mass

The relative formula mass (M_r) of a substance is the sum of the relative atomic masses of all the atoms shown in the formula. It is often just called *formula mass*.

▲ **Figure 3.1** The elephant has a relative mass of 500 compared to the child. This means that the elephant is 500 times heavier than the child.

For example, the formula of water is H_2O and so the relative formula mass is the sum of the relative atomic mass of two hydrogen atoms (2 × 1) and one oxygen atom (16) which adds up to 18. This and other examples are shown in Table 3.1.

Table 3.1

Name	Formula	A_r values	Sum	M_r
Water	H_2O	H = 1, O = 16	2(1) + 16 =	18
Copper	Cu	Cu = 63.5		63.5
Sodium chloride	NaCl	Na = 23, Cl = 35.5	23 + 35.5 =	58.5
Sulfuric acid	H_2SO_4	H = 1, S = 32, O = 16	2(1) + 32 + 4(16) =	98
Magnesium nitrate	$Mg(NO_3)_2$	Mg = 24, N = 14, O = 16	24 + 2(14) + 6(16) =	148
Ammonium sulfate	$(NH_4)_2SO_4$	N = 14, H = 1, S = 32, O = 16	2(14) + 8(1) + 32 + 4(16) =	132

Show you can...

Many compounds, for example carbon dioxide and calcium nitrate contain the element oxygen.

a) The formula of carbon dioxide is CO_2. How many oxygen atoms are present in one molecule? What is the relative formula mass of carbon dioxide?

b) Calcium nitrate is an ionic compound. Use ion charges to write the formula of calcium nitrate. How many oxygen atoms are shown in the formula? What is the relative formula mass of calcium nitrate?

c) A compound containing oxygen has the formula OX_2 and a relative formula mass of 54. Calculate the relative atomic mass of X and use your periodic table to identify X.

Test yourself

1 Calculate the relative formula mass of the following substances. You can find relative atomic masses on the periodic table at the back of this book.
a) ammonia, NH_3
b) nickel, Ni
c) butane, C_4H_{10}
d) calcium hydroxide, $Ca(OH)_2$
e) aluminium nitrate, $Al(NO_3)_3$

○ The mole

What is a mole?

Amounts of chemicals are often measured in moles. This makes it much easier to work out how much of a chemical is needed in a reaction.

One atom of ^{12}C has a relative mass of 12. The mass of 602 000 000 000 000 000 000 000 (6.02 × 10^{23}) atoms of ^{12}C is exactly 12 g. The number 6.02 × 10^{23} is a very special number and is known as the Avogadro constant.

When you have a pair of ^{12}C atoms, you have two of them. When you have a dozen ^{12}C atoms, you have 12 of them. When you have a mole of ^{12}C atoms, you have 6.02 × 10^{23} of them.

One mole of a substance contains the same number of the stated particles (atoms, molecules or ions) as one mole of any other substance. For example, the number of carbon atoms in one mole of carbon atoms (6.02 × 10^{23} atoms) is the same number of particles as there are molecules in one mole of water molecules (6.02 × 10^{23} molecules).

The value of the Avogadro constant was chosen so that the mass of one mole of that substance is equal to the relative formula mass (M_r) in grams. Table 3.2 shows some examples.

Table 3.2

Name	Formula	Relative formula mass (M_r)	Mass of 1 mole of that substance
Water	H_2O	18	18 g
Copper	Cu	63.5	63.5 g
Sodium chloride	NaCl	58.5	58.5 g
Sulfuric acid	H_2SO_4	98	98 g
Magnesium nitrate	$Mg(NO_3)_2$	148	148 g
Ammonium sulfate	$(NH_4)_2SO_4$	132	132 g

How many moles?

If one mole of ^{12}C atoms has a mass of 12 g, then it follows that the mass of two moles of ^{12}C atoms will be 24 g. There is a simple equation linking the mass of a substance to the number of moles (Figure 3.2).

When using this equation, the mass must be in grams. Table 3.3 shows conversion factors if the masses are not given in grams.

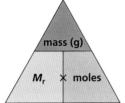

$$\text{mass (g)} = M_r \times \text{moles}$$

This formula triangle may help you. Cover up the quantity you want to show the equation you need to use.

Thinking of **Mr Moles** with a mass on his head may help you remember this triangle

▲ Figure 3.2

Table 3.3

Conversion factor	Example
1 tonne = 1 000 000 g	3 tonnes = 3 × 1 000 000 = 3 000 000 g
1 kg = 1000 g	0.5 kg = 0.5 × 1000 = 500 g
1 mg = 0.001 g	20 mg = 20 × 0.001 = 0.020 g

Table 3.4 gives some examples using the equation that links mass and moles. In calculations, the units of moles is usually abbreviated to mol.

Table 3.4

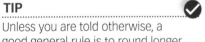

How many moles in each of the following?	180 g of H_2O	M_r of H_2O = 2(1) + 16 = 18 Moles $H_2O = \dfrac{\text{mass}}{M_r} = \dfrac{180}{18} = 10$ mol
	4 g of CH_4	M_r of CH_4 = 12 + 4(1) = 16 Moles $CH_4 = \dfrac{\text{mass}}{M_r} = \dfrac{4}{16} = 0.25$ mol
	2 kg of Fe_2O_3	M_r of Fe_2O_3 = 2(56) + 3(16) = 160 Mass of Fe_2O_3 = 2 kg = 2 × 1000 g = 2000 g Moles $Fe_2O_3 = \dfrac{\text{mass}}{M_r} = \dfrac{2000}{160} = 12.5$ mol
	50 mg of NaOH	M_r of NaOH = 23 + 16 + 1 = 40 Mass of NaOH = 50 mg = $\dfrac{50}{1000}$ g = 0.050 g Moles NaOH = $\dfrac{\text{mass}}{M_r} = \dfrac{0.050}{40} = 0.00125$ mol
What is the mass of each of the following?	20 moles of CO_2	M_r of CO_2 = 12 + 2(16) = 44 Moles $CO_2 = M_r \times$ moles = 44 × 20 = 880 g
	0.025 moles of Cl_2	M_r of Cl_2 = 2(35.5) = 71 Moles $Cl_2 = M_r \times$ moles = 71 × 0.025 = 1.78 g
What is the M_r of the following?	3.6 g of a substance is found to contain 0.020 mol	$M_r = \dfrac{\text{mass}}{\text{moles}} = \dfrac{3.6}{0.020} = 180$

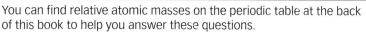

Show you can...

Use Figure 3.2 to write three different equations linking mass, moles and M_r. Use these equations to answer the questions below.

a) Calculate the number of moles in 9.8 g of H_2SO_4.
b) Calculate the mass in grams of 0.5 moles of $Ca(OH)_2$.
c) 6.9 g of a substance Y_2CO_3 contains 0.05 moles. What is the relative formula mass of the substance? Identify Y.

Test yourself

You can find relative atomic masses on the periodic table at the back of this book to help you answer these questions.

2 What is the mass of one mole of the following substances?
   a) iron, Fe
   b) oxygen, O_2
   c) ethane, C_2H_6
   d) potassium chloride, KCl
   e) calcium nitrate, $Ca(NO_3)_2$

3 Which one of the substances in each of the following pairs of substances contains the most particles or are they the same?
   a) 2 moles of water (H_2O) molecules and 2 moles of carbon dioxide (CO_2) molecules
   b) 10 moles of methane (CH_4) molecules and 10 moles of argon atoms (Ar)
   c) 10 moles of helium atoms (He) and 5 moles of oxygen molecules (O_2)

4 What are the following masses in grams?
   a) 20 kg
   b) 5 mg
   c) 0.3 tonnes

5 What is the mass of each of the following?
   a) 3 moles of oxygen, O_2
   b) 0.10 moles of ethanol, C_2H_5OH

6 How many moles are there in each of the following?
   a) 50 g of hydrogen, H_2
   b) 4 kg of calcium carbonate, $CaCO_3$
   c) 80 mg of bromine, Br_2
   d) 2 tonnes of calcium oxide, CaO

7 0.300 g of a substance was analysed and found to contain 0.0050 moles. Calculate the M_r of the substance.

○ Significant figures

We often quote answers to calculations to a certain number of significant figures. In chemistry, we usually give values to 2, 3 or 4 significant figures (sf), but it can be more or less than this. Table 3.5 shows some numbers given to 2, 3 and 4 significant figures.

We quote values to a limited number of significant figures because we cannot be sure of the exact value to a greater number of significant figures.

For example, if we measure the temperature rise in a reaction three times and find the values to be 21 °C, 21 °C and 22 °C, then the mean temperature rise shown on a calculator would be 21.333 333 33 °C. However, it is impossible for us to say that the temperature rise is exactly 21.333 333 33 °C as the thermometer could only measure to ±1 °C. Therefore, we should quote the temperature rise to 2 significant figures, i.e. 21 °C, which is the same number of significant figures as the values we measured.

Table 3.5

Number	2 sf	3 sf	4 sf
2.7358	2.7	2.74	2.736
604531	600000	605000	604500
0.108 36	0.11	0.108	0.108 4
0.004 298 1	0.0043	0.004 30	0.004 298

Test yourself

8 Give the following numbers to 2, 3 and 4 significant figures.
   a) 34.822 6
   b) 28 554 210
   c) 0.023 187 6
   d) 0.000 631 947

9 Find the mean of these measurements and give your answer to 3 significant figures.
   a) 25.4 cm³, 25.1 cm³, 25.3 cm³
   b) 162 s, 175 s, 169 s, 173 s
   c) 1.65 g, 1.70 g, 1.69 g, 1.64 g, 1.71 g

Conservation of mass

N₂ + 3H₂ → 2NH₃

1 molecule of N₂	3 molecules of H₂	2 molecules of NH₃
2 N atoms	6 H atoms	2 N atoms & 6 H atoms

▲ **Figure 3.3** There are 2 N and 6 H atoms on both sides of this equation.

KEY TERM ⭐

Conservation of mass In a reaction, the total mass of the reactants must equal the total mass of the products.

$$N_2 \quad + \quad H_2 \quad M_r = 2 \quad \to \quad NH_3 \quad M_r = 17$$
$$M_r = 28 \qquad H_2 \quad M_r = 2 \qquad\qquad NH_3 \quad M_r = 17$$
$$\qquad\qquad\qquad H_2 \quad M_r = 2$$

Sum of M_r of all reactants = 28 + 2 + 2 + 2 = 34	Sum of M_r of all products = 17 + 17 = 34

▲ **Figure 3.4** Conservation of mass: the mass of the products must equal the mass of the reactants.

○ Balanced equations

In a balanced chemical equation, the number of atoms of each element is the same on both sides of the equation. This is because atoms cannot be created or destroyed in chemical reactions. This means that you have the same atoms before and after the reaction, although how they are bonded to each other changes during the reaction.

For example, in Figure 3.3 one molecule of nitrogen (N_2) reacts with three molecules of hydrogen (H_2) to make two molecules of ammonia (NH_3). There are two N atoms and six H atoms on both sides of the equation.

○ Conservation of mass

As there are the same atoms present at the start and end of a chemical reaction, mass must be conserved in the reaction. In other words, the total mass of the reactants must equal the total mass of the products. This is known as the law of **conservation of mass**.

One way to look at this is using relative formula masses. The total of the relative formula masses of all the reactants in the quantities shown in the equation will add up to the total of the relative formula masses of all the products in the quantities shown in the equation (Figure 3.4).

All chemical reactions obey the law of conservation of mass. There are some reactions that may appear to break this law, but they do not as shown below.

○ Reaction of metals with oxygen

When metals react with oxygen the mass of the product is greater than the mass of the original metal. However, this does not break the law of conservation of mass. The 'extra' mass is the mass of the oxygen from the air that has combined with the metal to form a metal oxide.

In the example shown in Figure 3.5, some magnesium has been heated in a crucible. It may appear as though the products are 0.16 g heavier than the reactants, but 0.16 g of oxygen has reacted with the 0.24 g magnesium to make 0.40 g magnesium oxide.

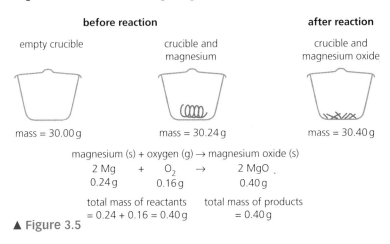

before reaction		after reaction
empty crucible	crucible and magnesium	crucible and magnesium oxide
mass = 30.00 g	mass = 30.24 g	mass = 30.40 g

magnesium (s) + oxygen (g) → magnesium oxide (s)

2 Mg	+	O₂	→	2 MgO
0.24 g		0.16 g		0.40 g

total mass of reactants = 0.24 + 0.16 = 0.40 g

total mass of products = 0.40 g

▲ **Figure 3.5**

○ Thermal decomposition reactions

TIP ✓
Metal carbonates decompose into a metal oxide and carbon dioxide when heated.

A thermal decomposition reaction is one where heat causes a substance to break down into simpler substances. In decomposition reactions, one or more of the products may escape from the reaction container into the air as a gas. These reactions may appear to lose mass, but the "missing" mass is the mass of the gas that has escaped into the air.

In the example shown in Figure 3.6, some copper carbonate is heated in a crucible. It may appear as though the products are 0.22 g lighter than the reactants, but 0.22 g of carbon dioxide has been released into the air.

before reaction

empty crucible

mass = 30.00 g

crucible and copper carbonate

mass = 30.62 g

after reaction

crucible and copper oxide

mass = 30.40 g

copper carbonate (s) → copper oxide (s) + carbon dioxide (g)

$$CuCO_3 \rightarrow CuO + CO_2$$
$$0.62\,g \quad\quad 0.40\,g \quad 0.22\,g$$

total mass of reactants = 0.62 g

total mass of products = 0.40 + 0.22 = 0.62 g

▲ Figure 3.6

Practical

Oxidation of titanium

The metal titanium reacts with oxygen to form an oxide of titanium. In an experiment a sample of titanium metal was heated in a crucible with a lid. During heating the lid was lifted from time to time.

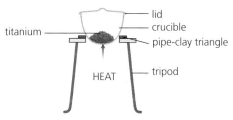

The following results were obtained:
Mass of crucible = 16.34 g
Mass of crucible + titanium metal = 17.36 g
Mass of crucible + titanium oxide = 18.04 g

Questions
1 Use the results to calculate
  a) the mass of titanium used in this experiment
  b) the mass of titanium oxide formed in this experiment
  c) the mass of oxygen used in this experiment.
2 The equation for the reaction is Ti + O_2 → TiO_2
  Explain using the masses calculated in (1) how this reaction follows the law of conservation of mass.
3 Suggest why it was necessary to lift the crucible lid during heating.

In a different experiment titanium metal was heated in a stream of oxygen:

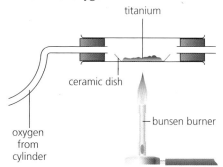

Questions
4 Describe a test which could be carried out, before the cylinder was used, to prove that the gas in it was oxygen.
5 What masses should be found before heating to determine the mass of titanium used in the experiment?
6 The ceramic dish and its contents are repeatedly weighed, heated, reweighed and heated until the mass is constant. State and explain if the mass increases or decreases during this experiment.
7 What safety precautions should be taken in this experiment?
8 How would the reliability of this experiment be checked?

Test yourself

You can find relative atomic masses on the periodic table at the back of this book to help you answer these questions.

10 Hydrogen reacts with oxygen to make water as shown in this equation below.

$$2H_2 + O_2 \rightarrow 2H_2O$$

a) Describe in words what this tells you about the reaction of hydrogen with oxygen in terms of how many molecules are involved in the reaction.

b) Show that the sum of the relative formula masses of all the reactants equals the sum of the relative formula masses of all the products.

11 A piece of copper was heated in air. After a few minutes it was reweighed and found to be heavier.

a) Explain why the copper gets heavier.

b) What is the law of conservation of mass?

c) Explain why this reaction does not break the law of conservation of mass.

12 When 1.19 g of solid nickel carbonate is heated for several minutes, only 0.75 g of solid remains. Explain clearly why the mass decreases and what happens to the remaining mass.

Reacting masses

○ Molar ratios in equations

Chemical equations can be interpreted in terms of moles. For example, in the equation in Figure 3.7 one mole of nitrogen (N_2) molecules reacts with three moles of hydrogen (H_2) molecules to form two moles of ammonia (NH_3) molecules.

N_2	+	$3H_2$	→	$2NH_3$
1 molecule of N_2	reacts with	3 molecules of H_2	to make	2 molecules of NH_3
100 molecules of N_2	reacts with	300 molecules of H_2	to make	200 molecules of NH_3
602 molecules of N_2	reacts with	1806 molecules of H_2	to make	1204 molecules of NH_3
6.02×10^{23} molecules of N_2	reacts with	18.06×10^{23} molecules of H_2	to make	12.04×10^{23} molecules of NH_3
1 mole of molecules of N_2	reacts with	3 moles of molecules of H_2	to make	2 moles of molecules of NH_3

▶ **Figure 3.7** Balanced equations give ratios in which substances react.

These ratios can be used to calculate how many moles would react and be produced in reactions (Figures 3.8 and 3.9).

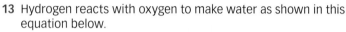

N_2	+	$3H_2$	→	$2NH_3$		C_3H_8	+	$5O_2$	→	$3CO_2$	+	$4H_2O$
1 mole N_2	reacts with	3 moles H_2	to make	2 moles NH_3		1 mole C_3H_8	reacts with	5 moles O_2	to make	3 moles CO_2	and	4 moles H_2O
10 moles N_2	reacts with	30 moles H_2	to make	20 moles NH_3		10 moles C_3H_8	reacts with	50 moles O_2	to make	30 moles CO_2	and	40 moles H_2O
2 moles N_2	reacts with	6 moles H_2	to make	4 moles NH_3		2 moles C_3H_8	reacts with	10 moles O_2	to make	6 moles CO_2	and	8 moles H_2O
0.5 moles N_2	reacts with	1.5 moles H_2	to make	1.0 moles NH_3		0.5 moles C_3H_8	reacts with	2.5 moles O_2	to make	1.5 moles CO_2	and	2.0 moles H_2O

▲ **Figure 3.8** Examples of molar quantities for the reaction of nitrogen with hydrogen.

▲ **Figure 3.9** Examples of molar quantities for the reaction of propane with oxygen.

Show you can...

a) **Select the correct words to complete the sentence below.**

In the equation C + O_2 → CO_2 one mole of C *atoms/molecules* reacts with one mole of O_2 *atoms/molecules* to form one mole of CO_2 *atoms/molecules*.

b) **Write a similar sentence about each equation below**
   i) $C_2H_4 + 2O_2 → CO_2 + 2H_2O$
   ii) $Be + Cl_2 → BeCl_2$
   iii) $C_xH_y + 3O_2 → 2CO_2 + 2H_2O$

c) **What is the value of x and y in equation (ii)?**

Test yourself

13 Hydrogen reacts with oxygen to make water as shown in this equation below.
   $2H_2 + O_2 → 2H_2O$
   a) How many moles of oxygen react with 10 moles of hydrogen?
   b) How many moles of hydrogen react with 3 moles of oxygen?
   c) How many moles of oxygen react with 0.3 moles of hydrogen?
   d) How many moles of hydrogen react with 2.5 moles of oxygen?

14 Titanium is made when titanium chloride reacts with sodium as shown in the equation below.
   $TiCl_4 + 4Na → Ti + 4NaCl$
   a) How many moles of sodium react with 4 moles of titanium chloride?
   b) How many moles of titanium are made from 2.5 moles of titanium chloride?
   c) How many moles of sodium react with 0.5 moles of titanium chloride?
   d) How many moles of titanium are made from 0.65 moles of titanium chloride?

15 Potassium chlorate ($KClO_3$) decomposes to form potassium chloride and oxygen as shown below.
   $2KClO_3 → 2KCl + 3O_2$
   a) How many moles of potassium chloride are formed when 10 moles of potassium chlorate decomposes?
   b) How many moles of oxygen are formed when 4 moles of potassium chlorate decomposes?
   c) How many moles of potassium chloride are formed when 0.5 moles of potassium chlorate decomposes?
   d) How many moles of oxygen are formed when 3 moles of potassium chlorate decomposes?

TIP

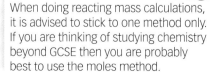

When doing reacting mass calculations, it is advised to stick to one method only. If you are thinking of studying chemistry beyond GCSE then you are probably best to use the moles method.

Calculating reacting masses

Scientists need to be able to calculate how much of each substance to use in a chemical reaction. There are two common ways to do these calculations. One method uses moles while the other uses ratios and relative formula masses.

Examples

Using moles

1. Work out the relative formula mass (M_r) of the substance whose mass is given and the one you are finding the mass of. (Remember that the balancing numbers in the equation are not part of the formulae).
2. Calculate the moles of the substance whose mass is given (using moles = mass / M_r).
3. Use molar ratios from the balanced equation to work out the moles of the substance the question asks about.
4. Calculate the mass of that substance (using mass = M_r × moles).

Example 1

Calculate the mass of hydrogen needed to react with 140 g of nitrogen to make ammonia.

$N_2 + 3H_2 \rightarrow 2NH_3$

M_r $N_2 = 2(14) = 28$, $H_2 = 2(1) = 2$

Moles $N_2 = \dfrac{mass}{M_r} = \dfrac{140}{28} = 5$

Moles H_2 = moles of N_2 × 3 = 5 × 3 = 15

Mass $H_2 = M_r$ × moles = 2 × 15 = 30 g

Example 2

What mass of oxygen reacts with 4.6 g of sodium?

$4Na + O_2 \rightarrow 2Na_2O$

M_r $Na = 23$, $O_2 = 2(16) = 32$

moles $Na = \dfrac{mass}{M_r} = \dfrac{4.6}{23} = 0.20$

Moles O_2 = moles of Na ÷ 4 = 0.20 ÷ 4 = 0.05

Mass $O_2 = M_r$ × moles = 32 × 0.05 = 1.6 g

Example 3

What mass of iron is produced when 32 kg of iron oxide is heated with carbon monoxide?

$Fe_2O_3 + 3CO \rightarrow 2Fe + 3CO_2$

M_r $Fe_2O_3 = 2(56) + 3(16) = 160$, $Fe = 56$

moles $Fe_2O_3 = \dfrac{mass}{M_r} = \dfrac{32\,000}{160} = 200$

Moles Fe = moles of Fe_2O_3 × 2 = 200 × 2 = 400

Mass Fe = M_r × moles = 56 × 400 = 22 400 g

Using ratios

1. Work out the relative formula mass (M_r) of the substance whose mass is given and the one you are finding the mass of. (Remember that the balancing numbers in the equation are not part of the formulae).
2. Find the reacting mass ratio for these substances using the M_r values and the balancing numbers in the equation.
3. Scale this to find what happens with 1 g of the substance given.
4. Scale this up/down to the mass you were actually given.

Example 1

Calculate the mass of hydrogen needed to react with 140 g of nitrogen to make ammonia.

$N_2 + 3H_2 \rightarrow 2NH_3$

M_r $N_2 = 2(14) = 28$, $H_2 = 2(1) = 2$

N_2 reacts with $3H_2$

28 g of N_2 reacts with 6 g (3 × 2) of H_2

1 g of N_2 reacts with $\dfrac{6}{28}$ g of H_2

140 g of N_2 reacts with $140 \times \dfrac{6}{28} = 30$ g of H_2

Example 2

What mass of oxygen reacts with 4.6 g of sodium?

$4Na + O_2 \rightarrow 2Na_2O$

M_r $Na = 23$, $O_2 = 2(16) = 32$

$4Na$ reacts with O_2

92 g (4 × 23) of Na reacts with 32 g of O_2

1 g of Na reacts with $\dfrac{32}{92}$ g of O_2

4.6 g of Na reacts with $4.6 \times \dfrac{32}{92} = 1.6$ g of O_2

Example 3

What mass of iron is produced when 32 kg of iron oxide is heated with carbon monoxide?

$Fe_2O_3 + 3CO \rightarrow 2Fe + 3CO_2$

M_r $Fe_2O_3 = 2(56) + 3(16) = 160$, $Fe = 56$

Fe_2O_3 makes $2Fe$

160 g of Fe_2O_3 makes 112 g (2 × 56) of Fe

1 g of Fe_2O_3 makes $\dfrac{112}{160}$ g of Fe

32 kg of Fe_2O_3 reacts with $32 \times \dfrac{112}{160} = 22.4$ kg of Fe

Test yourself

16. Hydrogen reacts with oxygen to make water as shown in the equation below. What mass of oxygen is needed to react with 10 g of hydrogen?

    $2H_2 + O_2 \rightarrow 2H_2O$

17. Calcium carbonate decomposes when heated as shown in the equation below. What mass of calcium oxide is formed from 25 g of calcium carbonate?

    $CaCO_3 \rightarrow CaO + CO_2$

18. Magnesium burns in oxygen to form magnesium oxide as shown in the equation below. What mass of magnesium oxide would be made from 3 g of magnesium?

    $2Mg + O_2 \rightarrow 2MgO$

19. Lithium reacts with water as shown in the equation below. What mass of hydrogen would be made from 1.4 g of lithium?

    $2Li + 2H_2O \rightarrow 2LiOH + H_2$

20. Aluminium is made by the electrolysis of aluminium oxide as shown in the equation below. What mass of aluminium can be formed from 1 kg of aluminium oxide? Give your answer to 3 significant figures.

    $2Al_2O_3 \rightarrow 4Al + 3O_2$

21. Hot molten copper can be produced for welding electrical connections in circuits by the reaction of copper oxide with aluminium powder shown below. What mass of aluminium is needed to react with 10 g of copper oxide? Give your answer to 3 significant figures.

    $3CuO + 2Al \rightarrow 3Cu + Al_2O_3$

○ Deducing the balancing numbers in an equation from reacting masses Ⓗ

The balancing numbers in a chemical equation can be calculated by calculating the moles of the substances in the reaction. In order to do this:

1 Calculate the moles of each substance (using moles = $\dfrac{Mass}{M_r}$)

2 Find the simplest whole number ratio of these mole values by dividing all the mole values by the smallest mole value.

3 If this does not give a whole number ratio, multiply up by a factor of 2 (where there is a value ending in approximately 0.5), of 3 (where there is a value ending in approximately 0.33 or 0.67), of 4 (where there is a value ending in approximately 0.25 or 0.75), etc.

Example

1.2 g of magnesium (Mg) reacts with 0.8 g of oxygen (O_2) to make 2.0 g of magnesium oxide (MgO). Use this information to deduce the equation for this reaction.

Answer

Substance	Magnesium (Mg)	Oxygen (O_2)	Magnesium oxide (MgO)
Calculate the moles of each substance	moles = $\dfrac{mass}{M_r}$ $= \dfrac{1.2}{24} = 0.05$	moles = $\dfrac{mass}{M_r}$ $= \dfrac{0.8}{32} = 0.025$	moles = $\dfrac{mass}{M_r}$ $= \dfrac{2.0}{40} = 0.05$
Find the simplest whole number ratio	$\dfrac{0.05}{0.025} = 2$	$\dfrac{0.025}{0.025} = 1$	$\dfrac{0.05}{0.025} = 2$

Therefore the reacting ratio is 2 : 1 : 2, and so the balanced equation is
$2Mg + O_2 \rightarrow 2MgO$

Example

0.81 g of aluminium (Al) reacts with 3.20 g of chlorine (Cl_2) to make 4.01 g of aluminium chloride ($AlCl_3$). Use this information to deduce the equation for this reaction.

Answer

Substance	Aluminium (Al)	Chlorine (Cl_2)	Aluminium chloride ($AlCl_3$)
Calculate the moles of each substance	moles = $\dfrac{mass}{M_r}$ $= \dfrac{0.81}{27} = 0.030$	moles = $\dfrac{mass}{M_r}$ $= \dfrac{3.2}{71} = 0.045$	moles = $\dfrac{mass}{M_r}$ $= \dfrac{4.01}{133.5} = 0.030$
Find the simplest whole number ratio	$\dfrac{0.030}{0.030} = 1$ × 2 to get rid of 0.5 values $1 \times 2 = 2$	$\dfrac{0.045}{0.030} = 1.5$ × 2 to get rid of 0.5 values $1.5 \times 2 = 3$	$\dfrac{0.030}{0.030} = 1$ × 2 to get rid of 0.5 values $1 \times 2 = 2$

Therefore the reacting ratio is 2 : 3 : 2, and so the balanced equation is
$2Al + 3Cl_2 \rightarrow 2AlCl_3$

▲ **Figure 3.10** Some camping stoves use propane gas.

Example

2.2 g of propane (C_3H_8) reacts with 8.0 g of oxygen (O_2). Calculate the molar ratio in which propane and oxygen react here.

Substance	propane (C_3H_8)	oxygen (O_2)
Calculate the moles of each substance	moles = $\dfrac{mass}{M_r} = \dfrac{2.2}{44} = 0.050$	moles = $\dfrac{mass}{M_r} = \dfrac{8.0}{32} = 0.25$
Find the simplest whole number ratio	$\dfrac{0.050}{0.050} = 1$	$\dfrac{0.25}{0.050} = 5$

Therefore the reacting ratio is 1:5, and so the ratio they react in is **$C_3H_8 + 5O_2$**

Test yourself

22 6.8 g of hydrogen peroxide (H_2O_2) decomposes into 3.6 g of water (H_2O) and 3.2 g of oxygen (O_2). By calculating molar ratios, deduce the balanced equation for this reaction.

23 2.0 g of calcium (Ca) reacts with 1.9 g of fluorine (F_2) to form 3.9 g of calcium fluoride (CaF_2). By calculating molar ratios, deduce the balanced equation for this reaction.

24 1.0 g of nickel (Ni) reacts with 0.27 g of oxygen (O_2) to form 1.27 g of nickel oxide (NiO). By calculating molar ratios, deduce the balanced equation for this reaction.

25 When 4.1 g of calcium nitrate ($Ca(NO_3)_2$) is heated, 1.4 g of calcium oxide (CaO), 2.3 g of nitrogen dioxide (NO_2) and 0.4 g of oxygen (O_2) are formed. By calculating molar ratios, deduce the balanced equation for this reaction.

26 11.7 g of potassium (K) reacts with 24 g of bromine (Br_2). Calculate the molar ratio in which potassium reacts with bromine.

27 11.4 g of titanium chloride ($TiCl_4$) reacts with 5.52 g of sodium (Na). Calculate the molar ratio in which titanium chloride reacts with sodium.

Experiment to find the equation for the action of heat on sodium hydrogencarbonate

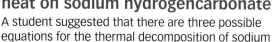

A student suggested that there are three possible equations for the thermal decomposition of sodium hydrogencarbonate.

Equation 1: $NaHCO_3 \rightarrow NaOH + CO_2$
Equation 2: $2NaHCO_3 \rightarrow Na_2CO_3 + H_2O + CO_2$
Equation 3: $2NaHCO_3 \rightarrow Na_2O + H_2O + 2CO_2$

In order to find out which is correct, she carried out the following experiment and recorded her results.

• Find the mass of an empty evaporating basin.
• Add approximately 8 g of sodium hydrogencarbonate to the basin and find the mass.
• Heat gently for about 5 minutes.
• Allow to cool and then find the mass.
• Reheat, cool and find the mass.
• Repeat the heating and measurement of mass until constant mass is obtained.

Results

Mass of evaporating basin = 21.05 g
Mass of basin and sodium hydrogencarbonate = 29.06 g
Mass of sodium hydrogencarbonate = 8.01 g
Mass of basin and residue after heating to constant mass = 26.10 g
Mass of residue = 5.05 g

Questions

1 Draw a labelled diagram of the assembled apparatus for this experiment.

2 Calculate the number of moles of sodium hydrogencarbonate used.

3 From the possible equations, and your answer to question 2, calculate:
   a) The number of moles of NaOH that would be formed in equation 1.
   b) The number of moles of Na_2CO_3 that would be formed in equation 2.
   c) The number of moles of Na_2O that would be formed in equation 3.

4 From your answers to question 3, calculate
   a) The mass of NaOH that would be formed in equation 1.
   b) The mass of Na_2CO_3 that would be formed in equation 2.
   c) The mass of Na_2O that would be formed in equation 3.

5 By comparing your answers in question 4 with the experimental mass of residue, deduce which is the correct equation for the decomposition of sodium hydrogencarbonate.

6 Why did the mass decrease?

Excess When the amount of a reactant is greater than the amount that can react.

Limiting reactant The reactant in a reaction that determines the amount of products formed. Any other reagents are in excess and will not all react.

Using an excess

In many reactions involving two reactants, it is very common for an excess of one of the reactants to be used to ensure that all of the other reactant is used up. This is often done if one of the reactants is readily available but the other one is expensive or is in limited supply. For example, when many fuels are burned an excess of oxygen is used. Fuels are expensive and in limited supply. The oxygen is readily available from the air and using an excess of oxygen ensures that all the fuel burns.

When one of the reactants is in excess, the other reactant is a limiting reactant that is completely used up. This is because it is the amount of this substance that determines the amount of product formed in a reaction, in other words it limits the amount of product made.

Example

Magnesium reacts with sulfuric acid as shown below:
5 moles of magnesium (Mg) is reacted with 7 moles of sulfuric acid (H_2SO_4). One of the reagents is in excess. Calculate the moles of the products formed.

$$Mg + H_2SO_4 \rightarrow MgSO_4 + H_2$$

Answer

As magnesium and sulfuric acid react with the ratio 1 to 1, only 5 moles of the sulfuric acid can react as there are only 5 moles of magnesium to react with. The rest of the sulfuric acid is in excess (2 moles). Therefore, the magnesium is the limiting reactant and determines how much of the products are made. In this case, 5 moles of $MgSO_4$ and 5 moles of H_2 would be made.

	Mg	+	H_2SO_4	$\rightarrow$	$MgSO_4$	+	H_2
Reacting ratio from the equation	1 mole of Mg	*reacts with*	1 mole of H_2SO_4	*to make*	1 mole of $MgSO_4$	*and*	1 mole of H_2
Amount provided	5 moles of Mg		7 moles of H_2SO_4				
	limiting reactant		in excess so it does not all react				
Reaction that takes place	5 moles of Mg	*reacts with*	5 moles of H_2SO_4	*to make*	5 moles of $MgSO_4$	*and*	5 moles of H_2

Example

Iron oxide reacts with carbon monoxide as shown below to produce iron. 10 moles of iron oxide (Fe_2O_3) is reacted with 50 moles of carbon monoxide (CO). Calculate the moles of products formed.

$$Fe_2O_3 + 3CO \rightarrow 2Fe + 3CO_2$$

Answer

In this reaction, the reacting ratio is one mole of iron oxide for every three moles of carbon monoxide. Therefore, 10 moles of iron oxide reacts with 30 moles of carbon monoxide. The rest of the carbon monoxide is in excess (20 moles). Therefore, the iron oxide is the limiting reactant and determines how much of the products are made. In this case, 20 moles of iron and 30 moles of carbon dioxide would be made.

	Fe_2O_3	+	3CO	$\rightarrow$	2Fe	+	$3CO_2$
Reacting ratio from the equation	1 mole of Fe_2O_3	*reacts with*	3 moles of CO	*to make*	2 moles of Fe	*and*	3 moles of CO_2
Amount provided	10 moles of Fe_2O_3		50 moles of CO				
	limiting reactant		in excess so it does not all react				
Reaction that takes place	10 moles of Fe_2O_3	*reacts with*	30 moles of CO	*to make*	20 moles of Fe	*and*	30 moles of CO_2

Example

Magnesium reacts with oxygen as shown below. 0.30 moles of magnesium (Mg) is reacted with 0.20 moles of oxygen (O_2). Calculate the moles of products formed.

$$2Mg + O_2 \rightarrow 2MgO$$

Answer

In this reaction, the reacting ratio is two moles of magnesium for every mole of oxygen. Therefore, 0.30 moles of magnesium can only react with 0.15 moles of oxygen. The rest of the oxygen is in excess (0.05 moles). Therefore, the magnesium is the limiting reactant and determines how much of the products is made. In this case, 0.30 moles of MgO would be made.

	2Mg	+	O_2	$\rightarrow$	2MgO
Reacting ratio from the equation	2 moles of Mg	reacts with	1 mole of O_2	to make	2 moles of MgO
Amount provided	0.30 moles of Mg limiting reactant		0.20 moles of O_2 in excess so it does not all react		
Reaction that takes place	0.30 moles of Mg	reacts with	0.15 moles of O_2	to make	0.30 moles of MgO

Example

Tungsten oxide reacts with hydrogen as shown below to produce tungsten (Figure 3.11). 23.2 g of tungsten oxide (WO_3, M_r = 232) is reacted with 20.0 g of hydrogen (H_2, M_r = 2). Calculate the mass of tungsten (M_r = 184) formed.

$$WO_3 + 3H_2 \rightarrow W + 3H_2O$$

▲ **Figure 3.11** Tungsten.

Answer

In this reaction, the reacting ratio is one mole of tungsten oxide for every three moles of hydrogen. Therefore, 0.10 moles (calculated from 23.2 g) of tungsten oxide reacts with 0.30 moles of hydrogen, and so the hydrogen is in excess (there are 10 moles of hydrogen calculated from 20 g). Therefore, the tungsten oxide is the limiting reactant and determines how much of the products are made. In this case, 0.10 moles of tungsten and 0.30 moles of water would be made.

	WO_3	+	$3H_2$	$\rightarrow$	W	+	$3H_2O$
Reacting ratio from the equation	1 mole of WO_3	reacts with	3 moles of H_2	to make	1 mole of W	and	3 moles of H_2O
Amount provided	Moles = $\frac{23.2}{232}$ = 0.10		Moles = $\frac{20}{10}$ = 10				
	0.10 moles of WO_3 limiting reactant		10 moles of H_2 in excess so it does not all react				
Reaction that takes place	0.10 moles of WO_3	reacts with	0.30 moles of H_2	to make	0.10 moles of W	and	0.30 moles of H_2O
					Mass of W = 184 × 0.10 = 18.4 g		

Test yourself

28 Copper can be made by reacting copper oxide with hydrogen as shown below.
$$CuO + H_2 \rightarrow Cu + H_2O$$
How many moles of copper would be made if:
  a) 5 moles of copper oxide were reacted with 10 moles of hydrogen?
  b) 2 moles of copper oxide were reacted with 2 moles of hydrogen?
  c) 0.4 moles of copper oxide were reacted with 0.3 moles of hydrogen?

29 Copper can be also be made by reacting copper oxide with methane as shown below.
$$4CuO + CH_4 \rightarrow 4Cu + 2H_2O + CO_2$$
How many moles of copper would be made if:
  a) 2 moles of copper oxide were reacted with 2 moles of methane?
  b) 2 moles of copper oxide were reacted with 1 mole of methane?
  c) 10 moles of copper oxide were reacted with 2 moles of methane?

30 Hydrogen gas is formed when magnesium reacts with hydrochloric acid as shown below.
$$Mg + 2HCl \rightarrow MgCl_2 + H_2$$
How many moles of hydrogen would be made if:
  a) 3 moles of magnesium were reacted with 3 moles of hydrochloric acid?
  b) 0.2 moles of magnesium were reacted with 0.3 moles of hydrochloric acid?
  c) 0.5 moles of magnesium were reacted with 1.5 moles of hydrochloric acid?

31 Calcium reacts with sulfur as shown in the equation below. What mass of calcium sulfide can be made when 8 g of calcium reacts with 8 g of sulfur?
$$Ca + S \rightarrow CaS$$

32 Titanium is made when titanium chloride reacts with magnesium as shown in the equation below. What mass of titanium can be made when 1.9 g of titanium chloride reacts with 6 g of magnesium?
$$TiCl_4 + 2Mg \rightarrow Ti + 2MgCl_2$$

33 Aluminium bromide is made when aluminium reacts with bromine as shown in the equation below. What mass of aluminium bromide can be made when 0.81 g of aluminium reacts with 6.4 g of bromine?
$$2Al + 3Br_2 \rightarrow 2AlBr_3$$

Show you can...

Calculate the mass of calcium carbide (CaC_2) formed when 84 g of calcium oxide (CaO) is reacted with 48 g of coke (C).

$$CaO + 3C \rightarrow CaC_2 + CO$$

Use the following headings.

- Number of moles of calcium oxide
- Number of moles of coke
- The reactant in excess is
- Number of moles of calcium carbide formed
- Mass of calcium carbide formed in grams.

Yield and atom economy

○ Percentage yield

When someone bakes a cake and starts with 500 g of ingredients, the final cake is likely to have a mass of less than 500 g (Figure 3.12). This will be because some of the ingredients will be left on the bowl, on the spoon or on the mixer, and some crumbs of cake will be lost when it is taken out of the cake tin.

In a similar way, when carrying out chemical reactions we are unlikely to produce all that we expect to. There are many reasons for this.

1 Some reactions do not go to completion (i.e. they do not completely finish) – sometimes this is because they are reversible and some of the products may turn back into reactants.

2 Some of the product may be lost when it is separated from the reaction mixture – for example, some may be left on the apparatus.

3 Some of the reactants may react in ways different to the desired reaction – in other words some of the reactants may take part in other reactions as well.

The amount of product made in a reaction is called the yield. The percentage yield is a measure of the amount produced in a reaction compared to the maximum theoretical amount that is expected as a percentage. For example, if a reaction was expected to form a maximum theoretical 20 g of product but only 10 g was made, then the yield is 10 g and the percentage yield is 50%.

▲ **Figure 3.12** When you bake a cake you do not get a 100% yield.

KEY TERMS

Yield The amount of product formed in a reaction.

Percentage yield The amount of product formed in a reaction compared to the maximum theoretical mass that could be produced as a percentage.

$$\text{Percentage yield} = 100 \times \frac{\text{mass of product actually made}}{\text{maximum theoretical mass of product}}$$

Example

In a reaction where the maximum theoretical mass of product was 40 g, the yield produced was 15 g. Calculate the percentage yield.

Answer

$$\text{Percentage yield} = 100 \times \frac{\text{mass of product actually made}}{\text{maximum theoretical mass of product}}$$

$$= 100 \times \frac{15}{40} = 37.5\%$$

When making products in industrial processes it is important that as high a percentage yield as possible is achieved. The higher the percentage yield, the more efficient the reaction, the more product formed, the less waste and the higher the profit for the manufacturer and/or the lower the cost to the consumer.

Show you can...

Butanol (C_4H_9OH) is used to prepare the chemical bromobutane (C_4H_9Br) according to the following equation:

$C_4H_9OH + HBr \rightarrow C_4H_9Br + H_2O$

a) Give an equation to explain the term percentage yield.
b) Assuming a 40% yield, what mass of butanol would be required to produce a yield of 5.48 g of bromobutane?

Test yourself

34 In a reaction to manufacture the medicine paracetamol, the maximum theoretical mass of paracetamol that could be produced was 200 g, but the yield was actually 150 g.
   a) Calculate the percentage yield.
   b) Give three possible reasons for the percentage being less than 100%.

35 Ammonia is made when nitrogen reacts with hydrogen. When 28 g of nitrogen reacts with 6 g of ammonia, the maximum theoretical mass of ammonia formed is 34 g. In a reaction the actual yield was 8 g.
   a) Calculate the percentage yield for this reaction. Give your answer to 3 significant figures.
   b) The percentage yield for this reaction is low because it is a reversible reaction. What is a reversible reaction?

36 Iron is made by reaction of iron oxide with carbon monoxide. In a reaction the maximum theoretical mass of iron that could be produced is 1 kg, but the yield was actually 800 g. Calculate the percentage yield.

H

○ Atom economy

Chemical reactions are used to make specific products that we want. In some reactions there is only one product and so all the atoms in the reactants end up in that product. For example, when ethene (C_2H_4) reacts with steam to make ethanol (C_2H_5OH), all of the atoms in the reactants end up in the desired product ethanol.

> ethene + steam $\rightarrow$ ethanol
>
> $C_2H_4 + H_2O \rightarrow C_2H_5OH$

However, in many reactions, there are one or more other products as well as the desired product. In these reactions, not all of the atoms that we start with end up in the desired product. For example, when ethanol is made by fermentation from glucose, carbon dioxide is formed as well as the ethanol. This means that some of the atoms from the glucose do not end up in the ethanol.

> glucose $\rightarrow$ ethanol + carbon dioxide
>
> $C_6H_{12}O_6 \rightarrow 2C_2H_5OH + 2CO_2$

The atom economy (atom utilisation) of a reaction is a way of measuring what percentage of the mass of all the atoms in the reactants ends up in the desired product of that reaction.

KEY TERM

Atom economy (atom utilisation)
A way of measuring what percentage of the mass of all the atoms in the reactants ends up in the desired product.

> Atom economy = 100 $\times$
> $$\frac{\text{sum of relative formula mass of desired product from equation}}{\text{sum of relative masses of all reactants from equation}}$$

TIP

It is important not to confuse percentage yield and atom economy. The percentage yield is about how much product is formed compared to the maximum amount that could be formed theoretically. Atom economy is about what percentage of the mass of the reactants ends up in the desired product.

In a reaction with only one product, all the atoms from the reactants end up in the products and so the atom economy must be 100%. Here are some examples for reactions where there are other products besides the desired product.

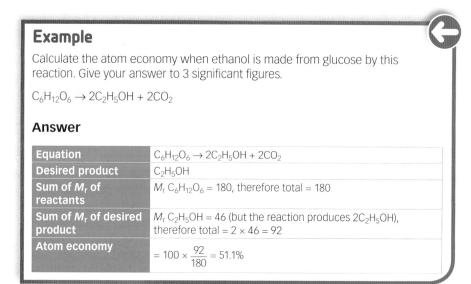

Example

Calculate the atom economy when ethanol is made from glucose by this reaction. Give your answer to 3 significant figures.

$C_6H_{12}O_6 \rightarrow 2C_2H_5OH + 2CO_2$

Answer

Equation	$C_6H_{12}O_6 \rightarrow 2C_2H_5OH + 2CO_2$
Desired product	C_2H_5OH
Sum of M_r of reactants	$M_r\ C_6H_{12}O_6 = 180$, therefore total = 180
Sum of M_r of desired product	$M_r\ C_2H_5OH = 46$ (but the reaction produces $2C_2H_5OH$), therefore total = $2 \times 46 = 92$
Atom economy	$= 100 \times \dfrac{92}{180} = 51.1\%$

Example

Calculate the atom economy when titanium is made from titanium chloride by this reaction. Give your answer to 3 significant figures.

$TiCl_4 + 2Mg \rightarrow Ti + 2MgCl_2$

Answer

Equation	$TiCl_4 + 2Mg \rightarrow Ti + 2MgCl_2$
Desired product	Ti
Sum of M_r of reactants	$M_r\ TiCl_4 = 190$; Mg = 24, therefore total = $190 + 2(24) = 238$
Sum of M_r of desired product	$M_r\ Ti = 48$, therefore total = 48
Atom economy	$= 100 \times \dfrac{48}{238} = 20.2\%$

The higher the atom economy of a reaction, the more of the mass of the reactants that ends up in the desired product. This means that there is less waste. For example, if 10 tonnes of reactants are used in a reaction with an atom economy of 80%, then 8 tonnes of the desired product are formed plus 2 tonnes of waste materials. However, if the same product could be made from 10 tonnes of starting materials in a reaction with an atom economy of 90%, then 9 tonnes of the desired product is formed along with only 1 tonne of waste material.

As the Earth's resources are precious, the less waste material produced in any process the better. In the long term, processes with higher atom economies are more sustainable. They can also be more economic as there is less waste to dispose of.

TIP

The higher the atom economy of a reaction, the less waste produced.

Test yourself

37 **a)** Calculate the mass of waste material in a reaction with a 60% atom economy if 1 kg of reactants is used.

   b) Calculate the mass of waste material in a reaction with a 80% atom economy if 1 kg of reactants is used.

   c) Explain why reactions with higher atom economies are more sustainable.

38 Calculate the atom economy to make calcium oxide by the thermal decomposition of calcium carbonate in the following reaction. Give your answer to 3 significant figures.

$$CaCO_3 \rightarrow CaO + CO_2$$

39 Calculate the atom economy to make iron by heating iron oxide with carbon monoxide in the following reaction. Give your answer to 3 significant figures.

$$Fe_2O_3 + 3CO \rightarrow 2Fe + 3CO_2$$

40 Calculate the atom economy to make sulfur trioxide by reaction of sulfur dioxide and oxygen in the following reaction.

$$2SO_2 + O_2 \rightarrow 2SO_3$$

41 Calculate the atom economy to make aluminium by electrolysis aluminium oxide in the following reaction. Give your answer to 3 significant figures.

$$2Al_2O_3 \rightarrow 4Al + 3O_2$$

42 Calculate the atom economy to make dichloromethane (CH_2Cl_2) by reaction of methane with chlorine in the following reaction. Give your answer to 3 significant figures.

$$CH_4 + 2Cl_2 \rightarrow CH_2Cl_2 + 2HCl$$

Show you can...

Titanium oxide is found in the naturally occurring ore rutile. It is possible to extract titanium from this ore by a displacement reaction with magnesium or by electrolysis of the ore.

Method 1: This uses a more reactive metal to displace the titanium:

$$TiO_2 + 2\,Mg \rightarrow Ti + 2\,MgO$$

Method 2: This is electrolysis of the ore. The overall reaction for this method is:

$$TiO_2 \rightarrow Ti + O_2$$

a) Calculate the atom economy for each reaction.

b) A 'green' method is one which produces little waste, uses little electricity and produces non-toxic, non-polluting products. Which method is 'greener'? What else might you want to know before making a final decision?

c) Oxygen is a useful product and can be sold. What is the atom economy of the electrolysis if the oxygen is collected and sold?

Gas volumes

○ The volume of gases

The volume of a gas varies with temperature and pressure:

- The higher the temperature of a gas, the greater its volume.
- The greater the pressure of a gas, the smaller its volume.

However, providing the temperature and pressure of gases are the same, equal numbers of moles of all gases have the same volume (Figure 3.13). In other words, the volume of a gas does not depend on which gas it is.

| argon (Ar) gas | oxygen (O_2) gas | carbon dioxide (CO_2) gas |

▶ **Figure 3.13** The same number of particles of different gases have the same volume (provided they are at the same temperature and pressure).

The volume of one mole of any gas at room temperature (20°C) and pressure (1 atmosphere) is 24 dm³; 24 dm³ is the same volume as 24 litres or 24 000 cm³. Therefore at room temperature and pressure, 1 mole of argon gas has a volume of 24 dm³, 1 mole of oxygen gas has a volume of 24 dm³, 1 mole of carbon dioxide gas has a volume of 24 dm³, and so does any other gas.

If the volume of 1 mole of a gas is 24 dm³ at room temperature and pressure, then the volume of 2 moles would be 48 dm³. There is a simple equation that links the volume of a gas to the number of moles (Figure 3.14).

volume (dm³) = 24 × moles

Volume (dm³)

24 × moles

This formula triangle may help you. Cover up the quantity you want to show the equation you need to use.

▲ **Figure 3.14**

Example

What is the volume of the following gases at room temperature and pressure?

Answer

10 moles of O_2: volume = 24 × moles = 24 × 10 = 240 dm³

0.2 moles of CH_4: volume = 24 × moles = 24 × 0.2 = 4.8 dm³

Example

How many moles are there of each of the following gases whose volumes are measured at room temperature and pressure?

Answer

48 dm³ of CO_2: $moles = \frac{volume}{24} = \frac{48}{24} = 2$ moles

1.2 dm³ of Ar: $moles = \frac{volume}{24} = \frac{1.2}{24} = 0.05$ moles

Masses can be converted to moles using the equation mass = M_r × moles. This can be used to find the volume of a gas from its mass and vice versa.

Example

What is the volume of 64 g of methane gas at room temperature and pressure?

Answer

64 g moles of CH_4:   moles $= \dfrac{\text{mass}}{M_r} = \dfrac{64}{16} = 4$ moles

volume = 24 × moles = 24 × 4 = 96 dm³

Example

What is the mass of 1.8 dm³ of nitrogen gas measured at room temperature and pressure?

Answer

1.8 dm³ of N_2:   moles $= \dfrac{\text{volume}}{24} = \dfrac{1.8}{24} = 0.075$ moles

mass = M_r × moles = 28 × 0.075 = 2.1 g

TIP

Remember that 1 dm³ = 1000 cm³.

Test yourself

43 Calculate the volume of the following gases at room temperature and pressure:
   a) 3 moles of hydrogen (H_2)
   b) 0.40 moles of oxygen (O_2)
   c) 11 g of carbon dioxide (CO_2)
   d) 40 g of helium (He)

44 Calculate the number of moles in each of the following gases at room temperature and pressure. Give your answer to 3 significant figures.
   a) 18.0 dm³ of carbon monoxide (CO)
   b) 1.00 dm³ of nitrogen (N_2)
   c) 600 cm³ of argon (Ar)

45 Calculate the mass of each of the following gases at room temperature and pressure. Give your answer to 3 significant figures.
   a) 72 dm³ of oxygen (O_2)
   b) 2 dm³ of methane (CH_4)
   c) 100 cm³ of fluorine (F_2)

Show you can...

Copy and complete the following table

	O_2	H_2	NH_3
Mass of 1 mole of gas in g			
Volume of 1 mole of gas in dm³ at room temperature and pressure			
Number of moles in 12 cm³ at room temperature and pressure			
Number of moles in 12 g			

○ Reacting volumes of gases

Due to equal amounts of moles of different gases having the same volume (at the same temperature and pressure), we can work out the volumes of gases involved in chemical reactions. Figures 3.15 and 3.16 show how the molar ratio from the equation can be used to do this.

$N_2(g)$	+	$3H_2(g)$	→	$2NH_3(g)$
1 mole of N_2 gas	reacts with	3 moles of H_2 gas	to make	2 moles of NH_3 gas
$24\,dm^3$	reacts with	$72\,dm^3$	to make	$48\,dm^3$
$12\,dm^3$	reacts with	$36\,dm^3$	to make	$24\,dm^3$
$1\,dm^3$	reacts with	$3\,dm^3$	to make	$2\,dm^3$
$100\,cm^3$	reacts with	$300\,cm^3$	to make	$200\,cm^3$

▲ **Figure 3.15** Examples of reacting gas volumes in the reaction of nitrogen with hydrogen.

$2SO_2(g)$	+	$O_2(g)$	→	$2SO_3(g)$
2 moles of SO_2 gas	reacts with	1 mole of O_2 gas	to make	2 moles of SO_3 gas
$48\,dm^3$	reacts with	$24\,dm^3$	to make	$48\,dm^3$
$12\,dm^3$	reacts with	$6\,dm^3$	to make	$12\,dm^3$
$2\,dm^3$	reacts with	$1\,dm^3$	to make	$2\,dm^3$
$100\,cm^3$	reacts with	$50\,cm^3$	to make	$100\,cm^3$

▲ **Figure 3.16** Examples of reacting gas volumes in the reaction of sulfur dioxide with oxygen.

TIP ✔

This method only works with gas volumes. It does NOT work with masses.

Example

What volume of oxygen reacts with $10\,dm^3$ of hydrogen with the volume of both gases measured at the same temperature and pressure?

Answer

$2H_2(g) + O_2(g) \rightarrow 2H_2O(l)$

2 moles of $H_2(g)$ reacts with 1 mole of $O_2(g)$

therefore volume of $O_2(g) = \frac{1}{2} \times$ volume of $H_2(g) = \frac{1}{2} \times 10 = 5\,dm^3$

Example

What volume of oxygen reacts with $600\,cm^3$ of propane with the volume of both gases measured at the same temperature and pressure?

Answer

$C_3H_8(g) + 5O_2(g) \rightarrow 3CO_2(g) + 4H_2O(l)$

1 mole of $C_3H_8(g)$ reacts with 5 moles of $O_2(g)$

therefore volume of $O_2(g) = 5 \times$ volume of $C_3H_8(g) = 5 \times 600 = 3000\,cm^3$

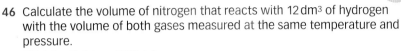

Test yourself

46 Calculate the volume of nitrogen that reacts with 12 dm³ of hydrogen with the volume of both gases measured at the same temperature and pressure.
$3H_2(g) + N_2(g) \rightarrow 2NH_3(g)$

47 Calculate the volume of nitrogen dioxide (NO_2) that is formed when 3 dm³ of dinitrogen tetroxide (N_2O_4) decomposes with the volume of both gases measured at the same temperature and pressure.
$N_2O_4(g) \rightarrow 2NO_2(g)$

48 Calculate the volume of oxygen that reacts with 1 dm³ of ethane with the volume of both gases measured at the same temperature and pressure.
$2C_2H_6(g) + 7O_2(g) \rightarrow 4CO_2(g) + 6H_2O(l)$

49 Calculate the volume of oxygen that reacts with 500 cm³ of hydrogen sulfide with the volume of both gases measured at the same temperature and pressure.
$2H_2S(g) + 3O_2(g) \rightarrow 2SO_2(g) + 2H_2O(l)$

Show you can...

12.5 g of zinc carbonate reacts with excess hydrochloric acid.

$ZnCO_3 + 2HCl \rightarrow ZnCl_2 + CO_2 + H_2O$

a) Calculate the number of moles of $ZnCO_3$ in 12.5 g of zinc carbonate.

b) Calculate the number of moles of carbon dioxide produced.

c) Calculate the volume of carbon dioxide gas produced at room temperature and pressure.

The concentration of solutions

▲ Figure 3.17 The darker copper sulfate solution is more concentrated than the lighter blue solution.

◯ Concentration of solutions in g/dm³

Figure 3.17 shows two solutions of copper sulfate. The one that is darker blue has much more copper sulfate dissolved in it. The darker blue one is more concentrated and the paler blue one is more dilute.

We can measure the concentration of a solution by considering what mass of solute is dissolved in the solution. This is usually found in g/dm³, which means the number of grams of solute dissolved in each dm³ of solution (Figure 3.18); 1 dm³ is the same volume as 1000 cm³ or 1 litre. For example, if 50 grams of copper sulfate is dissolved in 2 dm³ of solution, then the concentration is 25 g/dm³.

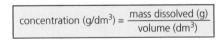

$$\text{concentration (g/dm}^3) = \frac{\text{mass dissolved (g)}}{\text{volume (dm}^3)}$$

mass

concentration (g/dm³) × volume (dm³)

This formula triangle may help you. Cover up the quantity you want to show the equation you need to use.

▲ Figure 3.18

In the laboratory, we often use volumes measured in cm³ rather than dm³. As there are 1000 cm³ in 1 dm³, we should divide the volume in cm³ by 1000 to find the volume in dm³. For example, 25 cm³ is $\frac{25}{1000} = 0.025$ dm³.

Examples

Find the concentration of the following solutions in g/dm³.

80 g dissolved in 4 dm³

$$\text{Concentration} = \frac{\text{mass}}{\text{volume}} = \frac{80}{4} = 20 \text{ g/dm}^3$$

12 g dissolved in 200 cm³

$$\text{volume of solution in dm}^3 = \frac{200}{1000} = 0.20 \text{ dm}^3$$

$$\text{concentration} = \frac{\text{mass}}{\text{volume}} = \frac{12}{0.20} = 60 \text{ g/dm}^3$$

Test yourself

50 What is the concentration in g/dm³ of a solution in which 90 g of solute are dissolved in 2 dm³ of solution?

51 What is the concentration in g/dm³ of a solution in which 18 g of solute are dissolved in 250 cm³ of solution?

52 What is the concentration in g/dm³ of a solution in which 3.50 g of solute are dissolved in 50 cm³ of solution?

53 What mass of solute is dissolved in 10 dm³ of a solution with concentration 36 g/dm³?

54 What mass of solute is dissolved in 25 cm³ of a solution with concentration 60 g/dm³?

Concentration of solutions in mol/dm³

We can measure how concentrated a solution is in mol/dm³. This is effectively the number of moles of solute dissolved in each 1 dm³ of solution (Figure 3.19); 1 dm³ is the same volume as 1 litre or 1000 cm³. If there were 6 moles of solute dissolved in 2 dm³ of solution, then the concentration would be 3 mol/dm³.

TIP

It can be useful to remember this equation as
moles = concentration × volume or
moles = cv.

$$\text{concentration (mol/dm}^3) = \frac{\text{moles}}{\text{volume (dm}^3)}$$

moles

concentration × volume (dm³)

This formula triangle may help you. Cover up the quantity you want to show the equation you need to use.

▲ Figure 3.19

Example

Find the concentration of each of the following solutions in mol/dm³.

10 moles dissolved in 4 dm³:

$$\text{concentration} = \frac{\text{moles}}{\text{volume}} = \frac{10}{4}$$
$$= 2.5 \text{ mol/dm}^3$$

0.2 moles dissolved in 100 cm³:

$$\text{volume of solution in dm}^3 = \frac{100}{1000}$$
$$= 0.10 \text{ dm}^3$$

$$\text{concentration} = \frac{\text{moles}}{\text{volume}} = \frac{0.20}{0.10}$$
$$= 2.0 \text{ mol/dm}^3$$

49 g of H_2SO_4 dissolved in 2 dm³:

$$\text{moles } H_2SO_4 = \frac{\text{mass}}{M_r} = \frac{49}{98} = 0.5$$

$$\text{concentration} = \frac{\text{moles}}{\text{volume}} = \frac{0.50}{2}$$
$$= 0.25 \text{ mol/dm}^3$$

Example

Find the number of moles of solute in each of the following solutions.

20 dm³ of 0.50 mol/dm³ solution:

moles = concentration × volume (dm³)
= 0.50 × 20 = 10 moles

25 cm³ of 1.50 mol/dm³ solution:

volume of solution in dm³ = $\frac{25}{1000}$ = 0.025 dm³

moles = concentration × volume (dm³) = 1.50 × 0.025 = 0.0375 moles

Test yourself

55 Calculate the concentration of the following solutions in mol/dm³:
   a) 4 moles dissolved in 8 dm³ of solution
   b) 30 moles dissolved in 10 dm³ of solution
   c) 0.36 moles dissolved in 250 cm³ of solution
   d) 0.10 moles dissolved in 50 cm³ of solution

56 Calculate the number of moles of solute in the following solutions:
   a) 4 dm³ of 0.30 mol/dm³ solution
   b) 1.5 dm³ of 2.0 mol/dm³ solution
   c) 100 cm³ of 1.50 mol/dm³ solution
   d) 25 cm³ of 0.50 mol/dm³ solution

57 Calculate the concentration of the following solutions in mol/dm³:
   a) 120 g of NaOH dissolved in 2 dm³ of solution
   b) 1.7 g of NH_3 dissolved in 200 cm³ of solution

58 What mass of sodium carbonate (Na_2CO_3) should be dissolved to make 500 cm³ a solution with a concentration of 0.80 mol/dm³?

We can easily convert concentrations in mol/dm³ to g/dm³ using this equation:

$$\text{concentration (g/dm}^3) = M_r \times \text{concentration (mol/dm}^3)$$

Example

What is the concentration in g/dm³ of a solution of NaOH of concentration 0.20 mol/dm³?

Answer
M_r NaOH = 40

concentration (g/dm³) = M_r × concentration (mol/dm³) = 40 × 0.20 = 8 g/dm³

Show you can...

A solution of nitric acid (HNO_3) contained 0.1 mole of solute dissolved in 250 cm³ of solution. What is the concentration of the solution in mol/dm³ and g/dm³?

Test yourself

59 What is the concentration in g/dm³ of a solution of HCl of concentration 0.60 mol/dm³?

60 What is the concentration in g/dm³ of a solution of $Ca(OH)_2$ of concentration 0.02 mol/dm³?

61 What is the concentration in mol/dm³ of a solution of H_2SO_4 of concentration 24.5 g/dm³?

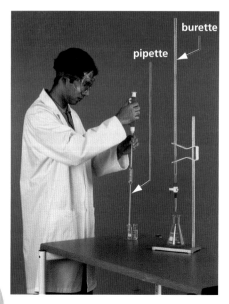

▲ **Figure 3.20** A titration.

Titrations

Titrations are a very accurate experimental technique that can be used to find the concentration of a solution by reacting it with a solution of known concentration. Titrations are often used to find the concentration of acids or alkalis.

Titrations use apparatus including a pipette, conical flask and burette (Figure 3.20). A pipette is a glass tube designed to measure a specific volume of a solution very accurately. A typical pipette measures out 25 cm³ within a margin of ±0.06 cm³. The pipette is filled using a pipette filler which is attached to the end of the pipette. A burette is a glass tube with a tap (to let out the liquid) with markings on to show the volume to the nearest 0.1 cm³.

The following steps are followed in a titration.

1 A known volume of a solution of an acid or alkali is measured out using a pipette and placed into a conical flask.

2 A few drops of a suitable indicator are added. For most acid-alkali titrations, methyl orange or phenolphthalein is suitable.

3 The other solution, the acid or alkali, is added to the conical flask from a burette.

4 The solution is added from the burette until the indicator changes colour (the end point). The solution is added dropwise around the point where the indicator changes colour to ensure the exact volume required is used.

5 The volume added from the burette is recorded.

6 The experiment is repeated until concordant results are achieved (i.e. results that are very close to each other). The mean volume is found using the concordant results.

Titrations are a very accurate technique and when done correctly often give a value within 1% of the true value.

Titration calculations

The concentration of a solution can be found using the results of a titration. The volume of both solutions has been measured and the concentration of one of the solutions will have been known at the start. This information is used as shown below.

1 Work out the moles of the solution whose concentration is known (using moles = concentration (mol/dm³) × volume (dm³))

2 Use molar ratios from the balanced equation to work out the moles of other solution.

3 Calculate the concentration of the other solution (using concentration (mol/dm³) = moles / volume (dm³)).

KEY TERMS

Pipette A glass tube used to measure volumes of liquids with a very small margin of error.

Burette A glass tube with a tap and scale for measuring liquids to the nearest 0.1 cm³.

End point The moment when the indicator changes colour in a titration showing that the moles of acid and alkali are equal.

Concordant Results that are very close together.

TIP

Remember that volumes in cm³ must be converted to dm³ in these calculations. Volume in dm³ is found by dividing volume in cm³ by 1000.

TIP

Results from titrations are usually given to three significant figures as a typical titration is accurate to within 1%.

Example

In a titration, 25.0 cm³ of 0.500 mol/dm³ sodium hydroxide solution reacted with 20.0 cm³ of a solution of nitric acid. Find the concentration of the nitric acid solution in mol/dm³.

Answer

$HNO_3 + NaOH \rightarrow NaNO_3 + H_2O$

Moles NaOH = concentration (mol/dm³) × volume (dm³)

$$= 0.500 \times \frac{25.0}{1000}$$

$$= 0.0125 \, mol$$

From the equation, 1 mole of NaOH reacts with 1 mole of HNO_3, so moles of HNO_3 = 0.0125

Concentration HNO_3 (mol/dm³) $= \dfrac{moles}{volume} = \dfrac{0.0125}{20.0/1000}$

$$= 0.625 \, mol/dm³$$

Example

In a titration, 25.0 cm³ of sodium hydroxide solution reacted with 22.9 cm³ of 0.600 mol/dm³ solution of sulfuric acid. Find the concentration of the sodium hydroxide solution in mol/dm³ and g/dm³. Give your answers to 3 significant figures.

Answer

$H_2SO_4 + 2NaOH \rightarrow Na_2SO_4 + 2H_2O$

Moles H_2SO_4 = concentration (mol/dm³) × volume (dm³)

$$= 0.600 \times \frac{22.9}{1000}$$

$$= 0.01374 \, mol$$

From the equation, 1 mole of H_2SO_4 reacts with 2 moles of NaOH, so moles of NaOH = 2 × 0.01374 = 0.02748

Concentration NaOH (mol/dm³) $= \dfrac{moles}{volume} = \dfrac{0.02748}{25.0/1000}$

$$= \textbf{1.10 mol/dm³}$$

Concentration NaOH (g/dm³) = M_r × concentration (mol/dm³)
$$= 40 \times 1.10 = 44.0$$

Test yourself

62 In a titration, 25.0 cm³ of 0.100 mol/dm³ sodium hydroxide solution reacted with 30.0 cm³ of a solution of hydrochloric acid. Find the concentration of the hydrochloric acid in mol/dm³. Give your answer to 3 significant figures.
$HCl + NaOH \rightarrow NaCl + H_2O$

63 In a titration, 25.0 cm³ of potassium hydroxide solution reacted with 27.5 cm³ of 0.500 mol/dm³ sulfuric acid solution. Find the concentration of the potassium hydroxide in mol/dm³. Give your answer to 3 significant figures.
$H_2SO_4 + 2KOH \rightarrow K_2SO_4 + 2H_2O$

64 In a titration, 20.0 cm³ of sodium carbonate solution reacted with 24.1 cm³ of 0.400 mol/dm³ hydrochloric acid solution. Find the concentration of the sodium carbonate solution in mol/dm³ and g/dm³. Give your answers to 3 significant figures.
$Na_2CO_3 + 2HCl \rightarrow 2NaCl + CO_2 + H_2O$

65 Citric acid is found in many fruits and fruit juices. A titration was carried out to find the concentration of a solution of citric acid. 25.0 cm³ of the citric acid solution reacted with 33.5 cm³ of 0.050 mol/dm³ sodium hydroxide solution. Calculate the concentration of the citric acid solution in mol/dm³. In the equation below, the citric acid is represented as H_3A. Give your answer to 3 significant figures.
$H_3A + 3NaOH \rightarrow Na_3A + 3H_2O$

66 Vinegar is a solution containing ethanoic acid (CH_3COOH). In a titration, 25.0 cm³ of vinegar reacted with 28.2 cm³ of 0.100 mol/dm³ sodium hydroxide solution. Find the concentration of the ethanoic acid in vinegar in mol/dm³. Give your answer to 3 significant figures.
$CH_3COOH + NaOH \rightarrow CH_3COONa + H_2O$

67 25.0 cm³ of limewater (calcium hydroxide solution) reacted with 22.8 cm³ of 0.0500 mol/dm³ hydrochloric acid solution in a titration. Find the concentration of the limewater in mol/dm³ and g/dm³. Give your answer to 3 significant figures.
$2HCl + Ca(OH)_2 \rightarrow CaCl_2 + 2H_2O$

Show you can...

a) A titration is a very accurate experimental technique. Choose from the statements below those which will help ensure accurate results in a titration to find the concentration of a solution of potassium hydroxide using hydrochloric acid.
   A   Use a burette to measure the volume of acid.
   B   Repeat the titration.
   C   Use a pipette to measure the volume of potassium hydroxide.
   D   Use a burette to add the acid as quickly as possible.
   E   Add the acid drop by drop at the end point.
   F   Swirl the flask during titration.
   G   Use a pipette filler.

b) In a titration, the first result is not usually used to calculate the average. Choose one statement that explains why the first result is not used.
   A   The first titration result is usually lower than the others.
   B   The first titration is done without indicator.
   C   It is only used to give a rough idea of the volume needed.

Required practical 2

Determination of the reacting volumes of solutions of a strong acid and a strong alkali by titration

A solution of a metal hydroxide XOH was made up by dissolving 3.92 g of solid in 250 cm³ of water.

Questions

1 Calculate the concentration of the metal hydroxide solution in g/dm³.

25.0 cm³ of XOH was placed in a conical flask with a few drops of bromothymol blue indicator. This indicator is yellow in acid and blue in alkali. The conical flask was placed on a white tile and titrated against a standard solution of hydrochloric acid of concentration 0.500 mol/dm³.

The balanced symbol equation for the reaction is $XOH + HCl \rightarrow XCl + H_2O$

Questions

2 Describe how you would safely and accurately measure out and place 25.0 cm³ of the metal hydroxide solution in the conical flask, naming all the apparatus you would use.

3 Why is a white tile used in this practical technique?

4 State the colour change of the bromothymol blue indicator at the end point.

5 Describe how the end point was accurately determined.

6 Why is the conical flask swirled during a titration?

The results obtained in this experiment are shown below.

Titration	1	2	3
Initial burette reading in cm³	0.0	14.9	28.9
Final burette reading in cm³	14.9	28.9	42.9
Volume of HCl used in cm³	14.9		

Questions

7 Calculate the volume of HCl used in titration 2 and titration 3.

8 Calculate the mean volume of HCl which is used in this titration. Which value should not be used?

9 Calculate the number of moles of HCl used in this titration.

10 Use the balanced symbol equation to calculate the number of moles of XOH which reacted with the HCl solution.

11 Calculate the concentration of the XOH in mol/dm³. Give your answer to 3 significant figures.

12 Using your answer to (1) and (11) calculate the relative formula mass (M_r) of XOH and identify the element X.

Another variation of this practical is found on page 112 in Chapter 4.

Chapter review questions

1 Calculate the relative formula mass (M_r) of the following substances.

    a) oxygen, O_2

    b) propane, C_3H_8

    c) magnesium sulfate, $MgSO_4$

    d) calcium nitrate, $Ca(NO_3)_2$

2 Sodium reacts with oxygen to make sodium oxide as shown in this equation:
$$4Na + O_2 \rightarrow 2Na_2O$$

    a) Show that the sum of the relative formula masses of all the reactants equals the sum of all the relative formula masses of the products.

    b) In a reaction, 11.5 g of sodium reacts with 4.0 g of oxygen. What mass of sodium oxide is formed?

3 When calcium carbonate is heated it decomposes. In a reaction 10 g of calcium carbonate was heated. At the end of the reaction, only 5.6 g of solid was left. What has happened to the other 4.4 g?

4 In a reaction to make iron metal, it was expected that 200 g of iron would be formed. However, only 180 g was produced.

    a) Calculate the percentage yield for this reaction.

    b) Give three possible reasons why the percentage yield was less than 100%.

5 a) Calculate the atom economy to make iron by heating iron oxide with carbon in the following reaction. Give your answer to 3 significant figures.

    $$Fe_2O_3 + 3C \rightarrow 2Fe + 3CO$$

    b) Calculate the atom economy to make ammonia by reacting hydrogen with nitrogen in the following reaction.

    $$3H_2 + N_2 \rightarrow 2NH_3$$

    c) Calculate the atom economy to make hydrogen by reacting of methane with steam in the following reaction. Give your answer to 3 significant figures.

    $$CH_4 + H_2O \rightarrow 3H_2 + CO$$

6 What is the mass of one mole of the following substances?

    a) potassium, K

    b) nitrogen, N_2

    c) sucrose, $C_{12}H_{22}O_{11}$

7 What is the mass of each of the following?

    a) 2.8 moles of chlorine, Cl_2

    b) 0.05 moles of methanol, CH_3OH

8 How many moles are there in each of the following?

    a) 2.4 g of oxygen, O_2

    b) 10 kg of iron oxide, Fe_2O_3

    c) 0.5 tonnes of ammonium nitrate, NH_4NO_3

    d) 25 mg of platinum, Pt

9 a) What mass of water molecules is the same number of molecules as the number of molecules in 88 g of carbon dioxide molecules?

    b) What mass of oxygen molecules is the same number of molecules as the number of atoms in 10 g of calcium atoms?

10 Hydrogen can be made by the reaction of methane with steam. What mass of hydrogen is formed from 80 g of methane?
$$CH_4 + H_2O \rightarrow 3H_2 + CO$$

11 Iron for welding railway lines together is produced from the reaction of iron oxide with aluminium. How much iron oxide will react with 1.00 kg of aluminium? Give your answer to 3 significant figures.
$$Fe_2O_3 + 2Al \rightarrow 2Fe + Al_2O_3$$

12 When 1.70 g of sodium nitrate ($NaNO_3$) is heated, 1.38 g of sodium nitrite ($NaNO_2$) and 0.32 g of oxygen (O_2) are formed. By calculating molar ratios, deduce the balanced equation for this reaction.

13 a) Calculate the concentration in mol/dm^3 of the solution formed when 0.25 moles of Na_2CO_3 is dissolved in 200 cm^3 of solution.

   b) Calculate the concentration in g/dm^3 of a solution of KOH with concentration 0.40 mol/dm^3.

14 In a titration, 25.0 cm^3 of 0.200 mol/dm^3 potassium hydroxide solution reacted with 28.9 cm^3 of a solution of nitric acid. Find the concentration of the nitric acid in mol/dm^3. Give your answer to 3 significant figures.
   $$HNO_3 + KOH \rightarrow KNO_3 + H_2O$$

15 In a titration, 20.0 cm^3 of a solution of sulfuric reacted with 16.4 cm^3 of 0.400 mol/dm^3 sodium hydroxide solution. Find the concentration of the sulfuric acid in mol/dm^3 and g/dm^3. Give your answer to 3 significant figures.
   $$H_2SO_4 + 2NaOH \rightarrow Na_2SO_4 + 2H_2O$$

16 Calculate the volume of the following gases at room temperature and pressure.

   a) 1.5 moles of methane, CH_4 b) 14 g of nitrogen, N_2

17 Calculate how many moles there are of each of the following gases given that their volumes are measured at room temperature and pressure. Give your answers to 3 significant figures.

   a) 8 dm^3 of argon, Ar   b) 1000 cm^3 of carbon dioxide, CO_2

18 Calculate the volume of oxygen that reacts with 2 dm^3 of ethene (C_2H_4) and the volume of carbon dioxide formed with the volume of all gases measured at the same temperature and pressure?
   $$C_2H_4(g) + 3O_2(g) \rightarrow 2CO_2(g) + 2H_2O(l)$$

19 The salt potassium sulfate can be made by reaction of sulfuric acid with potassium hydroxide.
   $$H_2SO_4 + 2KOH \rightarrow K_2SO_4 + 2H_2O$$
   How many moles of potassium sulfate would be made if:

   a) 10 moles of sulfuric acid was reacted with 10 moles of potassium hydroxide?

   b) 10 moles of sulfuric acid was reacted with 15 moles of potassium hydroxide?

   c) 10 moles of sulfuric acid was reacted with 25 moles of potassium hydroxide?

20 Chromium metal can be made when chromium oxide reacts with aluminium as shown in the equation below. What mass of chromium can be made when 30.4 g of chromium oxide reacts with 13.5 g of aluminium? One of the reagents is in excess. Give your answer to 3 significant figures.
   $$Cr_2O_3 + 2Al \rightarrow 2Cr + Al_2O_3$$

21 What volume of hydrogen gas is formed, measured at room temperature and pressure, when 5.75 g of sodium reacts with water?
   $$2Na(s) + 2H_2O(l) \rightarrow 2NaOH(aq) + H_2(g)$$

22 What mass of ethane burns in 600 cm^3 of oxygen, measured at room temperature and pressure, in the reaction below.
   $$2C_2H_6(g) + 7O_2(g) \rightarrow 4CO_2(g) + 6H_2O(l)$$

23 What volume of 0.100 mol/dm^3 hydrochloric acid solution reacts with 1 g of calcium?
   $$Ca + 2HCl \rightarrow CaCl_2 + H_2$$

24 What volume of carbon dioxide gas, measured at room temperature and pressure, will be formed when 10 g of sodium carbonate reacts with 100 cm^3 of 2.0 mol/dm^3 sulfuric acid solution? One of the reagents is in excess.
   $$Na_2CO_3(s) + H_2SO_4(aq) \rightarrow Na_2SO_4(aq) + H_2O(l) + CO_2(g)$$

Practice questions

1 Which one of the following would contain the same number of moles as 6 g of magnesium? [1 mark]

A 3 g of carbon   B 6 g of carbon

C 20 g of calcium   D 40 g of calcium

2 What is the mass of one mole of calcium nitrate, $Ca(NO_3)_2$? [1 mark]

A 82 g   B 164 g

C 204 g   D 220 g

3 Most metals are found naturally in rocks called ores. Some examples are shown in the table:

Metal ore	Formula of compound in ore
Galena	PbS
Haematite	Fe_2O_3
Dolomite	$CaMg(CO_3)_2$

a) Calculate the relative formula mass (M_r) of each compound in the metal ore. [3 marks]

b) The relative formula mass of another metal ore was calculated to be 102. The formula of this ore can be represented as X_2O_3. Use this information to calculate the relative atomic mass of metal X. Find the identity of metal X. [2 marks]

c) Iron is extracted from the Fe_2O_3 in haematite by reaction with carbon monoxide as shown in the equation below.

$Fe_2O_3 + 3CO \rightarrow 2Fe + 3CO_2$

   i) What is meant by the law of conservation of mass? [1 mark]

   ii) By working out the formula masses of the reactants and products in this equation show that the equation follows the law of conservation of mass. [2 marks]

4 Anaemia is a condition that occurs when the body has too few red blood cells. Anaemia often occurs in pregnant women. To prevent anaemia, iron(II) sulfate tablets can be taken to provide the iron needed by the body to produce red blood cells. One brand of iron(II) sulfate tablets contains 200 mg of iron(II) sulfate.

a) Calculate the relative formula mass (M_r) of iron(II) sulfate ($FeSO_4$). [1 mark]

b) What is the mass of one mole of iron(II) sulfate? [1 mark]

c) Calculate the number of moles of iron(II) sulfate present in one tablet. [2 marks]

5 a) In the upper part of the atmosphere, oxygen forms ozone gas O_3: $3O_2(g) \rightarrow 2O_3(g)$

Calculate the volume of ozone produced from 90 m³ of oxygen. [2 marks]

b) 1.92 g of sulfur dioxide SO_2 reacts completely with ozone to form 2.40 g of SO_3

   i) Calculate the number of moles of SO_2 used. [1 mark]

   ii) Calculate the mass of ozone which reacts. [1 mark]

   iii) Calculate the number of moles of ozone which reacts. [1 mark]

   iv) Calculate the number of moles of sulfur trioxide formed. [1 mark]

   v) Use your answers to (i) (iii) and (iv) to work out the mole ratio for this equation, and use it to balance the symbol equation for the reaction: $SO_2 + O_3 \rightarrow SO_3$ [2 marks]

6 Lead is extracted from the ore galena PbS.

a) The ore is roasted in air to produce lead(II) oxide PbO. Calculate the maximum mass of lead(II) oxide PbO produced from 4780 g of galena PbS. [3 marks]

$2PbS + 3O_2 \rightarrow 2PbO + 2SO_2$

b) The lead(II) oxide is reduced to lead by heating it with carbon in a blast furnace. The molten lead is tapped off from the bottom of the furnace.

$PbO + C \rightarrow Pb + CO$

Using your answer to part (a) calculate the maximum mass of lead that would eventually be produced. [2 marks]

c) Solid lead reacts with warm nitric acid according to equation:

$3Pb + 8HNO_3 \rightarrow 3Pb(NO_3)_2 + 2NO + 4H_2O$

20.0 cm³ of 2.4 mol/dm³ nitric acid reacted completely with solid lead. Some lead was left over.

   i) Calculate the amount, in moles, of HNO_3 used. [1 mark]

   ii) Calculate the amount, in moles, of NO gas formed. [1 mark]

   iii) Calculate the volume of NO gas in cm³ formed at room temperature and pressure. [1 mark]

7 Camping stoves use butane (C_4H_{10}) gas as a fuel. The equation for the complete combustion of butane is:

$2C_4H_{10}(g) + 13O_2(g) \rightarrow 8CO_2(g) + 10H_2O(g)$

a) 150 cm³ of butane are burned in an excess of oxygen. Calculate the volume of oxygen needed for complete combustion of butane. [1 mark]

b) A camping gas cylinder contains 0.454 kg of butane. Calculate the number of moles of butane (C_4H_{10}) in the cylinder. Give your answer to 3 significant figures. [2 marks]

8 Sodium hydrogencarbonate decomposes when it is heated. 3.36 g of sodium hydrogencarbonate were placed in a test tube and heated in a Bunsen flame for some time.

$$2NaHCO_3 \rightarrow Na_2CO_3 + H_2O + CO_2$$

a) Calculate the number of moles of sodium hydrogencarbonate used. [2 marks]

b) Calculate the number of moles of sodium carbonate formed. [1 mark]

c) Calculate the mass of sodium carbonate expected to be formed. [2 marks]

d) Calculate the volume of carbon dioxide (measured at room temperature and pressure) produced in this experiment. [2 marks]

9 a) Calcium nitrate is a Group 2 compound which can be produced in the laboratory by reacting calcium metal with nitric acid. In one experiment, 0.4 g of calcium metal was reacted with excess nitric acid.

$$Ca + 2HNO_3 \rightarrow Ca(NO_3)_2 + H_2$$

i) Calculate the volume of hydrogen gas which was produced in this reaction (1 mole of gas at room temperature and pressure has volume of 24 dm³). [3 marks]

ii) If the concentration of the nitric acid is 2.0 mol/dm³, calculate the volume of nitric acid needed to completely react with the 0.4 g of calcium. [2 marks]

iii) The calcium nitrate is often used in fertilisers. Calculate the relative formula mass of calcium nitrate.

b) Barium hydroxide is another Group 2 compound. It is often found in drain cleaners. To determine the concentration of a solution of barium hydroxide Ba(OH)₂, 25.0 cm³ of the solution was titrated with a solution of hydrochloric acid of concentration 0.200 mol/dm³. The balanced symbol equation for the reaction is

$$Ba(OH)_2 + 2HCl \rightarrow BaCl_2 + 2H_2O$$

i) Describe in detail how you would experimentally determine an accurate value for the volume of hydrochloric acid required to neutralise the 25.0 cm³ of barium hydroxide. [6 marks]

(H) ii) Which results should the student use to calculate the mean volume of acid added? [1 mark]

iii) Calculate the mean from the results. Give your answer to two decimal places. [1 mark]

iv) Calculate the number of moles of hydrochloric acid used in this titration. Give your answer to 3 significant figures. [1 mark]

v) Use the balanced symbol equation to work out the number of moles of barium hydroxide which reacted with the hydrochloric acid. [1 mark]

vi) Calculate the concentration of barium hydroxide in mol/dm³. [1 mark]

vii) Calculate the concentration of barium hydroxide in g/dm³. Give your answer to 3 significant figures. [1 mark]

10 Zinc sulfate crystals are prepared in the laboratory by reacting zinc carbonate with sulfuric acid, as shown in the equation below.

$$ZnCO_3 + H_2SO_4 \rightarrow ZnSO_4 + H_2O + CO_2$$

a) What is the maximum mass of zinc sulfate which could be formed when 2.5 g of zinc carbonate are reacted with sulfuric acid? [3 marks]

b) A student carried out this experiment and only obtained 2.8 g of zinc sulfate. Calculate the percentage yield. Give your answer to 2 significant figures. [1 mark]

c) Suggest two reasons why the percentage yield is not 100% in this reaction. [2 marks]

11 Willow bark contains salicylic acid and was once used as a painkiller. Salicylic acid is now used to manufacture aspirin. A student reacted 4.00 g of salicylic acid with 6.50 g of ethanoic anhydride.

$$C_6H_4(OH)COOH + (CH_3CO)_2O \rightarrow HOOCC_6H_4OCOCH_3 + CH_3COOH$$

| salicylic acid | ethanoic anhydride | aspirin |

a) How many moles of salicylic acid were used? [2 marks]

b) How many moles of ethanoic anhydride were present? [2 marks]

c) What is the maximum number of moles of aspirin which could be formed? [1 mark]

d) Calculate the maximum mass of aspirin which could be formed. [2 marks]

e) The student prepared 2.90 g of aspirin. Calculate the percentage yield of aspirin obtained by the student. Give your answer to 2 significant figures. [1 mark]

f) Calculate the atom economy of the reaction to prepare aspirin. Give your answer to 2 significant figures. [2 marks]

Titration	1	2	3	4
Final burette reading in cm³	23.10	45.50	22.45	22.45
Initial burette reading in cm³	0.00	23.10	0.00	0.05
Volume of HCl added in cm³	23.10	22.40	22.45	22.40

Working scientifically:
Interconverting units

Units

Many of the calculations used in chemistry will require different units. It is important that you can convert between units.

Volume

Volume is measured in cm^3 or dm^3 (cubic decimetres) or m^3.

$1000\,cm^3 = 1\,dm^3$

You need to be able to convert between these volume units, particularly for calculations on solution volume and concentration. The flow scheme in Figure 3.20 will help you to convert between volume units.

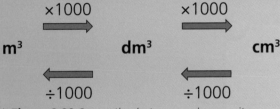

▲ **Figure 3.20** Converting between volume units.

Example

What is $15\,cm^3$ in dm^3?

Answer

To convert from cm^3 to dm^3 you need to divide by 1000

$15\,cm^3 = \dfrac{15}{1000} = 0.015\,dm^3$

Example

What is $0.4\,dm^3$ in cm^3?

Answer

To convert from dm^3 to cm^3 you need to multiply by 1000

$0.4\,dm^3 = 0.4 \times 1000 = 400\,cm^3$

Question

1 Convert the following volumes to the units shown.
   a) $25\,cm^3$ to dm^3
   b) $100\,cm^3$ to dm^3
   c) $10\,dm^3$ to cm^3
   d) $20\,dm^3$ to m^3
   e) $24\,000\,cm^3$ to dm^3

Mass

Mass can be measured in milligrams (mg), grams (g), kilograms (kg) and in tonnes.

1 tonne = 1000 kg

1 kilogram = 1000 g

1 gram = 1000 mg

The flow diagram in Figure 3.21 will help you to convert between mass units.

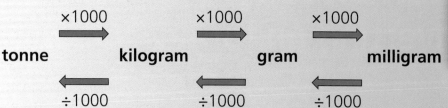

▲ **Figure 3.21** Converting between mass units.

Example

Convert 420 mg to grams.

Answer

To convert from mg to g you need to divide by 1000:

$$\frac{420}{1000} = 0.420\,g$$

Example

Convert 3.2 kg to grams.

Answer

To convert from kg to g you need to multiply by 1000:

$3.2 \times 1000 = 3200\,g$

Example

Convert 0.44 tonnes to grams.

Answer

First you need to convert from tonnes to kilograms by multiplying by 1000:

$0.44 \times 1000 = 440\,kg$

Then convert 440 kg to g by multiplying by 1000:

$440 \times 1000 = 440\,000\,g$

Example

Convert 250 mg to kg.

Answer

First you need to convert mg to g by dividing by 1000:

$$\frac{250}{1000} = 0.25\,g$$

Then convert 0.25 g to kg by dividing by 1000:

$$\frac{0.25}{1000} = 0.00025 = 2.5 \times 10^{-4}\,kg$$

Questions

2 Convert the following masses to the units shown.
   a) 25 g to kg
   b) 1 032 kg to tonnes
   c) 10 tonnes to kg
   d) 43 mg to g
   e) 6.13 tonnes to g
   f) 0.3 kg to g

3 Carry out the following unit conversions.
   a) 50 cm³ to dm³
   b) 32 000 g to tonnes
   c) 22 000 cm³ to dm³
   d) 0.7 kg to g
   e) 2.45 tonnes to g
   f) 12 cm³ to dm³

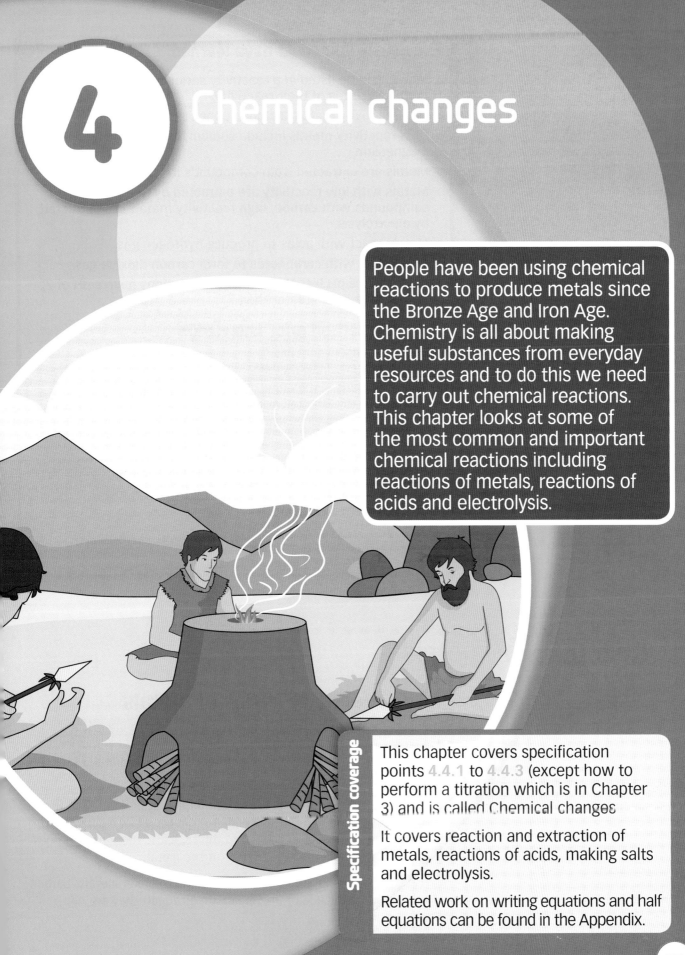

4 Chemical changes

People have been using chemical reactions to produce metals since the Bronze Age and Iron Age. Chemistry is all about making useful substances from everyday resources and to do this we need to carry out chemical reactions. This chapter looks at some of the most common and important chemical reactions including reactions of metals, reactions of acids and electrolysis.

Specification coverage

This chapter covers specification points 4.4.1 to 4.4.3 (except how to perform a titration which is in Chapter 3) and is called Chemical changes

It covers reaction and extraction of metals, reactions of acids, making salts and electrolysis.

Related work on writing equations and half equations can be found in the Appendix.

Previously you could have learned:

> Metals can be listed in a reactivity series which compares metals in terms of their reactivity.
> Low reactivity metals include copper, silver and gold.
> High reactivity metals include sodium, calcium and magnesium.
> Metals are extracted from compounds in ores.
> Metals with low reactivity are extracted by heating metal compounds with carbon; high reactivity metals are extracted by electrolysis.
> Metals react with acids to produce hydrogen gas.
> Acids react with carbonates to form carbon dioxide gas.
> Acids have a pH less than 7; neutral solutions have a pH of 7; alkalis have a pH greater than 7.

Test yourself on prior knowledge

1 Name two reactive metals.
2 Name two unreactive metals.
3 Sodium is found in the ore halite which contains sodium chloride. How is sodium extracted from this ore?
4 What gas is made when acids react with:
  a) metals
  b) carbonates?
5 State whether each of the following solutions is acidic, neutral or alkaline.
  a) pH 11
  b) pH 2
  c) pH 7
  d) pH 6

Reactions of metals

○ The reactivity series of metals

Metals have many uses. For example, they are used in electrical cables, cars, aeroplanes, buildings, mobile phones and computers. Some metals, such as gold, are very unreactive. Other metals, such as sodium, are very reactive.

The reactivity series of metals shows the metals in order of reactivity. This order can be worked out by comparing how metals react with substances such as oxygen, water and dilute acids. The more vigorous the reaction, the higher the reactivity of the metal.

The reactivity series in Figure 4.1 shows some common metals. Carbon and hydrogen are included for comparison although they are non-metals.

	Metal		Reaction with oxygen	Reaction with water	Reaction with acids
Most reactive ↑	Potassium	K	Burns to form oxide	Reacts and gives off $H_2(g)$	Reacts violently and gives off $H_2(g)$
	Sodium	Na			
	Lithium	Li			
	Calcium	Ca			Reacts and gives off $H_2(g)$
	Magnesium	Mg			
	Aluminium	Al			
	Carbon	C			
	Zinc	Zn	Forms oxide when heated (metal powder burns)	No reaction	Reacts slowly and gives off $H_2(g)$
	Iron	Fe			
	Tin	Sn			
	Lead	Pb			
	Hydrogen	H			
	Copper	Cu	Forms oxide when heated	No reaction	No reaction
	Silver	Ag	No reaction		
	Gold	Au			
Least reactive	Platinum	Pt			

▲ **Figure 4.1** The reactivity series of some common metals.

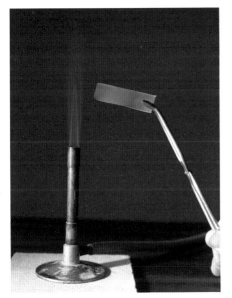

▲ **Figure 4.2** Copper reacts with oxygen but does not burn.

KEY TERMS ⭐

Oxidation A reaction where a substance gains oxygen and/or loses electrons.

Reduction A reaction where a substance loses oxygen and/or gains electrons.

Reaction with oxygen

Most metals react with oxygen. Reactive metals burn when they are heated as they react with oxygen in the air. The metals in the middle of the reactivity series react with oxygen when heated and can burn if the metal is powdered. Copper, a low reactivity metal, reacts with oxygen forming a layer of copper oxide on the surface of the copper but does not burn (Figure 4.2). Metals with very low reactivity, such as gold, do not react with oxygen at all.

When metals react with oxygen they form a metal oxide:

metal + oxygen → metal oxide

For example:

sodium + oxygen → sodium oxide
$4Na + O_2 → 2Na_2O$

copper + oxygen → copper oxide
$2Cu + O_2 → 2CuO$

These are examples of oxidation reactions. An oxidation reaction can be defined as a reaction where a substance gains oxygen. A reduction reaction can be defined as a reaction where a substance loses oxygen.

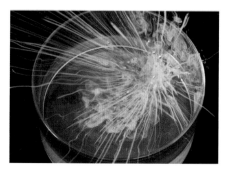

Reaction with water

Most metals do not react with cold water. However, metals with high reactivity react with cold water to form a metal hydroxide and hydrogen gas (Figure 4.3).

$$\text{metal} + \text{water} \rightarrow \text{metal hydroxide} + \text{hydrogen}$$

Table 4.1 shows how metals react with cold water.

◄ **Figure 4.3** Potassium reacts vigorously with water.

Table 4.1

Metal	Observations	Equation for reaction
Potassium (K)	Fizzes, melts, floats and moves on the surface of the water, lilac flame (Figure 4.3)	potassium + water → potassium hydroxide + hydrogen $2K + 2H_2O \rightarrow 2KOH + H_2$
Sodium (Na)	Fizzes, melts, floats and moves on the surface of the water (sometimes there is a yellow-orange flame)	sodium + water → sodium hydroxide + hydrogen $2Na + 2H_2O \rightarrow 2NaOH + H_2$
Lithium (Li)	Fizzes, floats and moves on the surface of the water	lithium + water → lithium hydroxide + hydrogen $2Li + 2H_2O \rightarrow 2LiOH + H_2$
Calcium (Ca)	fizzes, white solid forms	calcium + water → calcium hydroxide + hydrogen $Ca + 2H_2O \rightarrow Ca(OH)_2 + H_2$
Magnesium (Mg)	Very slow reaction	magnesium + water → magnesium hydroxide + hydrogen $Mg + 2H_2O \rightarrow Mg(OH)_2 + H_2$
Zinc (Zn)	No reaction	
Iron (Fe)	No reaction	
Copper (Cu)	No reaction	

▲ **Figure 4.4** Magnesium fizzing as it reacts with an acid giving off hydrogen gas.

Reaction with dilute acids

Metals that are more reactive than hydrogen react with dilute acids. When metals react with dilute acids they form a salt and hydrogen gas.

$$\text{metal} + \text{acid} \rightarrow \text{metal salt} + \text{hydrogen}$$

Hydrochloric acid makes chloride salts. Sulfuric acid makes sulfate salts. Nitric acid makes nitrate salts.

With high reactivity metals, the reaction with acids is explosive due to the hydrogen that is formed igniting. Metals that are less reactive than hydrogen do not react with dilute acids.

Table 4.2 shows how metals react with dilute hydrochloric acid.

Table 4.2

Metal	Observation	Equation for reaction
Potassium (K)	Explosive	potassium + hydrochloric acid → potassium chloride + hydrogen $2K + 2HCl \rightarrow 2KCl + H_2$
Sodium (Na)	Explosive	sodium + hydrochloric acid → sodium chloride + hydrogen $2Na + 2HCl \rightarrow 2NaCl + H_2$
Lithium (Li)	Explosive	lithium + hydrochloric acid → lithium chloride + hydrogen $2Li + 2HCl \rightarrow 2LiCl + H_2$
Calcium (Ca)	Fizzes	calcium + hydrochloric acid → calcium chloride + hydrogen $Ca + 2HCl \rightarrow CaCl_2 + H_2$
Magnesium (Mg)	Fizzes (Figure 4.4)	magnesium + hydrochloric acid → magnesium chloride + hydrogen $Mg + 2HCl \rightarrow MgCl_2 + H_2$
Zinc (Zn)	Fizzes slowly	zinc + hydrochloric acid → zinc chloride + hydrogen $Zn + 2HCl \rightarrow ZnCl_2 + H_2$
Iron (Fe)	Fizzes slowly	iron + hydrochloric acid → iron chloride + hydrogen $Fe + 2HCl \rightarrow FeCl_2 + H_2$
Copper (Cu)	No reaction	

What happens to metal atoms when they react?

When metal atoms react, they lose electrons to form positive ions. For example:

- When sodium atoms react with oxygen, the sodium atoms lose electrons and form sodium ions (Na^+) in the product sodium oxide.
- When calcium atoms react with water, the calcium atoms lose electrons and form calcium ions (Ca^{2+}) in the product calcium hydroxide.
- When zinc atoms react with hydrochloric acid, the zinc atoms lose electrons and form zinc ions (Zn^{2+}) in the product zinc chloride.

The greater the tendency of a metal to lose electrons to form ions, the more reactive it is. Reactive metals like potassium and sodium easily lose electrons to form ions, but metals like gold and platinum do not tend to form ions and so are unreactive.

Displacement reactions

In a displacement reaction, a more reactive metal will take the place of a less reactive metal in a compound. For example, aluminium will displace iron from iron oxide because aluminium is more reactive than iron.

$$\text{aluminium} + \text{iron oxide} \rightarrow \text{aluminium oxide} + \text{iron}$$
$$2Al + Fe_2O_3 \rightarrow Al_2O_3 + 2Fe$$

This reaction is used to weld railway lines together (Figure 4.5). A mixture of aluminium and iron oxide is placed over the gap between the railway lines and the reaction started. The reaction gets very hot and produces molten iron which flows into a mould, cools and solidifies to weld the lines together.

Displacement reactions also take place in solution. For example, copper will displace silver from silver nitrate solution because copper is more reactive than silver. Copper nitrate is blue and so the solution turns blue as the copper nitrate is formed (Figure 4.6).

$$\text{copper} + \text{silver nitrate} \rightarrow \text{copper nitrate} + \text{silver}$$
$$Cu + 2AgNO_3 \rightarrow Cu(NO_3)_2 + 2Ag$$

KEY TERM

Displacement reaction Reaction where a more reactive element takes the place of a less reactive element in a compound.

▲ **Figure 4.5** A displacement reaction is used to weld railway lines together.

▲ **Figure 4.6** Silver forms on the copper wire as silver is displaced from the silver nitrate solution.

Test yourself

1 Complete the following word equations, or write *no reaction* if no reaction would take place.

a) calcium + oxygen

b) gold + oxygen

c) copper + water

d) lithium + water

e) calcium + nitric acid

f) copper + sulfuric acid

g) zinc + hydrochloric acid

h) iron + sulfuric acid

i) tin + magnesium chloride

j) zinc + lead nitrate

k) magnesium + aluminium sulfate

2 Magnesium (Mg) reacts with oxygen (O_2) to form magnesium oxide (MgO).

a) Write a word equation for this reaction.

b) Write a balanced equation for this reaction.

c) What happens to the magnesium atoms in this reaction in terms of electrons?

3 Calcium (Ca) reacts with water to form calcium hydroxide ($Ca(OH)_2$) and hydrogen (H_2).

a) Describe what you would see in this reaction.

b) Write a word equation for this reaction.

c) Write a balanced equation for this reaction.

d) What happens to the calcium atoms in this reaction in terms of electrons?

4 Magnesium (Mg) and zinc (Zn) both react with sulfuric acid.

a) i) Which metal reacts more vigorously with sulfuric acid?

 ii) Explain, in terms of the tendency to form ions, why this metal reacts more vigorously.

b) For the reaction of magnesium with sulfuric acid:

 i) Write a word equation   ii) Write a balanced equation.

5 The metal chromium can be made in a displacement reaction between aluminium and chromium oxide.

aluminium + chromium oxide → aluminium oxide + chromium

$2Al + Cr_2O_3 \rightarrow Al_2O_3 + 2Cr$

a) Why does aluminium displace chromium in this reaction?

b) Which substance is oxidised in this reaction?

c) Which substance is reduced in this reaction?

Show you can...

To determine the order of reactivity of the metals copper, magnesium, nickel and zinc each metal was heated with the oxides of other metals and the results obtained recorded in the table below.

	Copper	Magnesium	Nickel	Zinc
Copper oxide		Reaction	Reaction	Reaction
Magnesium oxide	No reaction		No reaction	No reaction
Nickel oxide	No reaction	Reaction		Reaction
Zinc oxide	No reaction	Reaction	No reaction	

a) Determine the order of the four metals from the most reactive to the least reactive.

b) Write a balanced chemical equation for the reaction of nickel oxide (NiO) with magnesium.

c) From the list below, write word and balanced chemical equations for all reactions which occur (when nickel reacts, it forms Ni^{2+} ions).

i) nickel + hydrochloric acid   ii) zinc + water

iii) nickel + water   iv) zinc + sulfuric acid

v) magnesium + zinc oxide

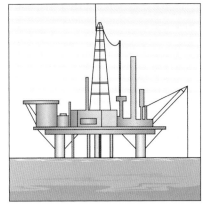

Oxidation and reduction in terms of electrons

Oxidation can be defined as a reaction where a substance gains oxygen. However, a better definition of oxidation is a reaction where a substance loses electrons.

Reduction can be defined as a reaction where a substance loses oxygen. However, a better definition of reduction is a reaction where a substance gains electrons.

One way to remember this is the phrase OIL RIG (Figure 4.7).

When metals react with oxygen, water and acids, the metal atoms lose electrons and form metal ions (Table 4.3). This means that each reaction can be defined as oxidation in terms of the loss of electrons. However, the reaction with oxygen is the only one that can be defined as oxidation in terms of gaining oxygen.

O xidation
I s
L oss
R eduction } of electrons
I s
G ain

▲ **Figure 4.7**

Table 4.3 Reactions of metals

Reaction	General equation	Oxidation in terms of gaining oxygen	Oxidation in terms of losing electrons
+ oxygen	metal + oxygen → metal oxide	Metal gains oxygen	Metal atoms lose electrons to form metal ions in the metal oxide
+ water	metal + water → metal hydroxide + hydrogen		Metal atoms lose electrons to form metal ions in the metal hydroxide
+ acids	metal + acid → metal salt + hydrogen		Metal atoms lose electrons to form metal ions in the metal salt

KEY TERM

Redox reaction A reaction where both reduction and oxidation take place.

Displacement reactions involve oxidation and reduction. In some reactions this can be explained in terms of oxygen and in terms of electrons. This is the case, for example, in the displacement of iron from iron oxide by aluminium (Figure 4.8).

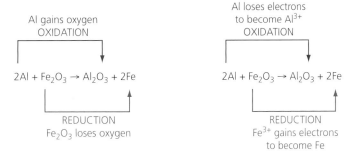

▲ **Figure 4.8** The displacement of iron from iron oxide by aluminium.

In other displacement reactions, only the definitions in terms of electrons can be used to define the reaction as involving oxidation and reduction. This is the case, for example, in the displacement of silver from silver nitrate by copper (Figure 4.9).

In a reaction in which one substance loses electrons, another substance gains those electrons. This means that both reduction and oxidation take place and these are called redox reactions (**red**uction–**ox**idation).

Cu loses electrons
to become Cu^{2+}
OXIDATION

$Cu + 2AgNO_3 \rightarrow Cu(NO_3)_2 + 2Ag$

REDUCTION
Ag^+ gains electrons to become Ag

▲ **Figure 4.9** The displacement of silver from silver nitrate by copper.

Writing ionic equations and/or half equations for displacement reactions

Ionic equations and/or half equations can be written for displacement reactions. (See pages 288–291 for more help on writing these equations.)

Example

Write two half equations for the displacement of iron from iron oxide by aluminium.

aluminium + iron oxide → aluminium oxide + iron
$2Al$ + Fe_2O_3 → Al_2O_3 + $2Fe$

Answer

In this reaction, the Al atoms become Al^{3+} ions in Al_2O_3 while the Fe^{3+} ions in Fe_2O_3 become Fe atoms.

The two half equations for this are:

Al atoms lose electrons to form Al^{3+} ions: $Al - 3e^- \rightarrow Al^{3+}$ (or $Al \rightarrow Al^{3+} + 3e^-$)

Fe^{3+} ions in Fe_2O_3 gain electrons to form Fe atoms: $Fe^{3+} + 3e^- \rightarrow Fe$

Example

Write an overall ionic equation and two half equations for the displacement of silver from silver nitrate by copper.

copper + silver nitrate → copper nitrate + silver
Cu + $2AgNO_3$ → $Cu(NO_3)_2$ + $2Ag$

Answer

In this reaction, the Cu atoms become Cu^{2+} ions in $Cu(NO_3)_2$ while the Ag^+ ions in $AgNO_3$ become Ag atoms. We can leave out the NO_3^- ions from the ionic equation as they do not change.

Therefore, the overall ionic equation is: $Cu + 2Ag^+ \rightarrow Cu^{2+} + 2Ag$

The two half equations for this are: Cu atoms lose electrons to form Cu^{2+} ions: $Cu - 2e^- \rightarrow Cu^{2+}$ (or $Cu \rightarrow Cu^{2+} + 2e^-$)

Ag^+ ions in $AgNO_3$ gain electrons to form Ag atoms: $Ag^+ + e^- \rightarrow Ag$

Example

Write an overall ionic equation and two half equations for the displacement of copper from copper sulfate by iron. Identify which half equation represents a reduction process and which represents an oxidation process.

iron + copper sulfate → iron sulfate + copper
Fe + $CuSO_4$ → $FeSO_4$ + Cu

Answer

In this reaction, the Fe atoms become Fe^{2+} ions in $FeSO_4$ while the Cu^{2+} ions in $CuSO_4$ become Cu atoms. We can leave out the SO_4^{2-} ions from the ionic equation as they do not change.

Therefore, the overall ionic equation is: $Fe + Cu^{2+} \rightarrow Fe^{2+} + Cu$

The two half equations for this are:

Fe atoms lose electrons to form Fe^{2+} ions: $Fe - 2e^- \rightarrow Fe^{2+}$ (or $Fe \rightarrow Fe^{2+} + 2e^-$)

Cu^{2+} ions in $CuSO_4$ gain electrons to form Cu atoms: $Cu^{2+} + 2e^- \rightarrow Cu$

The reduction half equation is: $Cu^{2+} + 2e^- \rightarrow Cu$

The oxidation half equation is: $Fe - 2e^- \rightarrow Fe^{2+}$ (or $Fe \rightarrow Fe^{2+} + 2e^-$)

Test yourself

6 Rubidium is a metal in Group 1 of the periodic table. It reacts vigorously with water.

   a) Write a word equation for this reaction.

   b) Write a balanced equation for this reaction.

   c) What happens to the rubidium atoms in this reaction in terms of electrons?

   d) Are the rubidium atoms oxidised or reduced in this reaction? Explain your answer.

7 Magnesium reacts with copper oxide in a displacement reaction to form copper.

   $Mg + CuO \rightarrow MgO + Cu$

   a) Explain, in terms of oxygen, why the magnesium is oxidised in this reaction.

   b) Write a half equation to show the oxidation of magnesium atoms to magnesium ions in this reaction.

   c) Explain, in terms of electrons, why the magnesium is oxidised in this reaction.

   d) Explain, in terms of oxygen, why the copper oxide is reduced in this reaction.

   e) Write a half equation to show the reduction of copper ions to copper atoms in this reaction.

   f) Explain, in terms of electrons, why the copper oxide is reduced in this reaction.

8 Write an overall ionic equation and two half equations for each of the following displacement reactions.

   a) Displacement of copper from copper sulfate by zinc.
      zinc + copper sulfate $\rightarrow$ zinc sulfate + copper
      $Zn + CuSO_4 \rightarrow ZnSO_4 + Cu$

   b) Displacement of silver from silver nitrate by zinc.
      zinc + silver nitrate $\rightarrow$ zinc nitrate + silver
      $Zn + 2AgNO_3 \rightarrow Zn(NO_3)_2 + 2Ag$

   c) Displacement of copper from copper sulfate by aluminium.
      aluminium + copper sulfate $\rightarrow$ aluminium sulfate + copper
      $2Al + 3CuSO_4 \rightarrow Al_2(SO_4)_3 + 3Cu$

9 Zinc displaces iron from a solution of iron(II) sulfate. For this reaction:

   a) Write a word equation for this reaction.

   b) Write a balanced equation for this reaction.

   c) Write an ionic equation for this reaction.

   d) Write the two half equations for this reaction.

   e) Identify which half equation is a reduction process.

   f) Identify which half equation is an oxidation process.

   g) Explain why this is a redox reaction.

10 Magnesium displaces silver from a solution of silver nitrate. For this reaction:

   a) Write a word equation for this reaction.

   b) Write a balanced equation for this reaction.

   c) Write an ionic equation for this reaction.

   d) Write the two half equations for this reaction.

   e) Identify which half equation is a reduction process.

   f) Identify which half equation is an oxidation process.

   g) Explain why this is a redox reaction.

Show you can...

Each of the following reactions can be classified as an oxidation or a reduction reaction.

Reaction 1: $CH_4 + 2O_2 \rightarrow CO_2 + 2H_2O$

Reaction 2: $CuO + Mg \rightarrow Cu + MgO$

Reaction 3: $ZnO + H_2 \rightarrow Zn + H_2O$

a) Write the formula of the substance which is oxidised in reaction 1.

b) Explain which substance in reaction 2 is reduced.

c) Explain which substance in reaction 3 is oxidised.

d) Write a balanced ionic equation for reaction 2 and state and use a half equation to explain which species is oxidised.

Extraction of metals

An ore is a compound from the ground that contains enough metal to make it economically viable to extract it.

○ Where do metals come from?

A few metals, such as gold and platinum, occur naturally on Earth as elements. These are metals with very low reactivity.

Most metals are only found on Earth in compounds. For example, iron is often found in the compound iron oxide and aluminium in the compound aluminium oxide. In order to extract the metal from these compounds, a chemical reaction is required.

The metal compounds are found in rocks called **ores**. An ore is a rock from which a metal can be extracted for profit (Figure 4.10).

▲ **Figure 4.10** Aluminium metal is extracted from the aluminium oxide (Al_2O_3) in the ore bauxite.

○ Methods of extraction

Most of the compounds from which metals are extracted are oxides. In order to extract the metal, the oxygen is removed in a reduction reaction. The way in which this is done depends on the reactivity of the metal (Figure 4.11).

Metals that are less reactive than carbon can be extracted by heating the metal oxide with carbon. For example, iron is extracted by heating iron oxide with carbon. The iron oxide is reduced in this reaction (Figure 4.12).

Metals that are more reactive than carbon can be extracted by electrolysis. This is studied in detail on page 117.

When metals are extracted from compounds, the metal ions in the compounds gain electrons. For example, when iron is extracted from iron oxide, Fe^{3+} ions in the iron oxide gain electrons to form Fe atoms. This is reduction because the iron oxide loses oxygen and also because the Fe^{3+} ions in the iron oxide gain electrons.

	Metal		Method of extraction
Most reactive	Potassium	K	Electrolysis
	Sodium	Na	
	Lithium	Li	
	Calcium	Ca	
	Magnesium	Mg	
	Aluminium	Al	
	Carbon	C	
	Zinc	Zn	Heat with carbon
	Iron	Fe	
	Tin	Sn	
	Lead	Pb	
	Copper	Cu	
	Silver	Ag	Metals found as elements
	Gold	Au	
Least reactive	Platinum	Pt	

▲ Figure 4.11

C gains oxygen
OXIDATION

$Fe_2O_3 + 3C \rightarrow 2Fe + 3CO$

REDUCTION
Fe_2O_3 loses oxygen

▲ Figure 4.12

Test yourself

11 Most metals are extracted from compounds found in rocks called ores. A few metals are found as elements.

   a) Why are some metals, such as gold, found as elements and not in compounds?

   b) What is an ore?

12 Which method is used to extract the following metals?

   a) iron

   b) aluminium

   c) magnesium

   d) platinum

   e) zinc

13 Tin is extracted from the ore tin oxide by heating with carbon.

   a) Write a word equation for this reaction.

   b) Explain why this is a reduction reaction.

Reactions of acids

KEY TERMS

Acid Solution with a pH less than 7; produces H^+ ions in water.

Aqueous Dissolved in water.

Alkali Solution with a pH more than 7; produces OH^- ions in water.

○ What are acids and alkalis?

An **acid** is a substance that produces hydrogen ions, **H^+**, in **aqueous** solution. For example, solutions of:

- hydrochloric acid (HCl) contain hydrogen (**H^+**) ions and chloride (Cl^-) ions
- sulfuric acid (H_2SO_4) contain hydrogen (**H^+**) ions and sulfate (SO_4^{2-}) ions
- nitric acid (HNO_3) contain hydrogen (**H^+**) ions and nitrate (NO_3^-) ions.

An **alkali** is a substance that produces hydroxide ions, **OH^-**, in aqueous solution. For example, solutions of

- sodium hydroxide (NaOH) contain sodium (Na^+) ions and hydroxide (**OH^-**) ions
- potassium hydroxide (KOH) contain potassium (K^+) ions and hydroxide (**OH^-**) ions
- calcium hydroxide ($Ca(OH)_2$) contain calcium (Ca^{2+}) ions and hydroxide (**OH^-**) ions.

○ The pH scale

The pH scale is a measure of how acidic or alkaline a solution is. A solution with a pH of 7 is neutral, whereas a solution with a pH below 7 is acidic and one with a pH above 7 is alkaline. The further away from 7 the pH is, the more acidic or alkaline the solution is (Figure 4.13).

The scale is often shown as running from 0 to 14, but it does go further in both directions. For example, it is common for the solutions of acids in school laboratories to have a pH that is less than 0 (typically about −0.3).

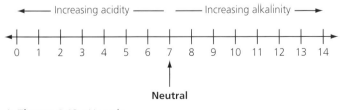

▲ **Figure 4.13** pH scale.

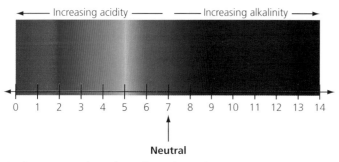

← Increasing acidity ——— | ——— Increasing alkalinity →

0 1 2 3 4 5 6 7 8 9 10 11 12 13 14

↑
Neutral

▲ **Figure 4.14** The colours for universal indicator.

7.0

▲ **Figure 4.15** pH probe.

The approximate pH of a solution can be measured using universal indicator solution. A few drops of the indicator is added to the solution. The colour is compared to a colour chart to give the approximate pH of the solution (Figure 4.14).

A more accurate way of finding the pH of a solution is to use a pH probe (Figure 4.15). There are different types but the probe is dipped into the solution and the pH shown on the display, often to 1 or 2 decimal places.

The pH of a solution is based on the concentration of H^+ ions in the solution. The higher the concentration of H^+ ions the lower the pH.

As the pH decreases by one unit, the concentration of hydrogen ions increases by a factor of 10. For example, a solution with a pH of 2 has a concentration of H^+ ions that is 10 times greater than one with a pH of 3. A solution with a pH of 1 has a concentration of H^+ ions that is 100 times greater than one with a pH of 3.

In a neutral solution, the concentration of H^+ ions equals the concentration of OH^- ions.

Show you can...

In an experiment a sample of human saliva was removed from the mouth every five minutes after a meal and the pH value determined. The graph shows how the pH value of the saliva changed.

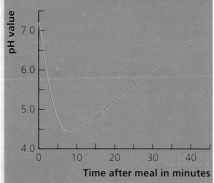

pH value

7.0

6.0

5.0

4.0

0   10   20   30   40

Time after meal in minutes

a) How were the pH values of the saliva likely to have been determined in this experiment?
b) When the pH in the mouth is 5.0 or less tooth decay occurs. Use the graph to find the time after the meal at which teeth would start to decay.
c) At what time is the pH of the saliva most acidic?

Test yourself

14 Classify each of the following solutions as acidic, neutral or alkaline.
  a) A solution with pH 9.
  b) A solution with pH 3.
  c) A solution with pH 0.
  d) A solution with pH 7.
15 a) Three solutions had pH values of 8, 11 and 13. Which one was the most alkaline?
  b) Three solutions had pH values of –1, 2 and 5. Which one was the most acidic?
16 a) Give two ways in which the pH of a solution can be measured.
  b) Which method will give the most accurate value?
17 a) Which ion do aqueous solutions of acids all contain?
  b) Which ion do aqueous solutions of alkalis all contain?
18 The table gives some information about three solutions.

Solution	A	B	C
pH	4	3	1

  a) Which solution has the highest concentration of H^+ ions?
  b) By what factor is concentration of H^+ ions in solution **B** bigger or smaller than solution **A**?
  c) By what factor is concentration of H^+ ions in solution **C** bigger or smaller than solution **A**?
19 A solution has a pH of 7. Comment on the concentration of H^+ ions compared to the concentration of OH^- ions.

○ Strong and weak acids

Hydrogen chloride (HCl), hydrogen sulfate (H_2SO_4) and hydrogen nitrate (HNO_3) are molecules in their pure state. When they are added to water all of their molecules break down into ions forming hydrochloric acid (HCl), sulfuric acid (H_2SO_4) and nitric acid (HNO_3). They are **strong acids** because their molecules are completely ionised in water – this means that all their molecules break into ions in water (Figure 4.16).

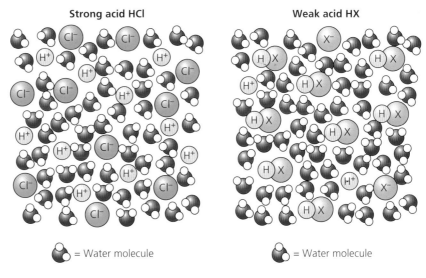

$$H-Cl\,(g) \xrightarrow{H_2O} H^+(aq) + Cl^-(aq)$$

(a) $\quad HCl(g) \xrightarrow{H_2O} H^+(aq) + Cl^-(aq)$

(b) $\quad H_2SO_4\,(l) \xrightarrow{H_2O} 2H^+(aq) + SO_4^{2-}\,(aq)$

▶ **Figure 4.16 (a)** Hydrogen chloride reacts with water and breaks up into ions. **(b)** Hydrogen sulfate reacts with water and breaks up into ions.

In **weak acids** the molecules are only partially ionised in water – this means that only a small fraction of the molecules break into ions when added to water. Figure 4.17 shows the difference between the strong acid HCl and a weak acid HX in water.

Strong acid HCl   **Weak acid HX**

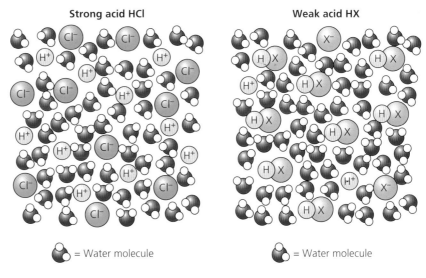

= Water molecule   = Water molecule

Every HCl molecule breaks down to form H^+ and Cl^- ions   Only a small fraction of the HX molecules break down to form H^+ and X^- ions.

▶ **Figure 4.17** The difference between strong acid (HCl) and a weak acid (HX) in water.

If solutions of equal concentration of a strong acid and weak acid are compared, there will be more H^+ ions in the strong acid solution. This means that the solution of the strong acid will have a lower pH.

There are many weak acids in food and drink. They tend to have a sour taste and are not dangerous because there is only a low concentration of H^+ ions. Ethanoic acid in vinegar, citric acid in citrus fruits (Figure 4.18) and carbonic acid in fizzy drinks are examples of weak acids in food and drink.

▲ **Figure 4.18** Citrus fruits contain citric acid.

The terms strong and weak refer to the degree of ionisation in water of acids. The terms **dilute** and **concentrated** refer to the amount of acid dissolved in the solution. This is summarised in Figure 4.19.

Strong
(all of the molecules break down into ions in water)

Weak
(a small fraction of the molecules break down into ions in water)

Concentrated
(a lot of acid is dissolved)

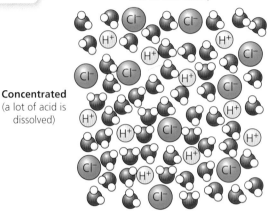
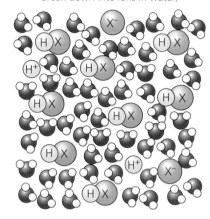

Dilute
(small amount of acid dissolved)

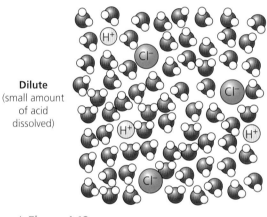
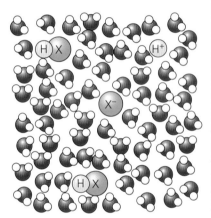

▲ Figure 4.19

Test yourself

20 a) Sulfuric acid is a strong acid. Explain what this means.

   b) Citric acid is a weak acid. Explain what this means.

21 a) What is the difference between a concentrated and dilute solution of an acid?

   b) Which of these two acids is more concentrated: 0.4 mol/dm³ ethanoic acid or 0.4 mol/dm³ nitric acid?

   c) Which of the two acids in (b) has the lowest pH? Explain your answer.

Example

Compare the following two solutions of acids:

1.0 mol/dm³ solution of ethanoic acid and 0.5 mol/dm³ solution of hydrochloric acid

1 Which is more concentrated? Explain your answer.

2 Which one is a strong acid? Explain your answer.

3 Which one has a lower pH? Explain your answer.

Answers

1 Ethanoic acid is more concentrated as it has a concentration of 1.0 mol/dm³ whereas the hydrochloric acid is only 0.5 mol/dm³.

2 Hydrochloric acid is a strong acid because all the molecules break into ions when added to water. Ethanoic acid is a weak acid as only a small fraction of the molecules break into ions when added to water.

3 Hydrochloric acid will contain more H⁺ ions because all the molecules break into ions while in ethanoic acid, even though it is twice as concentrated, only a small fraction of the molecules break into ions. Therefore, as hydrochloric acid contains more H⁺ ions it will have a lower pH.

Show you can...

Two solutions A and B were tested using a pH meter, red litmus paper, blue litmus paper and universal indicator paper. The results are shown in the table.

Test	Result for solution A	Result for solution B
pH meter	1.82	3.85
Red litmus	Red	Red
Blue litmus	Red	Red
Universal indicator paper	Red	Orange

a) Describe how the results with universal indicator may be converted into a pH value.

b) Are solutions A and B acidic, neutral or alkaline?

c) If the experiment was repeated using a more concentrated solution of A would the results be different? Explain your answer.

d) Explain why universal indicator gives more information than litmus.

○ Reaction of acids with metals

Metals that are more reactive than hydrogen react with dilute acids. When metals react with dilute acids they form a salt and hydrogen gas. The reaction fizzes as hydrogen gas is produced.

$$\text{metal} + \text{acid} \rightarrow \text{metal salt} + \text{hydrogen}$$

Table 4.4 shows how different acids form different types of salts.

Table 4.5 shows how magnesium, zinc and iron react with dilute hydrochloric and sulfuric acids.

When a metal reacts with an acid, a redox reaction takes place (Figure 4.20). The metal atoms lose electrons and so are oxidised. The H^+ ions in the acid gain electrons and are reduced.

▲ **Figure 4.20** A redox reaction takes place when a metal reacts with an acid.

Table 4.4

Acid	Reaction	Type of salt formed
Hydrochloric acid HCl	magnesium + hydrochloric acid → magnesium chloride + hydrogen Mg + 2HCl → $MgCl_2$ + H_2	Chloride (contains Cl^- ions)
Sulfuric acid H_2SO_4	magnesium + sulfuric acid → magnesium sulfate + hydrogen Mg + H_2SO_4 → $MgSO_4$ + H_2	Sulfate (contains SO_4^{2-} ions)
Nitric acid HNO_3	magnesium + nitric acid → magnesium nitrate + hydrogen Mg + $2HNO_3$ → $Mg(NO_3)_2$ + H_2	Nitrate (contains NO_3^- ions)

Table 4.5

Metal	Reaction with hydrochloric acid	Reaction with sulfuric acid
Magnesium	Fizzes vigorously magnesium + hydrochloric acid → magnesium chloride + hydrogen Mg + 2HCl → $MgCl_2$ + H_2	Fizzes vigorously magnesium + sulfuric acid → magnesium sulfate + hydrogen Mg + H_2SO_4 → $MgSO_4$ + H_2
Zinc	Fizzes gently zinc + hydrochloric acid → zinc chloride + hydrogen Zn + 2HCl → $ZnCl_2$ + H_2	Fizzes gently zinc + sulfuric acid → zinc sulfate + hydrogen Zn + H_2SO_4 → $ZnSO_4$ + H_2
Iron	Fizzes very slowly iron + hydrochloric acid → iron chloride + hydrogen Fe + 2HCl → $FeCl_2$ + H_2	Fizzes very slowly iron + sulfuric acid → iron sulfate + hydrogen Fe + H_2SO_4 → $FeSO_4$ + H_2

○ Reaction of acids with metal hydroxides

Acids react with metal hydroxides to form a salt and water. Some examples of this reaction are shown in the Table 4.6.

$$\text{metal hydroxide} + \text{acid} \rightarrow \text{metal salt} + \text{water}$$

Table 4.6

Full equation	Ionic equation
sodium hydroxide + nitric acid → sodium nitrate + water $NaOH$ + HNO_3 → $NaNO_3$ + H_2O	$H^+(aq) + OH^-(aq) \longrightarrow H_2O(l)$
potassium hydroxide + sulfuric acid → potassium sulfate + water $2KOH$ + H_2SO_4 → K_2SO_4 + $2H_2O$	$H^+(aq) + OH^-(aq) \longrightarrow H_2O(l)$
calcium hydroxide + hydrochloric acid → calcium chloride + water $Ca(OH)_2$ + $2HCl$ → $CaCl_2$ + $2H_2O$	$H^+(aq) + OH^-(aq) \longrightarrow H_2O(l)$

The ionic equation for each of these reactions is the same. In each reaction, H^+ ions from the acid are reacting with OH^- ions from the metal hydroxide. This produces water in each case.

Metal hydroxides that dissolve in water are alkalis because they release OH^- ions into the water. Metal hydroxides that are insoluble in water do not release OH^- ions into the water and so are not alkalis. However, they still react with acids to produce a salt and water.

In these reactions, the H^+ ions from the acid are being used up and so they are examples of neutralisation reactions.

Practical

Investigation of pH changes when a strong acid neutralises a strong alkali

In an experiment a student slowly added solution Y in $0.5\,cm^3$ portions to solution X in a conical flask and swirled the solution. The apparatus for the experiment is shown.

The pH after each addition was measured and recorded and a graph of pH against volume of Y added drawn.

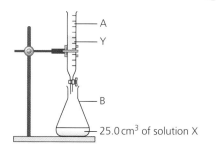

Questions

1 What is the piece of apparatus labelled A?

2 What is the piece of apparatus labelled B?

3 What piece of apparatus could have been used to measure out $25.0\,cm^3$ of solution X?

4 How could the pH of the solution be measured?

5 Why was the flask swirled after each addition?

6 Is solution X acidic, alkaline or neutral? Use the graph to explain your answer.

7 Is solution Y acidic, alkaline or neutral? Use the graph to explain your answer.

8 Use the graph to describe what happens to the pH of the mixture in the conical as solution Y is slowly added.

9 Describe what would happen to the shape of the graph if solution X was in the burette and solution Y in the conical.

10 What volume of Y is needed to react with all of solution X?

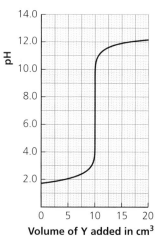

◯ Reaction of acids with metal oxides

Acids react with metal oxides to form a salt and water. Most metal oxides are insoluble in water and the reactions usually need to be heated. Some examples of this reaction are shown in Table 4.7.

metal oxide + acid → metal salt + water

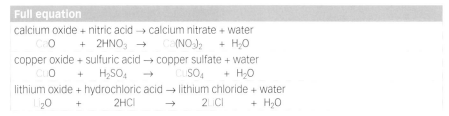

Table 4.7

Full equation
calcium oxide + nitric acid → calcium nitrate + water
CaO + $2HNO_3$ → $Ca(NO_3)_2$ + H_2O
copper oxide + sulfuric acid → copper sulfate + water
CuO + H_2SO_4 → $CuSO_4$ + H_2O
lithium oxide + hydrochloric acid → lithium chloride + water
Li_2O + $2HCl$ → $2LiCl$ + H_2O

In these reactions, the H⁺ ions from the acid are being used up and so they are examples of neutralisation reactions.

◯ Reaction of acids with metal carbonates

Acids react with metal carbonates to form a salt, water and carbon dioxide. These reactions usually take place readily. The reaction fizzes as carbon dioxide gas is produced. Some examples of this reaction are shown in the Table 4.8.

metal carbonate + acid → metal salt + water + carbon dioxide

Table 4.8

Full equation
sodium carbonate + nitric acid → sodium nitrate + water + carbon dioxide
Na_2CO_3 + $2HNO_3$ → $2NaNO_3$ + H_2O + CO_2
copper carbonate + sulfuric acid → copper sulfate + water + carbon dioxide
$CuCO_3$ + H_2SO_4 → $CuSO_4$ + H_2O + CO_2
calcium carbonate + hydrochloric acid → calcium chloride + water + carbon dioxide
$CaCO_3$ + $2HCl$ → $CaCl_2$ + H_2O + CO_2

In these reactions, the H⁺ ions from the acid are being used up and so they are examples of neutralisation reactions.

Test yourself

22 Complete the following word equations.
   a) calcium + hydrochloric acid
   b) tin + sulfuric acid
   c) barium hydroxide + nitric acid
   d) lithium hydroxide + hydrochloric acid
   e) nickel oxide + nitric acid
   f) magnesium oxide + sulfuric acid
   g) potassium carbonate + nitric acid
   h) zinc carbonate + hydrochloric acid

23 a) There is fizzing when hydrochloric acid reacts with calcium. What gas causes this?
   b) There is fizzing when hydrochloric acid reacts with calcium carbonate. What gas causes this?

24 Hydrochloric acid reacts with sodium hydroxide as shown in this equation:
   HCl(aq) + NaOH(aq) → NaCl(aq) + H₂O(l)
   a) What does the symbol (aq) mean?
   b) Write an ionic equation for this reaction.
   c) Explain why sodium hydroxide is an alkali.

25 Write a balanced equation for each of the reactions in question 22(a), (d), (f) and (g).

Making salts

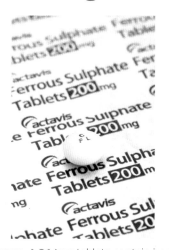

▲ Figure 4.21 Iron tablets contain iron(ɪɪ) sulfate (sometimes called ferrous sulfate).

○ What are salts?

Salts are substances made when acids react with metals, metal hydroxides, metal oxides and metal carbonates.

Salts are very useful substances. For example:

● many medicines are salts (Figure 4.21)
● fertilisers are salts
● toothpaste contains salts
● many food additives are salts.

Salts are made up of a metal ion combined with the ion left over from the acid when the H^+ ions react. The salt produced depends on which acid is used and what it reacts with. Table 4.9 below gives some examples of the salts formed when acids react.

Table 4.9

	Hydrochloric acid HCl	Sulfuric acid H_2SO_4	Nitric acid HNO_3
Magnesium Mg	Magnesium chloride (+ H_2) $MgCl_2$	Magnesium sulfate (+ H_2) $MgSO_4$	Magnesium nitrate (+ H_2) $Mg(NO_3)_2$
Sodium hydroxide NaOH	Sodium chloride (+ H_2O) NaCl	Sodium sulfate (+ H_2O) Na_2SO_4	Sodium nitrate (+ H_2O) $NaNO_3$
Copper oxide CuO	copper chloride (+ H_2O) $CuCl_2$	copper sulfate (+ H_2O) $CuSO_4$	Copper nitrate (+ H_2O) $Cu(NO_3)_2$
Zinc carbonate $ZnCO_3$	Zinc chloride (+ $H_2O + CO_2$) $ZnCl_2$	Zinc sulfate (+ $H_2O + CO_2$) $ZnSO_4$	Zinc nitrate (+ $H_2O + CO_2$) $Zn(NO_3)_2$

Stage 1 THE REACTION

- React the acid with an insoluble substance (e.g. a metal, metal oxide, metal carbonate, metal hydroxide) to produce the desired salt
- Add this substance until it no longer reacts
- This reaction may need to be heated

↓

Stage 2 FILTER OFF THE EXCESS

- Filter off the left over metal/metal oxide/metal carbonate/metal hydroxide

↓

Stage 3 CRYSTALLISE THE SALT

- Heat the solution to evaporate some water until crystals start to form
- Leave the solution to cool down – more crystals will form
- Filter off the crystals of the salt
- Allow the crystals to dry

▲ **Figure 4.22** Making salts that are soluble in water.

TIP ✓

Crystals of many salts contain water within the crystal structure, e.g. blue crystals of copper(II) sulfate have the formula, $CuSO_4.5H_2O$. These are hydrated salts.

TIP ✓

If an aqueous solution of a soluble salt is heated to evaporate all the water, then it is not possible to form hydrated crystals.

TIP ✓

Some salts only form anhydrous crystals (crystals containing no water). Sodium chloride (NaCl) is a good example.

◯ Making soluble salts

It is important to be able to make pure samples of salts, especially if they are being used in medicines or food.

Some salts are soluble in water and some are not. Figure 4.22 shows the method for making salts that are soluble in water.

It is easier to make a pure salt by reacting an acid with a substance that is insoluble in water rather than one that is soluble. Suitable substances to use include some metals, metal oxides, metal hydroxides or metal carbonates. As it is insoluble, you can add an excess of that substance to ensure that all the acid is used up. The excess can then be filtered off. In this way, there is no left over acid or the substance it reacts with mixed in with the salt that is formed.

After filtration, you are left with a solution of the salt in water. If some of the water is boiled off, a hot saturated solution is formed. As this hot, saturated solution cools, crystals form as the salt is less soluble at lower temperatures and so cannot all stay dissolved.

Test yourself

26 Suggest two chemicals that could be reacted together to make the following salts.
   a) calcium chloride
   b) copper nitrate
   c) aluminium sulfate

27 The salt iron(II) sulfate is used in iron tablets. It can be made by reacting an excess of iron with sulfuric acid.
   a) Why is it important that the iron(II) sulfate used in iron tablets is pure?
   b) Why is an excess of iron used?
   c) How is the excess iron removed?
   d) Write a word equation for the reaction between iron and sulfuric acid.
   e) Write a balanced equation for the reaction between iron and sulfuric acid.

28 Crystals of the salt nickel nitrate form as a hot, saturated solution of nickel nitrate cools down.
   a) What is a saturated solution?
   b) Explain why crystals of nickel nitrate form as the hot, saturated solution cools down.

29 The salt copper chloride can be made by reacting hydrochloric acid with copper oxide, copper carbonate or copper hydroxide. It cannot be made by reacting copper with hydrochloric acid.
   a) Explain why copper chloride cannot be made by reacting copper with hydrochloric acid.
   b) For each of the reactions of hydrochloric acid with copper oxide, copper carbonate and copper hydroxide:
      i) Write a word equation.
      ii) Write a balanced equation.

Show you can...

A solution of the salt magnesium chloride can be prepared by any of reactions A to D in the diagram.

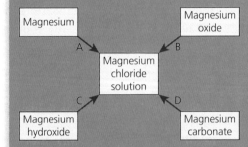

a) Write word equations for each of the reactions A to D.
b) State two observations that you would make during reaction D.

Preparation of a pure dry sample of a soluble salt from an insoluble oxide or carbonate (magnesium sulfate from magnesium carbonate)

Bath crystals are a mixture of water soluble solids which are added to bathwater for health benefits. Bath crystals contain Epsom salts (Figure 4.23) (hydrated magnesium sulfate) which relax muscles, reduce inflammation and help muscle function. The name Epsom salts comes from the town of Epsom, which has mineral springs from which hydrated magnesium sulfate was extracted.

To produce pure dry crystals of hydrated magnesium sulfate ($MgSO_4.7H_2O$) in the laboratory by reacting magnesium carbonate with sulfuric acid, the following method was followed.

- Measure $25\,cm^3$ of dilute sulfuric acid and place in a conical flask.
- Warm the dilute sulfuric acid using a Bunsen burner and add magnesium carbonate, stirring until it is in excess.
- Filter the solution.
- Heat the filtered solution to make it more concentrated.
- Cool and crystallise.
- Filter the crystals from the solution.
- Dry the crystals.

▲ **Figure 4.23** Epsom Salts.

Questions

1 What piece of apparatus would you use to measure $25\,cm^3$ of sulfuric acid?
2 Draw a labelled diagram of the apparatus used for the second step of the method.
3 Explain why the magnesium carbonate is added until it is in excess.
4 State one way in which you would know that the magnesium carbonate is in excess.
5 What is the general name given to the solid trapped by the filter paper?
6 Why was the filtered solution evaporated using a water bath or electric heater?
7 Why is the solution not evaporated to dryness?
8 Why do crystals form as the solution is cooled?
9 State two methods of drying the crystals.
10 Write a word and balanced symbol equation for the reaction between magnesium carbonate and sulfuric acid.

Electrolysis

Electrolysis

KEY TERMS

Electrolysis Decomposition of ionic compounds using electricity.

Electrolyte A liquid that conducts electricity.

Discharge Gain or lose electrons to become electrically neutral.

Anode An electrode where oxidation takes place (oxidation is the loss of electrons) – in electrolysis it is the positive electrode.

Cathode An electrode where reduction takes place (reduction is the gain of electrons) – in electrolysis it is the negative electrode.

◯ What is electrolysis?

Electrolysis is the decomposition of ionic compounds using electricity. Ionic compounds contain metals combined with non-metals. Examples include

- sodium chloride (NaCl) – a combination of the metal sodium with the non-metal chlorine
- copper sulfate ($CuSO_4$) – a combination of the metal copper with the non-metals sulfur and oxygen.

Ionic compounds are made up of positive and negative ions. As solids, ionic compounds cannot conduct electricity because the ions cannot move around. However, when ionic compounds are melted or dissolved, the ions are free to move and conduct electricity. These liquids or solutions are called **electrolytes** because they are able to conduct electricity.

If two electrodes connected to a supply of electricity are put into the electrolyte, the negative ions are attracted to the positive electrode and the positive ions are attracted to the negative electrode (Figure 4.24). This happens because opposite charges attract each other.

When the ions reach the electrodes they are **discharged**. This means that they gain or lose electrons so that they lose their charge and become neutral. Positive ions gain electrons. Negative ions lose electrons.

The negative ions are discharged by losing electrons at the positive electrode. These electrons move around the circuit through the wires to the negative electrode. The positive ions are discharged by gaining electrons at the negative electrode.

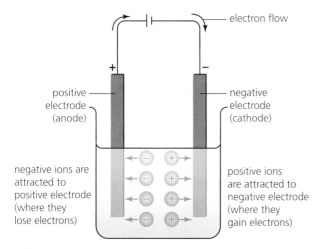

electron flow

positive electrode (anode)

negative electrode (cathode)

negative ions are attracted to positive electrode (where they lose electrons)

positive ions are attracted to negative electrode (where they gain electrons)

▲ Figure 4.24

TIP

Remember that electrons have a negative charge.

TIP

In the circuit as a whole, the current is carried by ions through the electrolyte and by electrons through the wires.

TIP

In electrolysis the **anode** is the positive electrode and the **cathode** is the negative electrode. In electrochemical cells (see Chapter 5) this is the opposite way round.

Table 4.10

Electrode	+ electrode (anode)	– electrode (cathode)
Which ions are attracted to the electrode	– ions	+ ions
What happens at the electrodes	– ions are discharged by losing electrons	+ ions are discharged by gaining electrons
Oxidation or reduction	Oxidation (loss of electrons) 🄷	Reduction (gain of electrons) 🄷

◯ Electrolysis of molten ionic compounds

Binary ionic compounds are ones made from one metal combined with one non-metal. Examples include lead bromide ($PbBr_2$) (Figure 4.25), sodium chloride (NaCl) and aluminium oxide (Al_2O_3) (Table 4.11). The electrolysis of these compounds when molten produces the metal and non-metal.

Table 4.11 Products from the electrolysis of aqueous solutions of ionic compounds

Electrode	Ions	+ Electrode (anode)	– Electrode (cathode)
Lead bromide, $PbBr_2$	lead ions, Pb^{2+} bromide ions, Br^-	bromide ions (Br^-) lose electrons to form bromine (Br_2) $2Br^- - 2e^- \rightarrow Br_2$	lead ions (Pb^{2+}) gain electrons to form lead (Pb) $Pb^{2+} + 2e^- \rightarrow Pb$
Sodium chloride, NaCl	sodium ions, Na^+ chloride ions, Cl^-	chloride ions (Cl^-) lose electrons to form chlorine (Cl_2) $2Cl^- - 2e^- \rightarrow Cl_2$	sodium ions (Na^+) gain electrons to form sodium (Na) $Na^+ + e^- \rightarrow Na$
Aluminium oxide, Al_2O_3	aluminium ions, Al^{3+} oxide ions, O^{2-}	oxide ions (O^{2-}) lose electrons to form oxygen (O_2) $2O^{2-} - 4e^- \rightarrow O_2$	aluminium ions (Al^{3+}) gain electrons to form aluminium (Al) $Al^{3+} + 3e^- \rightarrow Al$

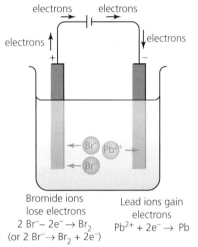

▲ Figure 4.25 The electrolysis of molten lead bromide.

Bromide ions lose electrons
$2\,Br^- - 2e^- \rightarrow Br_2$
(or $2\,Br^- \rightarrow Br_2 + 2e^-$)

Lead ions gain electrons
$Pb^{2+} + 2e^- \rightarrow Pb$

TIP ✔

When writing half equations, remember that the non-metals produced at the positive electrode are often made of molecules containing two atoms, e.g. F_2, Cl_2, Br_2, I_2 and O_2.

Show you can...

Copy and complete the table below for the electrolysis of molten lithium chloride.

	Anode	Cathode
Product		
Observation		
Half equation		
Oxidation or reduction		

Test yourself

30 a) Why do ionic compounds conduct electricity when molten?
  b) Why do ionic compounds not conduct electricity when solid?
31 Copy and complete the table to show the products of electrolysis of some molten ionic compounds.

Ionic compound (molten)	Product at the negative electrode (cathode)	Product at the positive electrode (anode)
Potassium iodide (KI)		
Zinc bromide ($ZnBr_2$)		
Magnesium oxide (MgO)		

32 a) What happens to negative ions at the positive electrode?
  b) What happens to positive ions at the negative electrode?
  c) What carries the electric charge through the electrolyte?
  d) What carries the electric charge through the wires?
33 a) What is an electrolyte?
  b) What is an anode?
  c) What is a cathode?
34 Balance each of the following half equations and state if each one is a reduction or oxidation process.
  a) $Cu^{2+} + e^- \rightarrow Cu$ **b)** $I^- - e^- \rightarrow I_2$
  c) $F^- \rightarrow F_2 + e^-$ **d)** $Fe^{3+} + e^- \rightarrow Fe$
35 Write a half equation for each of the following.
  a) The process at the positive electrode in the electrolysis of molten calcium oxide.
  b) The process at the negative electrode in the electrolysis of molten potassium bromide.

○ Metal extraction

Many metals are extracted from metal compounds in ores by heating with carbon in a reduction reaction. However, some metals cannot be extracted this way. This is because

● some metals are more reactive than carbon and/or
● some metals would react with carbon in the process.

Electrolysis is usually used to extract metals that cannot be extracted by heating with carbon. However, metals produced this way are expensive because of

● the high cost of heat energy to melt the metal compounds and
● the high cost of the electricity for the process.

Extraction of aluminium

Aluminium is the second-most commonly used metal after iron/steel. It is too reactive to be extracted by heating with carbon and so is extracted by electrolysis.

The main ore of aluminium is bauxite which contains aluminium oxide. This has a very high melting point of 2072°C. The cost of the thermal energy to melt aluminium oxide for electrolysis is very high. However, if the aluminium oxide is mixed with a substance called cryolite, the mixture melts at about 950°C and so the cost of thermal energy to melt this mixture is lower.

The electrodes for the process are made of graphite, a form of carbon (Figure 4.26). Aluminium ions (Al^{3+}) are attracted to the negative electrode where they gain electrons and form aluminium metal. As it is so hot, this is produced as a liquid and is run off at the bottom. Oxide ions (O^{2-}) are attracted to the positive electrode where they lose electrons and form oxygen. This oxygen reacts with the graphite anode and so the anode burns to produce carbon dioxide. This means that the anode has to be replaced regularly.

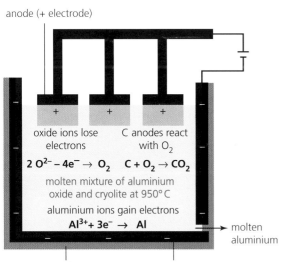

anode (+ electrode)

oxide ions lose electrons

$2\,O^{2-} - 4e^- \rightarrow O_2$

C anodes react with O_2

$C + O_2 \rightarrow CO_2$

molten mixture of aluminium oxide and cryolite at 950°C

aluminium ions gain electrons

$Al^{3+} + 3e^- \rightarrow Al$

molten aluminium

outer casing of electrolysis cell   cathode (- electrode)

▲ **Figure 4.26** The extraction of aluminium by electrolysis.

Show you can...

Complete the table below for the electrolysis of molten aluminium oxide

	Anode	Cathode
Product		
Half equation		
Oxidation or reduction		
Conditions for the electrolysis		

Test yourself

36 **a)** Give two reasons why some metals cannot be extracted by heating with carbon.
   b) Name two metals other than aluminium that are extracted by electrolysis.
   c) Give two reasons why the energy cost of extracting metals by electrolysis is so high.

37 This question is about the extraction of aluminium by electrolysis.
   a) Identify the compound from which aluminium is extracted.
   b) Why is this compound mixed with cryolite?
   c) What happens at the positive electrode?
   d) What happens at the negative electrode?
   e) What are the electrodes made of?
   f) Why does the positive electrode have to be replaced regularly?

○ **Electrolysis of aqueous ionic compounds**

When an ionic compound is dissolved in water, the products of electrolysis are often different to those when the compound is molten. In water, a small fraction of the molecules break down into hydrogen ions (H^+) and hydroxide ions (OH^-). These ions can be discharged instead of the ions in the ionic compound. More water molecules can break down if these ions are used up.

At each electrode there are two ions that could discharge, one from the ionic compound and one from the water. The one that is easier to discharge is the one that is discharged. Table 4.12 shows which ions are discharged at each electrode when inert electrodes are used. Inert electrodes are electrodes that will allow the electrolysis to take place but do not react themselves. Graphite electrodes are the most common inert electrodes used.

KEY TERMS

Inert electrodes Electrodes that allow electrolysis to take place but do not react themselves.

4 Chemical changes

TIP ✓

The halide ions include chloride (Cl^-), bromide (Br^-) and iodide (I^-). The halogens include chlorine (Cl_2), bromine (Br_2) and iodine (I_2).

TIP ✓ Ⓗ

Remember that oxidation is the loss of electrons, reduction is the gain of electrons.

Table 4.12 Products from the electrolysis of aqueous solutions of ionic compounds

Electrode	Positive electrode	Negative electrode
Which ions are discharged	Negative ions discharged: Oxygen is produced from the discharge of hydroxide ions, unless the ionic compound contains halide ions when these are discharged producing a halogen.	Positive ions discharged: Hydrogen is produced from the discharge of hydrogen ions, unless the ionic compound contains metal ions from a metal that is less reactive than hydrogen when the metal ions are discharged producing the metal.

Table 4.13 gives some examples to show which ions are discharged in the electrolysis of some aqueous solutions using inert electrodes.

Table 4.13

		Sodium chloride NaCl(aq)	Copper(II) bromide CuBr$_2$(aq)	Silver nitrate AgNO$_3$(aq)	Potassium sulfate K$_2$SO$_4$(aq)
Negative electrode **CATHODE CATIONS ATTRACTED**	Positive ions present	Na^+ and H^+	Cu^{2+} and H^+	Ag^+ and H^+	K^+ and H^+
	Positive ion discharged	H^+	Cu^{2+}	Ag^+	H^+
	Product	Hydrogen, H_2	Copper, Cu	Silver, Ag	Hydrogen, H_2
	Notes	Sodium is more reactive than hydrogen	Copper is less reactive than hydrogen	Silver is less reactive than hydrogen	Potassium is more reactive than hydrogen
	Half equation Ⓗ	$2H^+ + 2e^- \rightarrow H_2$ Ⓗ	$Cu^{2+} + 2e^- \rightarrow Cu$ Ⓗ	$Ag^+ + e^- \rightarrow Ag$ Ⓗ	$2H^+ + 2e^- \rightarrow H_2$ Ⓗ
	Process Ⓗ	Reduction Ⓗ	Reduction Ⓗ	Reduction Ⓗ	Reduction Ⓗ
Positive electrode **ANODE ANIONS ATTRACTED**	Negative ions present	Cl^- and OH^-	Br^- and OH^-	NO_3^- and OH^-	SO_4^{2-} and OH^-
	Negative ion discharged	Cl^-	Br^-	OH^-	OH^-
	Product	Chlorine , Cl_2	Bromine , Br_2	Oxygen, O_2	Oxygen, O_2
	Notes	Cl^- is a halide ion	Br^- is a halide ion	NO_3^- is not a halide ion	SO_4^{2-} is not a halide ion
	Half equation Ⓗ	$2Cl^- - 2e^- \rightarrow Cl_2$ Ⓗ	$2Br^- - 2e^- \rightarrow Br_2$ Ⓗ	$4OH^- - 4e^- \rightarrow 2H_2O + O_2$ Ⓗ	$4OH^- - 4e^- \rightarrow 2H_2O + O_2$ Ⓗ
	Process Ⓗ	Oxidation Ⓗ	Oxidation Ⓗ	Oxidation Ⓗ	Oxidation Ⓗ

Test yourself

38 Copy and complete the table to show the products of electrolysis of some aqueous solutions of ionic compounds with inert electrodes.

Ionic compound (aqueous)	Product at the negative electrode (cathode)	Product at the positive electrode (anode)
Potassium iodide (KI)		
Copper(II) chloride (CuCl$_2$)		
Magnesium sulfate (MgSO$_4$)		
Copper(II) nitrate (Cu(NO$_3$)$_2$)		
Zinc bromide (ZnBr$_2$)		

39 For the electrolysis of aqueous sodium bromide solution using inert electrodes, write a half equation for the process at the electrode shown and state whether it is an oxidation or reduction process:

a) at the positive electrode

b) at the negative electrode.

40 For the electrolysis of aqueous copper(II) sulfate solution using inert electrodes, write a half equation for the process at the electrode shown and state whether it is an oxidation or reduction process:

a) at the positive electrode

b) at the negative electrode.

At the anode, oxygen is produced unless the solution contains a halide ↓ group 7.

120

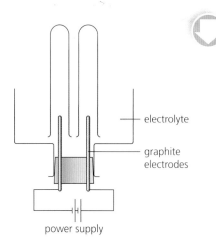

electrolyte

graphite electrodes

power supply

Required practical

Investigation of what happens when aqueous solutions are electrolysed using inert electrodes

A student made a hypothesis which stated, 'Oxygen is produced at the positive electrode in the electrolysis of an aqueous solution, unless the compound contains halide ions.' To test this hypothesis the apparatus was set up as shown. The electricity supply was switched on and the observations at each electrode recorded. The experiment was repeated using different aqueous solutions as electrolyte and the results obtained recorded in the table.

KEY TERM

Hypothesis A proposal intended to explain certain facts or observations.

		Copper(II) chloride	Calcium nitrate	Silver sulfate	Potassium bromide	Sodium iodide	Sulfuric acid
Cathode (negative electrode)	Observations	Red brown solid	Colourless gas	Grey solid	Colourless gas	Colourless gas	Colourless gas
	Test used for product	Appearance	Squeaky pop when a burning splint is inserted	Appearance	Squeaky pop when a burning splint is inserted	Squeaky pop when a burning splint is inserted	Squeaky pop when a burning splint is inserted
	Identity of product						
Anode (positive electrode)	Observations	Colourless gas	Colourless gas	Colourless gas	Yellow-orange solution	Brown solution	Colourless gas
	Test used for product	Bleaches damp litmus paper	Relights a glowing splint	Relights a glowing splint	Bleaches damp litmus paper	Goes blue-black when starch is added	Relights a glowing splint
	Identity of product						

Questions

1 Why are graphite electrodes used?

2 Why are the aqueous solutions made up in distilled water?

3 What would you add to the circuit to show that a current is flowing?

4 Suggest why the experiment was only carried out long enough to make the necessary observations.

5 Write down the identity of the missing products at the cathode to complete the table.

6 Summarise the rules for deciding what is formed at the cathode when aqueous solutions of ionic compounds are electrolysed using inert electrodes.

7 Write down the identity of the missing products at the anode, to complete the table.

8 Using the results of this experiment, state if the hypothesis made by the student was correct.

Practice questions

1 Which one of the following substances does not react with dilute hydrochloric acid at room temperature? [1 mark]

A calcium carbonate

B copper

C magnesium

D potassium hydroxide

2 The electrical conductivity of an electrolyte in electrolysis is due to the movement of particles through the substance when it is molten or in solution. Which of the following are the particles that move in the electrolyte? [1 mark]

A atoms

B electrons

C ions

D protons

3 Strips of four different metals were placed in solution of the nitrates of the same metals. Any reactions which occur are represented by a ✓ in the table. Which of the following is the correct order of reactivity with the most reactive first? [1 mark]

Metal	P nitrate	Q nitrate	R nitrate	S nitrate
P	X	X	X	✓
Q	✓	X	✓	✓
R	✓	X	X	✓
S	X	X	X	X

A P Q R S

B Q R S P

C R S P Q

D Q R P S

4 Which one of the following could neutralise a solution of pH 10? [1 mark]

A ammonia solution

B hydrochloric acid

C sodium hydroxide solution

D water

5 Which one of the following pairs of substances could be used for the preparation of magnesium sulfate crystals? [1 mark]

A dilute hydrochloric acid and magnesium carbonate

B dilute nitric acid and magnesium oxide

C dilute sulfuric acid and magnesium chloride

D dilute sulfuric acid and magnesium carbonate.

6 What is the product formed at the anode in the electrolysis of aqueous sodium chloride solution? [1 mark]

A chlorine

B hydrogen

C oxygen

D sodium

7 Neutralisation occurs when an acid and an alkali react to form a salt and water.

a) i) Copy and complete the table below to give the names and formulae of the ions present in all acids and alkalis. [4 marks]

	Ion present in all acids	Ion present in all alkalis
Name		
Formula		

ii) Write an ionic equation for neutralisation, including state symbols. [2 marks]

b) Sulfuric acid solution can be neutralised using an alkali such as sodium hydroxide or adding a solid oxide such as copper(II) oxide.

i) Write a balanced equation for the reaction between sodium hydroxide and sulfuric acid. [2 marks]

ii) Write a balanced equation for the reaction between copper(II) oxide and sulfuric acid. [2 marks]

c) In the preparation of hydrated copper sulfate(II) crystals, an excess of copper(II) oxide was added to warm dilute sulfuric acid. The excess copper(II) oxide was removed by filtration. Describe how you would obtain pure dry crystals of hydrated copper(II) sulfate from the filtrate collected. [4 marks]

8 The description that follows is about a metal, other than zinc, which belongs to the reactivity series.

At room temperature the metal is a silver coloured solid. It reacts rapidly with cold dilute sulfuric acid to produce hydrogen. It conducts electricity. On heating in air the metal burns with a very bright flame leaving a white powder.

a) State one property from the description above that is common to all metals. [1 mark]

b) Give two reasons why the metal is NOT copper. [2 marks]

c) What part of the air reacts with the metal? [1 mark]

d) Suggest one metal which fits the description. [1 mark]

e) Write a word equation for the reaction of your chosen metal with sulfuric acid. [1 mark]

9 Hydrochloric acid can react with calcium hydroxide and with calcium. **Compare and contrast** the reaction of hydrochloric acid with calcium hydroxide with the reaction of hydrochloric acid with calcium. In your answer you must include the names of all products for each reaction and the observations for each reaction. [6 marks]

10 Aluminium is the most abundant metal in the Earth's crust. Aluminium ore is first purified to give aluminium oxide and the metal is then extracted from the aluminium oxide by electrolysis.

a) What is meant by the term electrolysis?

b) Name the ore from which aluminium is extracted.

c) The electrolysis of the purified ore is carried out in the Hall-Héroult cell. The diagram below shows the cell used.

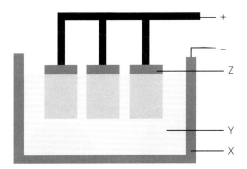

   i) Name X and Z. [2 marks]

   ii) Y is the electrolyte. Name the substances in the electrolyte. [2 marks]

   iii) Why is the electrolysis cell kept at about 950°C? [2 marks]

   iv) Name the products formed at the positive and negative electrodes. [2 marks]

   v) Write half equations for the reactions taking place at each electrode. [2 marks]

   vi) Which electrode must be replaced regularly? Write a balanced symbol equation to explain your answer. [2 marks]

d) Give a reason in terms of electrons why the extraction of aluminium in this process is a reduction reaction. [1 mark]

11 Some substances, for example molten lead bromide and aqueous sodium chloride, are described as electrolytes. Other substances for example copper metal, are conductors.

An experiment, to investigate the electrolysis of the electrolyte molten lead bromide, was set up as shown in the diagram below.

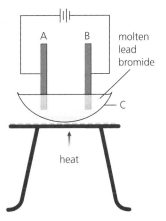

a) Some pieces of apparatus in the diagram are labelled A–C. State the correct name for each piece of apparatus. [3 marks]

b) Name a piece of apparatus which could be connected in the circuit to show that an electric current is flowing through the molten lead bromide. [1 mark]

c) Copy and complete the table to state the names of the products, and the half equations for the electrodes. [4 marks]

Electrode	Name of product	Half equation
A		☰ Ⓗ
B		☰ Ⓗ

d) Why does this electrolysis need to be carried out in a fume cupboard? [1 mark]

e) Describe the differences in conduction of electricity by copper metal and molten lead bromide. [2 marks]

f) Name the product at each electrode when aqueous sodium chloride is electrolysed. [2 marks]

Working scientifically:
Measurements and uncertainties

We often use apparatus to make measurements in chemistry. For example, we often measure the mass, volume or temperature of a substance.

Measurement instruments which you should be able to use correctly in chemistry include

▶ measuring cylinder (Figure 4.27)

▶ pipette (Figure 4.28)

▶ burette

▶ thermometer

▶ balance.

Measuring cylinders, pipettes and burettes are used to measure out volumes of liquids. Pipettes and burettes, which are used in titrations (see Chapter 3), measure volumes more accurately than measuring cylinders.

A meniscus is the curve seen at the top of a liquid in response to its container. When reading the volume of the liquid, the measurement at the bottom of the meniscus curve is read. This should be read at eye level (Figure 4.29).

▲ Figure 4.27 Measuring cylinders.

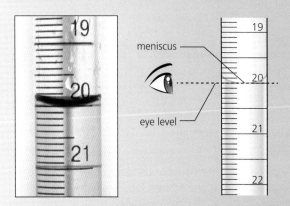

▲ Figure 4.29 The meniscus should be read at eye level.

▲ Figure 4.28 Pipette.

The resolution of a piece of apparatus is the smallest change it can measure. For example, in Figure 4.29, the resolution of the burette is ±0.1 cm^3.

When a measurement is made, there is always some doubt or uncertainty about its value. Uncertainty is often recorded after a measurement as a ± value.

The uncertainty can be estimated from the range of results that are obtained when an experiment is repeated several times.

> **KEY TERMS** ⭐
>
> **Meniscus** The curve at the surface of a liquid in a container.
>
> **Resolution** The smallest change a piece of apparatus can measure.
>
> **Uncertainty** The range of measurements within which the true value can be expected to lie.

For example, the volume of a gas produced in a reaction was measured five times. The results are 82, 77, 78, 96 and 80 cm^3.

The mean value is found after excluding any anomalous results (96 cm^3 is anomalous here as it is significantly different from all the others):

Mean volume $= \dfrac{82 + 77 + 78 + 80}{4} = 79 \pm 3$ cm^3

The mean is quoted to the nearest unit as all the values are measured to the nearest unit. The uncertainty is ±3 cm^3 as the highest and lowest values are within 3 cm^3 of the mean.

Questions

1 Record the volume of liquid in each measuring cylinder A–D. All scales are shown in cm^3.

2 Record the volume of liquid in each of the burettes E–G shown below. All scales are shown in cm^3.

3 Record the temperatures shown on each thermometer H–M. All scales are shown in °C.

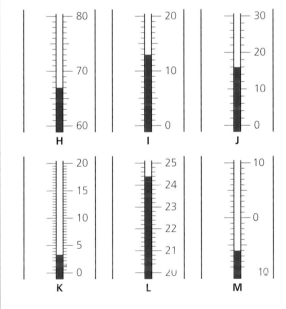

4 In each pair of diagrams, find the difference in mass between the first reading and the second reading. All scales are shown in grams.

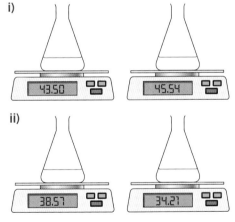

5 Find the mean value and its uncertainty for the boiling point of a substance which was measured several times. Values measured were 124, 126, 123, 125 and 123°C.

6 Find the mean value and its uncertainty for the mass of gas produced in a reaction which was measured several times. Values measured were 0.36, 0.18, 0.33 and 0.40 g.

7 Find the mean value and its uncertainty for the volume of acid needed to neutralise an alkali which was measured several times. Values measured were 25, 27, 26, 20 and 25 cm^3.

Many people have a gas fire to keep warm at home. A chemical reaction takes place when the gas burns. This chemical reaction releases a lot of thermal energy that keeps us warm. This chapter looks at why some chemical reactions release thermal energy and increase the temperature while other reactions remove thermal energy and lower the temperature.

This chapter covers specification points 4.5.1 to 4.5.2 and is called Energy changes.

It covers exothermic and endothermic reactions, as well as chemical cells and fuel cells.

Prior knowledge

Previously you could have learned:

> Energy cannot be made or destroyed – it can only be transferred from one form to another (this is the law of conservation of energy).
> An energy change is a sign that a chemical reaction has taken place.
> Some chemical reactions lead to an increase in temperature and some to a decrease in temperature.
> Fuels are substances that are good stores of energy.

Test yourself on prior knowledge

1 Chemical energy and thermal energy are two forms of energy. Write down three forms of energy besides these two.
2 What is the law of conservation of energy?
3 What is a fuel?
4 Name three common fuels.

Exothermic and endothermic reactions

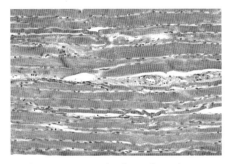

▲ Figure 5.1 Respiration takes place in all cells, including the cells in this muscle tissue.

KEY TERM

Exothermic reactions Reaction where thermal energy is transferred from the chemicals to the surroundings and so the temperature increases.

▲ Figure 5.2 An exothermic reaction takes place inside hand warmers.

○ Energy changes in reactions

When chemical reactions take place, thermal energy is transferred to or from the surroundings. Some reactions make their surroundings hotter and some make them colder.

Exothermic reactions

In exothermic reactions, thermal energy is transferred from the chemicals to the surroundings. As there is more thermal energy, the temperature increases and so it gets hotter. Most chemical reactions are exothermic.

In some exothermic reactions, only a small amount of thermal energy is transferred and the temperature may only rise by a few degrees. In some reactions a lot of thermal energy is transferred and the surroundings get very hot. Sometimes there is so much thermal energy transferred that the reactants catch fire (Table 5.1).

Applications of exothermic reactions

Hand warmers are used by people in cold places to keep their hands warm (Figure 5.2). There are many different types, but they all work by using an exothermic reaction that transfers thermal energy to the surroundings.

Exothermic reactions are reactions where the heat energy is transferred to the surroundings which increases the temperature of the surroundings

Endothermic reactions are reactions where heat energy is absorbed from the surroundings decreasing the temperature of the surroundings.

Table 5.1 Examples of exothermic reactions

Example	Comments
Oxidation reactions e.g. respiration (Figure 5.1)	Oxidation reactions take place when substances react with oxygen. Many oxidation reactions are exothermic. Respiration takes place in the cells of living creatures and is the reaction of glucose with oxygen. Thermal energy is released in this reaction. For example: glucose + oxygen → carbon dioxide + water $C_6H_{12}O_6 + 6O_2 → 6CO_2 + 6H_2O$
Combustion reactions e.g. burning fuels	Combustion reactions take place when substances react with oxygen and catch fire. They are a type of oxidation reaction. Fuels burning, such as methane (CH_4) in natural gas, are combustion reactions. These reactions are very exothermic, release a lot of thermal energy and catch fire. For example: methane + oxygen → carbon dioxide + water $CH_4 + 2O_2 → CO_2 + 2H_2O$
Neutralisation reactions e.g. acids reacting with alkalis	Neutralisation reactions take place when acids react with bases. Many neutralisation reactions are exothermic. For example, when hydrochloric acid reacts with sodium hydroxide some thermal energy is transferred to the surroundings and the temperature rises by a few degrees. For example: hydrochloric acid + sodium hydroxide → sodium chloride + water $HCl + NaOH → NaCl + H_2O$

▲ **Figure 5.3** Self-heating cans of foods and drinks.

> **KEY TERM** ★
>
> **Endothermic reactions** Reaction where thermal energy is transferred from the surroundings to the chemicals and so the temperature decreases.

> **TIP** ✓
>
> Remember that en**d**othermic reactions get col**d**er.

Self-heating cans can also be very useful (Figure 5.3). They can be used to provide hot drinks, such as coffee and hot chocolate, or even foods. Inside these cans, the food or drink is separated from a layer containing the chemicals used for heating. This is often calcium oxide and a bag of water. When a button is pressed by the user, the bag of water is punctured and the water mixes with the calcium oxide. The calcium oxide reacts with the water in an exothermic reaction. The thermal energy released heats up the food or drink.

Endothermic reactions

In **endothermic** reactions, thermal energy is transferred from the surroundings to the chemicals. As there is less thermal energy, the temperature decreases and so it gets colder (Table 5.2). Endothermic reactions are less common than exothermic reactions.

Applications of endothermic reactions

Some sports injury packs act as a cold pack to put on an injury to prevent swelling (Figure 5.4). Inside the pack is a bag of water and a substance such as ammonium nitrate. When the pack is squeezed' the water bag bursts and the ammonium nitrate dissolves in the water in an endothermic process.

Table 5.2 Examples of endothermic reactions

Example			Comments
Decomposition, e.g. thermal decomposition of metal carbonates	Green copper carbonate decomposes into black copper oxide when heated		Decomposition reactions occur when substances break down into simpler substances. For example when metal carbonates are heated they break down into a metal oxide and carbon dioxide. Decomposition reactions are always endothermic. For example: copper carbonate → copper oxide + carbon dioxide $CuCO_3 → CuO + CO_2$
Reaction of acids with metal hydrogencarbonates	Citric acid and sodium hydrogencarbonate are in sherbet sweets and react in your mouth in an endothermic reaction		When metal hydrogencarbonates (e.g. $NaHCO_3$) react with acids (e.g. citric acid), the temperature drops as this is an endothermic reaction. For example: citric acid + sodium hydrogencarbonate → sodium citrate + carbon dioxide + water

▲ **Figure 5.4** Sports injury packs use an endothermic reaction to keep the injury cold.

Test yourself

1 Is each of the following reactions endothermic or exothermic?
   a) the temperature started at 21°C and finished at 46°C *–Exothermic*
   b) the temperature started at 18°C and finished at 14°C *–Endothermic*
   c) the temperature started at 19°C and finished at 25°C *– Exothermic*

2 Is each of the following reactions endothermic or exothermic?
   a) burning alcohol
   b) thermal decomposition of iron carbonate
   c) reaction of vinegar (containing ethanoic acid) with baking powder (sodium hydrogencarbonate)
   d) reaction of vinegar (containing ethanoic acid) with sodium hydroxide

3 a) Why does the temperature increase when an exothermic reaction takes place in solution?
   b) Why does the temperature decrease when an endothermic reaction takes place in solution?

Show you can...

The reactions P and Q can be classified in different ways.

P calcium carbonate → calcium oxide + carbon dioxide
Q sodium hydroxide + nitric acid → sodium nitrate + water

Complete the table by placing one or more ticks (✔) in each row for reactions P and Q to indicate the terms which apply to each reaction. More than one term can apply to each reaction.

	Combustion	Decomposition	Neutralisation	Oxidation	Respiration	Exothermic	Endothermic
P							
Q							

○ Reaction profiles

Chemical reactions can only occur when particles collide with each other with enough energy to react. The *minimum* energy particles must have to react is called the activation energy.

We can show the relative energy of reactants and products in a reaction profile (Figure 5.5). This can also show the activation energy.

KEY TERM

Activation energy The minimum energy particles must have to react.

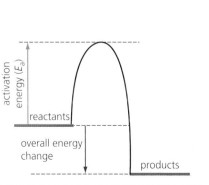

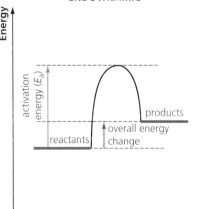

▲ **Figure 5.5** Reaction profiles for exothermic and endothermic reactions.

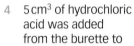

Investigating the variables that affect temperature change in reacting solutions – the temperature change in a neutralisation reaction

In an experiment the following method was followed:

1 25 cm³ of 1 mol/dm³ sodium hydroxide was measured out and placed in a polystyrene cup.

2 A burette was filled with 1 mol/dm³ hydrochloric acid.

3 The temperature of the sodium hydroxide was measured and recorded.

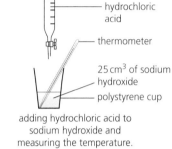

adding hydrochloric acid to sodium hydroxide and measuring the temperature.

4 5 cm³ of hydrochloric acid was added from the burette to the plastic beaker, and the temperature recorded. Additional hydrochloric acid was added 5 cm³ at a time and the temperature recorded until the total volume of hydrochloric acid added was 40 cm³.

5 The experiment was repeated.

6 The results are shown in Table 5.3.

Table 5.3 The temperature recorded when increasing volumes of 1 mol/dm³ HCl was added to 25 cm³ of 1 mol/dm³ NaOH.

Volume of acid added in cm³	Temperature in °C		
	Experiment 1	Experiment 2	Average
0	19.4	19.6	19.5
5	21.6	21.8	21.7
10	24.0	24.0	24.0
15	25.2	25.0	25.1
20	25.4	25.6	25.5
25	26.2	26.4	26.3
30	25.4	25.4	25.4
35	25.2	25.0	25.1
40	25.1	24.9	25.0

Questions

1 In this experiment identify the
   a) independent variable
   b) dependent variable
   c) key control variables

2 Why was the experiment repeated?

3 Why was a polystyrene cup used rather than a glass beaker?

4 Why should the solution in the polystyrene cup be stirred after each addition of acid?

5 Plot a graph of average temperature (y-axis) against volume of acid added (x-axis).

6 Describe the trend shown by the results plotted.

7 It is thought that the highest temperature is reached when complete neutralisation has occurred. Suggest how you would experimentally confirm that the highest temperature reached is the point at which neutralisation has occurred.

A different experiment was carried out by adding 1 mol/dm³ solutions of different acids to 25 cm³ of 1 mol/dm³ sodium hydroxide in a plastic beaker. The highest temperature reached was recorded and presented in Table 5.4.

Table 5.4 The highest temperature recorded when 40 cm³ of different types of 1 mol/dm³ acid of was added to 25 cm³ of 1 mol/dm³ NaOH.

Type of acid	hydrochloric acid	sulfuric acid	ethanoic acid
Highest temp in °C	26.2	26.5	25.2
Repeat highest temperature in °C	26.4	26.3	25.4
Average highest temperature in °C	26.3	26.4	25.3

Questions

8 In this experiment identify the
   a) independent variable
   b) dependent variable
   c) key control variables

9 State two conclusions you can draw from the results.

Example

Draw a reaction profile diagram for the reaction below which is exothermic.

$CH_4 + 2O_2 \rightarrow CO_2 + 2H_2O$

Answer

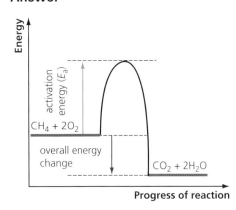

Show you can...

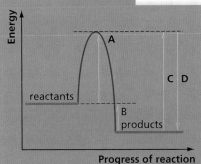

The reaction profile for a reaction is shown.

a) **Is it an exothermic reaction or an endothermic reaction?**
b) **Which letter represents the activation energy for the conversion of reactants to products?**

Test yourself

4 Look at the following reaction profiles.

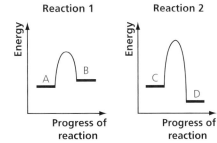

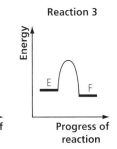

a) Which reaction(s) is/are exothermic?
b) Which reaction(s) is/are endothermic?

5 a) Sketch the reaction profile for the following reaction which is endothermic:
$CuCO_3 \rightarrow CuO + CO_2$

b) Draw an arrow to show the overall energy change for the reaction and label it **O**.

c) Draw an arrow to show the activation energy for the reaction and label it **A**.

○ Bond energies

▲ Figure 5.6

Breaking a chemical bond takes energy. For example, 436 kJ of energy is needed to break one mole of H—H covalent bonds. Due to the law of conservation of energy, 436 kJ of energy must be *released* when making one mole of H—H covalent bonds (Figure 5.6).

During a chemical reaction:

- energy must be supplied to break bonds in the reactants
- energy is released when bonds in the products are made.

The overall energy change for a reaction equals the difference between the energy needed to break the bonds in the reactants

Bond energy is the energy needed to break bonds or the energy released by making bonds

and the energy released when bonds are made in the products (Table 5.5).

Energy change = energy needed breaking bonds in reactants
  − energy released making bonds in products

$$\Delta H = B_b - B_f$$

Table 5.5

	Exothermic reaction	Endothermic reaction
Comparison of bond energies	More energy is released making new bonds than is needed to break bonds	More energy is needed to break bonds than is released making new bonds
Sign of energy change	−	+

Example

Find the energy change in the following reaction using the bond energies given.

Bond energies: C—H 412 kJ, O=O 496 kJ, C=O 743 kJ,
    O—H 463 kJ

H—C—H (with H above and below) + 2 O=O ⟶ O=C=O + 2 H—O—H

Explain why the reaction is exothermic or endothermic using bond energies.

Answer
Bonds broken: Bonds made:

4C—H = 4(412)   2C=O = 2(743)

2O=O = 2(496)   4O—H = 4(463)

Total   = 2640 kJ   Total   = 3338 kJ

Energy change = energy needed to break bonds
    − energy released making bonds

    = 2640 – 3338

    = –698 kJ

This reaction is exothermic because more energy is released making bonds than is needed to break bonds.

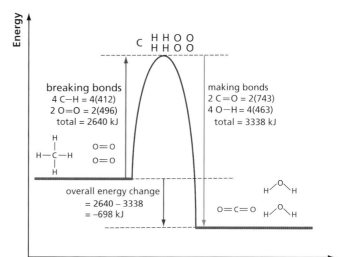

Example

Find the energy change in the following reaction using the bond energies given.

Bond energies: H—H 436 kJ, O=O 496 kJ, O—H 463 kJ

2 H—H   +   O=O ⟶ 2 H—O—H

Explain why the reaction is exothermic or endothermic using bond energies.

Answer
Bonds broken: Bonds made:

2H—H = 2(436)   4O—H = 4(463)

1O=O = 496   Total   = 1852 kJ

Total   = 1368 kJ

Energy change = energy needed to break bonds
    − energy released making bonds

    = 1368 – 1852

    = –484 kJ

This reaction is exothermic because more energy is released making bonds than is needed to break bonds.

Example

Find the energy change in the following reaction using the bond energies given.

Bond energies: C—C 348 kJ, C—H 412 kJ, C—O 360 kJ,
O—H 463 kJ, C=C 612 kJ

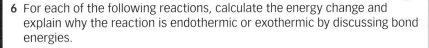

Explain why the reaction is exothermic or endothermic using bond energies.

Answer

Bonds broken:		Bonds made:
5C—H = 5(412)		4C—H = 4(412)
C—C = 348		C=C = 612
C—O = 360		2O—H = 2(463)
O—H = 463		Total = 3186 kJ
Total = 3231 kJ		

Energy change = energy needed to break bonds
– energy released making bonds

= 3231 – 3186

= +45 kJ

This reaction is endothermic because more energy is needed breaking bonds than is released making bonds.

TIP ✔

Remember that energy is *needed to break bonds* but is *released when bonds are made.*

Test yourself

6 For each of the following reactions, calculate the energy change and explain why the reaction is endothermic or exothermic by discussing bond energies.

Bond	C—C	C=C	C—H	C—Br	C=O	O=O	O—H	H—H	H—Br	N≡N	N—H	Br—Br
Bond energy in kJ	348	612	412	276	743	496	463	436	366	944	388	193

a) H—H + Br—Br ⟶ 2 H—Br

b) N≡N + 3 H—H ⟶ 2 H—N—H (with H above N)

c) H—C=C—H (each C with H) + Br—Br ⟶ H—C—C—H with Br Br below

d) H—C—C—C—H (propane) + 5 O=O ⟶ 3 O=C=O + 4 H—O—H

Chemical cells and fuel cells

○ Chemical cells

We use **cells** and **batteries** every day. They contain chemicals which react to supply electricity.

A very simple chemical cell can be made using a lemon and two pieces of different metals (Figure 5.7). The lemon juice acts as an **electrolyte**. An electrolyte is a liquid that conducts electricity.

Salt solution is another example of a good electrolyte. In most electrolytes, *ions* (not electrons) carry the electric charge through the liquid. For example, in sodium chloride solution it is the sodium ions and chloride ions that move through the solution carrying the electric charge.

Chemical cells can be made by placing two different metals in a beaker containing sodium chloride solution as electrolyte (Figure 5.8). There is a potential difference (voltage) between the two metals, so electrons flow through the wires from the more reactive metal to the less reactive metal, while ions from sodium chloride carry the electric charge through the solution.

There are several factors that affect the voltage produced by the cell. These include the identity of the metals and the electrolyte.

1 The metals: The identity of the metals used as electrodes in the cell makes a significant difference. The greater the difference in reactivity between the two metals, the greater the voltage.

2 The electrolyte: Changing the electrolyte and/or the concentration of the electrolyte can affect the voltage.

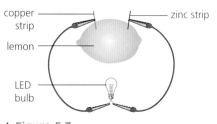

copper strip — zinc strip
lemon —
LED bulb —

▲ Figure 5.7

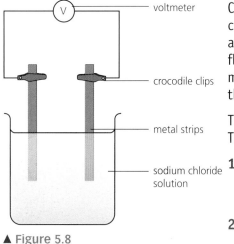

voltmeter
crocodile clips
metal strips
sodium chloride solution

▲ Figure 5.8

KEY TERMS ⭐

Cell Two electrodes in an electrolyte used to generate electricity.

Battery Two or more chemical cells connected together.

Electrolyte A liquid that conducts electricity.

Test yourself

7 A simple chemical cell can be made from two different metals placed in an electrolyte.
  a) What is an electrolyte?
  b) Name a good electrolyte.
  c) Explain why pure water is not an electrolyte.
8 In a simple chemical cell, how is the electric charge carried through
  a) the electrolyte
  b) the wire joining the two electrodes?
9 State two factors that affect the potential difference (voltage) of a cell.
10 Why would a cell made from two pieces of copper joined by a wire and placed in an electrolyte not produce a potential difference (voltage)?

Show you can...

The voltages between pairs of metals can be used to place them in a reactivity series. In an experiment each of the metals A to D was connected via a voltmeter to a piece of copper. The copper is connected to the positive terminal of the voltmeter. The metals were placed in a beaker containing an electrolyte. The experimental results are as follows:

Metal	Potential difference in V
A	+0.6
B	−0.2
C	+0.9
D	−0.4

a) Which is the most reactive metal?
b) Which is the least reactive metal?
c) Which pair of metals would give you the highest voltage if connected together?
d) What would the potential difference be for the pair of metals selected in (c)?

Batteries

The voltage produced by many chemical cells is typically 1.5 volts. In order to increase this voltage, several cells can be connected together. A battery is two or more cells connected together in series to provide a greater voltage. A 6 volt battery is made from four 1.5 volt cells connected together (Figure 5.9).

> **TIP**
>
> People often incorrectly refer to cells as batteries. A battery is made from two or more cells joined together.

(a)

(b)

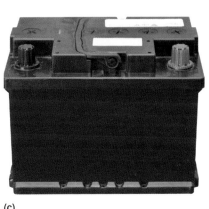
(c)

▲ Figure 5.9 **(a,b)** There are four 1.5 V cells are connected together inside a 6 V battery. **(c)** A 12 V car battery is made of six 2 V cells connected together.

As a current is drawn from a cell, the chemicals inside the cell are used up. In non-rechargeable cells and batteries, once one or more of the reactants has been used up the battery is 'flat' and no longer works. Alkaline cells and lithium button cells are non-rechargeable (Figure 5.10).

In rechargeable cells and batteries, the chemical reactions can be reversed to re-form these chemicals. This means that they can be used over and over again. The cells and batteries in mobile phones, tablets and laptops are rechargeable (Figure 5.11). The chemical reactions inside the cell can be reversed by passing an electric current through the cell.

Evaluating the use of cells and batteries

Cells and batteries are very useful but there are some problems with their use. Table 5.6 evaluates the use of cells and batteries.

▲ Figure 5.10 Alkaline cells and lithium button cells are non-rechargeable.

▲ Figure 5.11 Many cells and batteries are rechargeable (e.g. lithium ion cells).

Table 5.6 Evaluation of the use of cells and batteries.

Advantages of using cells and batteries	Disadvantages of using cells and batteries
Provide a very convenient portable source of electricity which can be used in many devices	Many cells contain toxic or harmful chemicals and so can harm the environment unless these chemicals are recycled
Relatively cheap	
Some are rechargeable and so can be used many times over and over again	

Show you can...

Create a table with two columns, one for rechargeable and one for non-rechargeable cells. Insert the phrases below in your table to show advantages and disadvantages of these cells.

- cheap to buy/more expensive to buy
- using these cells means disposal of fewer cells and less chemical pollution/using these cells means disposal of many cells and more chemical pollution
- can only be used once/can be used many times
- output falls gradually with time/output stays constant until flat

KEY TERM

Fuel cell A chemical cell with a continuous supply of chemicals to fuel the cell.

Test yourself

11 What is a battery?
12 What is the reason for making batteries from cells?
13 Why do cells and batteries go flat after use?
14 How does recharging a rechargeable cell or battery work?
15 One of the most common cells used is the alkaline AA 1.5 V cell. Is this non-rechargeable or rechargeable?
16 Why is it important that used cells and batteries are not thrown away in household rubbish?

○ Fuel cells

Non-rechargeable cells and batteries contain a limited amount of chemicals that are used up. However, in a fuel cell there is a supply of the chemicals into the cell replacing those that are used up and so the cell keeps working. The chemicals are usually oxygen (or air which contains oxygen) plus a fuel which is usually hydrogen.

In a hydrogen fuel cell, an electrochemical reaction takes place between hydrogen and oxygen producing a potential difference. The hydrogen is oxidised and the only product of the reaction is water. This means that hydrogen fuel cells do not produce any polluting gases.

The overall reaction in a hydrogen fuel cell is:

$$\text{hydrogen} + \text{oxygen} \rightarrow \text{water}$$
$$2H_2 + O_2 \rightarrow 2H_2O$$

This reaction is very efficient and the electrical energy supplied can be used to power electrical devices or even power motors in vehicles such as cars and buses (Figures 5.12 and 5.13).

▲ Figure 5.12 Some vehicles including cars and buses are powered by hydrogen fuel cells.

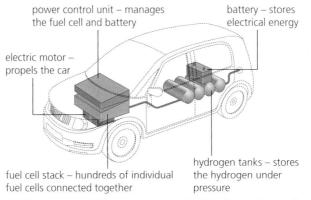

power control unit – manages the fuel cell and battery

battery – stores electrical energy

electric motor – propels the car

hydrogen tanks – stores the hydrogen under pressure

fuel cell stack – hundreds of individual fuel cells connected together

▲ Figure 5.13 Inside a car powered by hydrogen fuel cells.

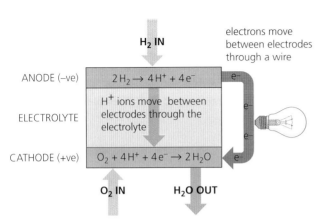

ANODE (−ve) $2H_2 \rightarrow 4H^+ + 4e^-$

ELECTROLYTE H^+ ions move between electrodes through the electrolyte

CATHODE (+ve) $O_2 + 4H^+ + 4e^- \rightarrow 2H_2O$

H_2 IN

O_2 IN H_2O OUT

electrons move between electrodes through a wire

▲ **Figure 5.14** How a fuel cell works.

What happens inside a fuel cell?

- At one of the electrodes, hydrogen gas loses electrons and forms hydrogen ions. The half equation for this is: $2H_2 \rightarrow 4H^+ + 4e^-$.
- The hydrogen ions move to the other electrode through the electrolyte.
- The electrons move to the other electrode through a wire. This current can be used to power a connected device.
- The hydrogen ions and electrons react with the oxygen gas at the other electrode making water. The half equation for this is: $O_2 + 4H^+ + 4e^- \rightarrow 2H_2O$ (Figure 5.14).

Evaluating the use of fuel cells

Fuel cells are very useful but there are some problems with their use. Table 5.7 evaluates the use of fuel cells compared to rechargeable cells and batteries.

Table 5.7

	Advantages	Disadvantages
Fuel cells	The only waste product from the fuel cell is water	Very expensive
		Contain some toxic chemicals in the electrodes and electrolyte which could harm the environment if not recycled
		Use hydrogen which is a flammable gas and so is difficult to transport and store
	Produce electricity continuously (so long as fuel is supplied)	The production of hydrogen may cause pollution
		Most hydrogen is made from the fossil fuel methane, also making the greenhouse gas carbon dioxide in the process
		Another method of making hydrogen is from the electrolysis of water which may use electricity generated from fossil fuels which also gives off the greenhouse gas carbon dioxide
Rechargeable cells and batteries	Much cheaper than fuel cells	Go 'flat' and need to be recharged regularly
		Contain harmful/toxic chemicals that could harm the environment if not recycled

Test yourself

17 What is the key difference between a fuel cell and other types of cell?

18 a) What is the only chemical product from a hydrogen fuel cell?

b) Write an overall equation for the reaction that takes place in a hydrogen fuel cell.

c) Hydrogen fuel cells are often said to have 'zero emissions'. Explain why this term is used but why it is not actually true.

19 What is the main reason that fuel cells are not used more than they are?

Show you can...

The table shows the reactions that occur at the anode and the cathode for a hydrogen fuel cell.

Cathode	Anode
$O_2 + 4H^+ + 4e^- \rightarrow 2H_2O$	$2H_2 \rightarrow 4H^+ + 4e^-$
Reduction	Oxidation

a) From the table select a species which is
   i) an ion
   ii) diatomic
   iii) a molecule
   iv) a compound
b) Explain why the reaction at the cathode is reduction.
c) Explain why the reaction at the anode is oxidation.
d) Copy the equation at the anode and the cathode and insert state symbols.

Chapter review questions

1 The following reactions all took place in solution in a beaker. The temperature before and after the chemicals were mixed was recorded in each case. Decide whether each reaction is exothermic or endothermic.

	Start temperature in °C	End temperature in °C
Reaction 1	21	15
Reaction 2	20	27
Reaction 3	22	67

2 Copy and complete the spaces in the following sentences.

In an exothermic reaction, thermal energy is transferred from the chemicals to their surroundings and so the temperature _____. In an _____ reaction, thermal energy is transferred away from the surroundings to the chemicals and so the temperature _____.

3 Copy and complete the spaces in the following sentences.

Chemical reactions can only take place when particles _____ with each other and have enough energy. The minimum energy particles need to react is called the _____ energy.

4 The reaction profile for a reaction is shown.

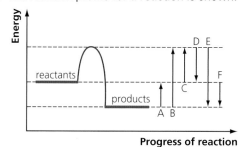

a) Give the letter of the arrow that shows the activation energy for the reaction.

b) Give the letter of the arrow that shows the overall energy change for the reaction.

c) Is this reaction endothermic or exothermic?

5 Decide whether each of the following reactions is likely to be endothermic or exothermic.

a) burning magnesium

b) decomposition of silver oxide

c) reaction inside a sports injury cold pack

d) reaction inside a self-heating food can

e) neutralisation of sulfuric acid by sodium hydroxide

f) neutralisation of sulfuric acid by sodium hydrogencarbonate

6 The table shows the potential difference (voltage) when chemical cells were set up by placing two strips of metal into a beaker of salt solution. In each case the copper was connected to the positive terminal of a voltmeter.

Metal 1	Metal 2	Potential difference (V)
Copper	Zinc	+1.10
Copper	Magnesium	+2.71
Copper	Silver	−0.46
Copper	Nickel	+0.59

a) What is an electrolyte?

b) What is the electrolyte in this experiment?

c) Place the five metals in the table in order of reactivity, clearly showing which is the most reactive.

d) What would be the potential difference if two pieces of copper metal were used as the electrodes?

e) What would be the potential difference if zinc and magnesium were used as the electrodes?

7 Cells and batteries are very useful portable sources of electricity. Some cells and batteries are rechargeable.

a) What is a battery?

b) Why do cells and batteries go flat?

c) What happens inside rechargeable cells and batteries when they are recharged?

d) Give one problem with the use of cells and batteries.

8 a) Calculate the energy change for the following reaction using these bond energies.

H—H = 436 kJ, Cl—Cl = 242 kJ, H—Cl = 431 kJ

H—H + Cl—Cl $\longrightarrow$ 2 H—Cl

b) Is this reaction endothermic or exothermic? Explain your answer by discussing bond energies.

9 Calculate the energy change for the following reaction using these bond energies.

N—H = 388 kJ, N—N = 158 kJ, N≡N = 944 kJ, O—H = 463 kJ, O=O = 496 kJ

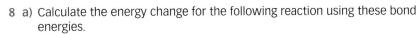

10 Many scientists believe that hydrogen fuel cells will become a major energy source in the future.

a) Give a major advantage of the use of fuel cells over rechargeable cells.

b) Write an equation for the overall reaction that takes place in a hydrogen fuel cell.

c) Write the half equation to show what happens to the hydrogen gas in a hydrogen fuel cell.

d) Write the half equation to show what happens to the oxygen gas in a hydrogen fuel cell.

e) The operation of a fuel cell to produce electrical energy is described as a 'clean' process. Explain why.

f) Explain why the production of hydrogen for a fuel cell may make the use of fuel cells less environmentally friendly.

Practice questions

1 Which one of the following is an endothermic reaction? [1 mark]

  A   the combustion of hydrogen

  B   the reaction of citric acid and sodium hydrogencarbonate

  C   the reaction of magnesium and hydrochloric acid

  D   the reaction of sodium hydroxide and hydrochloric acid

2 When potassium hydroxide dissolves in water the temperature of the solution rises. This is an example of

  A   an endothermic change

  B   an exothermic change

  C   a neutralisation reaction

  D   a thermal decomposition reaction [1 mark]

3 a) For each of the reactions A to E, choose the appropriate word from the list below to describe the type of reaction. Each word may be used once, more than once or not at all. [5 marks]

combustion   decomposition   neutralisation
oxidation   reduction

  A   copper carbonate → copper oxide + carbon dioxide

  B   ethanoic acid + sodium hydroxide → sodium ethanoate + water

  C   magnesium + oxygen → magnesium oxide

  D   methane + oxygen → carbon dioxide + water

  E   sodium hydroxide + hydrochloric acid → sodium chloride + water

  b) Explain the difference between an exothermic reaction and an endothermic reaction. [2 marks]

  c) For each of the reactions A to E above decide if it is an exothermic or an endothermic reaction. [5 marks]

  d) Describe how you would experimentally prove that the reaction between magnesium and hydrochloric acid is an exothermic reaction. [4 marks]

4 Photosynthesis is an endothermic process used by plants to produce carbohydrates.

$6CO_2 + 6H_2O \rightarrow C_6H_{12}O_6 + 6O_2$

  a) What is meant by the term endothermic? [1 mark]

  b) Describe what is meant by the term activation energy. [1 mark]

  c) Draw a labelled reaction profile for this reaction. You must show: the position of the reactants and products; the activation energy; and the energy change of the reaction. [4 marks]

  d) Explain in terms of bond making and breaking why this reaction is endothermic. [2 marks]

5 Sodium azide (NaN_3) is present in car airbags. During a car crash the airbag rapidly fills with nitrogen gas from the reaction shown below:

$2NaN_3(s) \rightarrow 2Na(s) + 3N_2(g)$

The energy change for the reaction is positive.

  a) What is the name given to reactions which have a positive enthalpy change? [1 mark]

  b) Draw a labelled reaction profile for the reaction. [4 marks]

  c) Calculate the mass of sodium azide needed to produce 72 dm³ of nitrogen gas at 20°C and one atmosphere pressure. (Remember from Chapter 3 that 1 mole of any gas has a volume of 24 dm³ at 20°C and one atmosphere pressure). [3 marks]

6 The diagram shows the profile for a reaction with a catalyst and without a catalyst.

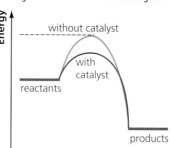

  a) Is this an exothermic or an endothermic reaction? [1 mark]

  b) Does a catalyst have an effect on the overall energy change of the reaction? [1 mark]

  c) On a copy of the diagram label the activation energy for the catalysed reaction as A and the activation energy for the uncatalysed reaction as B. [2 marks]

  d) From the information shown in the graph, state the effect of a catalyst on the activation energy of a reaction. [1 mark]

7 Hydrogen reacts with fluorine to form hydrogen fluoride.

$$H_2 + F_2 \rightarrow 2HF$$

Use the bond energies in the table to calculate the energy change for this reaction. Explain if the reaction is exothermic or endothermic. [4 marks]

Bond	Bond energy in kJ
H—H	436
F—F	158
H—F	568

8 Ethanol burns in oxygen.

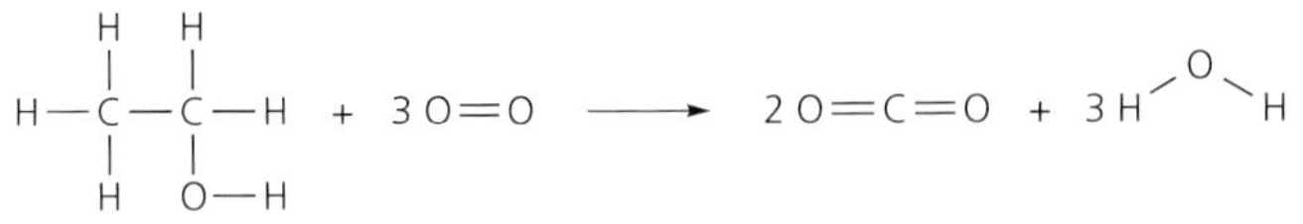

$$H-\underset{\underset{H}{|}}{\overset{\overset{H}{|}}{C}}-\underset{\underset{O-H}{|}}{\overset{\overset{H}{|}}{C}}-H \;+\; 3\,O{=}O \;\longrightarrow\; 2\,O{=}C{=}O \;+\; 3\,H{\overset{O}{\diagdown}}H$$

Use the bond energies below to calculate the energy change for this reaction. [3 marks]

Bond energies:
C—H 412 kJ
O=O 496 kJ
C=O 743 kJ
O—H 463 kJ
C—C 348 kJ
C—O 360 kJ

9 Hydrogen is used as a liquid propellant to launch rockets.

a) Explain using a balanced symbol equation, why hydrogen is a clean fuel. [2 marks]

b) Hydrogen is also used in fuel cells which are taken into space to supply power. Explain how a hydrogen fuel cell produces electricity. In your answer include two half equations. [5 marks]

H

Working scientifically: Identifying variables when planning experiments

When carrying out an experiment different variables are used. Variables are the things that can change. When we plan experiments, we choose to change some variables while keeping others the same.

A variable is a physical, chemical or biological quantity or characteristic that can have different values. It may be for example, temperature, mass, volume, pH or even the type of chemical used in an experiment.

There are different types of variables which you need to be familiar with.

A continuous variable has values that are numbers. Mass, temperature and volume are examples of continuous variables. The values of these variables can either be found by counting (e.g. the number of drops) or by measurement (e.g. the temperature).

A categoric variable is one which is best described by words. Variables such as the type of acid or the type of metal are categoric variables.

▲ Figure 5.15 The volume of carbon dioxide gas produced over time, from the reaction between calcium carbonate and 1.0 mol/dm HCl was recorded using this apparatus. The experiment was repeated using different concentrations of HCl. Can you identify the independent, dependent and controlled variables?

Questions

1 Decide if the following are categoric or continuous variables.
   a) temperature
   b) type of acid
   c) volume
   d) name of gas produced
   e) concentration of solution
   f) volume of gas produced
   g) mass
   h) name of metal
   i) colour of solution
   j) drops of acid

Scientists often plan experiments to investigate if there is a relationship between two variables, the independent and the dependent variable.

The independent variable is the variable for which values are changed or selected by the investigator (i.e. it is the one which you deliberately change during an experiment).

The dependent variable is one which may change as a result of changing the independent variable. It is the one which is measured for each and every change in the independent variable.

A control variable is one which may, in addition to the independent variable, affect the outcome of the investigation. Control variables must be kept constant during an experiment to make it a fair test.

As an example, the variables are shown below for an experiment to find the effect of temperature on the rate of the reaction between magnesium and excess hydrochloric acid. The rate was measured by timing how long it took for all the magnesium to react.

Independent variable	Dependent variable	Control variables
Temperature	Time taken for all the magnesium to react	The mass of magnesium The surface area of magnesium The volume of hydrochloric acid The concentration of hydrochloric acid

Questions

2 For the following experiments identify the
   (i)   independent variable,
   (ii) dependent variable and
   (iii) control variables.

   a) Some magnesium was added to hydrochloric acid and the temperature recorded. The experiment was repeated several times using different volumes of hydrochloric acid.

   b) In the reaction between copper carbonate and hydrochloric acid the time taken for a mass of copper carbonate to all be used up was recorded. The experiment was repeated using different masses of copper carbonate.

   c) 2 g of magnesium was added to copper sulfate solution and the highest temperature reached was recorded. The experiment was repeated using five different concentrations of copper sulfate.

   d) In an experiment to find the effect of stirring on speed of dissolving, the time taken to dissolve some copper sulfate in water was measured. This was repeated stirring the solution.

   e) The volume of carbon dioxide gas produced when calcium carbonate reacts with hydrochloric acid was measured, and the experiment repeated using different masses of calcium carbonate.

   f) The temperature of nitric acid was recorded before and after some sodium hydroxide was added. The experiment was repeated using sulfuric acid, ethanoic acid and methanoic acid.

6 The rate and extent of chemical change

The explosion of dynamite is a very fast reaction. The rusting of steel is a very slow reaction. Some reactions completely react, but in others the products can turn back into the reactants. Being able to control the speed of a reaction and reducing the amount of products turning back into reactants are very important and are studied in this chapter.

Specification coverage

This chapter covers specification points 4.6.1 to 4.6.2 and is called The rate and extent of chemical change

It covers the rate of reactions, reversible reactions and dynamic equilibrium.

Prior knowledge

Previously you could have learned:

- › Some reactions are fast and some reactions are slow.
- › Some chemical changes are reversible.
- › Exothermic reactions release thermal energy to the surroundings and so the temperature increases.
- › Endothermic reactions take in thermal energy from the surroundings and so the temperature decreases.

Test yourself on prior knowledge

1 Give two examples of chemical reactions that are
 a) very fast
 b) very slow
 c) exothermic
 d) endothermic

2 a) Explain why the temperature rises in an exothermic reaction.
 b) Explain why the temperature falls in an endothermic reaction.

Rate of reaction

We can make a chemical
action happen by:-
at (temperature)
the concentration of the
ubstance
talyst
urface area
essure

Measuring the rate of reaction

Some chemical reactions are fast while others are slow. Reactions that are fast have a high rate of reaction. Reactions that are slow have a low rate of reaction.

Measuring the mean rate of reaction

The mean rate of a reaction can be found by measuring the quantity of a reactant used or a product formed over the time taken:

$$\text{mean rate of reaction} = \frac{\text{quantity of reactant used}}{\text{time}}$$

or

$$\text{mean rate of reaction} = \frac{\text{quantity of product formed}}{\text{time}}$$

The time is typically measured in seconds.

The quantity of a chemical could be measured as a:

- mass in grams (g) or
- volume of gas in cubic centimetres (cm^3) or
- (H) amount in moles (mol).

The rate of a reaction changes during a reaction. Most reactions are fastest at the beginning, slow down and then eventually stop. As the rate is constantly changing, the rate at one moment is likely to be different to the rate at another. This means that when we calculate the rate of reaction over a period of time, we are actually working out the mean rate of reaction over that time.

Example

In a reaction where some calcium carbonate reacts with hydrochloric acid, 40 cm³ of carbon dioxide was produced in 10 seconds. Calculate the mean rate of reaction during this time in cm³/s of carbon dioxide produced.

Answer

$$\text{rate of reaction} = \frac{\text{quantity of product formed}}{\text{time}} = \frac{40}{10} = 4.0\,\text{cm}^3/\text{s}$$

Example

Magnesium reacts with sulfuric acid. In a reaction, 0.10 g of magnesium was used up in 20 seconds. Calculate the mean rate of reaction during this time in g/s of magnesium used.

Answer

$$\text{rate of reaction} = \frac{\text{quantity of reactant used}}{\text{time}} = \frac{0.10}{20} = 0.0050\,\text{g/s}$$

Example

Copper carbonate decomposes when it is heated. In a reaction, 0.024 moles of copper carbonate had decomposed in 30 seconds. Calculate the mean rate of reaction during this time in mol/s of copper carbonate decomposing.

Answer

$$\text{rate of reaction} = \frac{\text{quantity of reactant used}}{\text{time}} = \frac{0.024}{30} = 0.00080\,\text{mol/s}$$

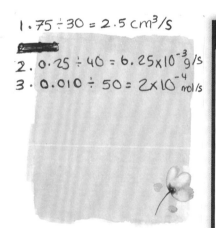

1. $75 \div 30 = 2.5\,\text{cm}^3/\text{s}$

2. $0.25 \div 40 = 6.25 \times 10^{-3}\,\text{g/s}$

3. $0.010 \div 50 = 2 \times 10^{-4}\,\text{mol/s}$

Test yourself

1 Carbon dioxide gas is formed when sodium carbonate reacts with ethanoic acid. In a reaction, 75 cm³ of carbon dioxide was produced in 30 seconds. Calculate the mean rate of reaction during this time in cm³/s of carbon dioxide produced.

2 When nickel carbonate is heated it decomposes. During an experiment it was found that 0.25 g of nickel carbonate decomposed in 40 seconds. Calculate the mean rate of reaction during this time in g/s of nickel carbonate decomposing.

3 When sodium thiosulfate solution reacts with hydrochloric acid, one of the products is solid sulfur. In a reaction, 0.010 moles of sulfur was formed in 50 seconds. Calculate the mean rate of reaction during this time in mol/s of sulfur formed.

Show you can...

Some copper carbonate was placed in a crucible and weighed. It was heated to decompose it for 40 seconds and reweighed, and the results recorded in the table below. Carbon dioxide is formed and escapes from the crucible as the copper carbonate decomposes.

Mass of copper carbonate and crucible before heating in g	18.1
Mass of copper carbonate and crucible after 40 seconds heating in g	16.9

Calculate the mean rate of reaction during this time in g/s of carbon dioxide produced.

Reaction rate graphs

Graphs can be drawn to show how the quantity of reactant used or product formed changes with time. The shapes of these graphs are usually very similar for most reactions. The slope of the line represents the rate of the reaction. The steeper the slope, the faster the reaction.

The relative slope can be judged by simply looking at the line but it can be useful to draw tangents to the curve (Figure 6.1). At any point on a curve, a straight line is drawn with a ruler that just touches the curve and that we judge to be at the same slope as the curve at that point. The steeper the tangent, the faster the rate of reaction.

In an individual reaction, the line is steepest at the start when the reaction is at its fastest. The line becomes less steep as the reaction slows down and eventually becomes horizontal when the reaction stops.

In reactions that produce a gas, it is easy to measure the volume or mass of gas formed over time (Table 6.1) and draw a graph of this type.

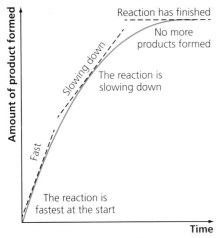

▲ **Figure 6.1** Graph showing the progress of a reaction.

[handwritten notes:] he rate of reaction can be
ilculated by :-
ne volume of gas produced
he mass of the solid
creasing.

Table 6.1

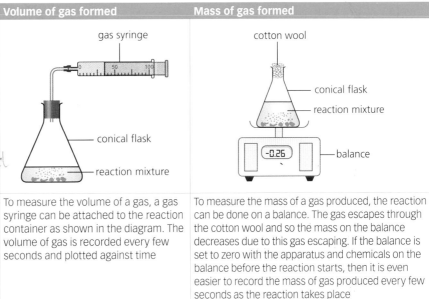

Volume of gas formed	Mass of gas formed
To measure the volume of a gas, a gas syringe can be attached to the reaction container as shown in the diagram. The volume of gas is recorded every few seconds and plotted against time	To measure the mass of a gas produced, the reaction can be done on a balance. The gas escapes through the cotton wool and so the mass on the balance decreases due to this gas escaping. If the balance is set to zero with the apparatus and chemicals on the balance before the reaction starts, then it is even easier to record the mass of gas produced every few seconds as the reaction takes place

The rate of one reaction can be compared to the rate of another using these graphs. The steeper the slope of the line, the greater the rate of reaction. For example, in the graph in Figure 6.2, reaction A is faster than reaction B. We can see that the line for reaction A is steeper, but this is confirmed if we draw a tangent to each curve near the start.

Finding the rate of reaction from the gradient of tangents on reaction rate curves

The rate of reaction at any given moment can be found by measuring the gradient of the tangent to the curve at that point.

To calculate the rate of reaction by drawing a tangent to a graph:

1 Select the time at which you want to measure the rate.

2 With a ruler, draw a tangent to the curve at that point (the tangent should have the same slope as the graph line at that point).

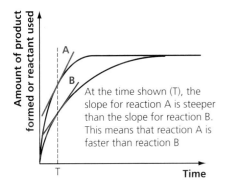

▲ **Figure 6.2** Comparing reaction rates with tangents.

(H)

3 Choose two points a good distance apart on the tangent line and draw lines until they meet the axes.

4 Find the slope using this equation: slope = $\dfrac{\text{change in } y \text{ axis}}{\text{change in } x \text{ axis}}$

In Figure 6.3, the volume of carbon dioxide produced when calcium carbonate reacts with acid was plotted against time. The rate of reaction was found at 18 and 40 seconds by drawing tangents to the curve and calculating the slope. The rate of reaction was 1.11 cm³/s at 18 seconds. The rate was lower at 40 seconds where it was 0.53 cm³/s.

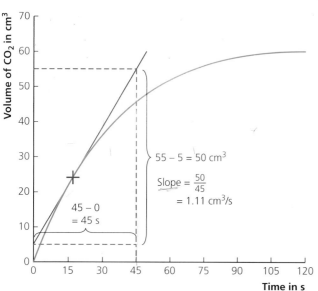
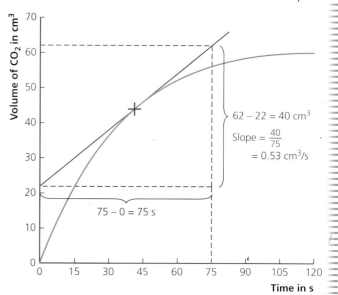

▲ **Figure 6.3** Using tangents to work out the rate of reaction.

Test yourself

4 The graph shows how the mass of carbon dioxide formed when sodium carbonate reacts with hydrochloric acid varies over time.

   a) At what point on the graph is the rate of reaction greatest? Explain your answer.

   b) At what point on the graph does the reaction stop? Explain your answer.

   c) Is the rate of reaction faster at point B or E? Explain your answer.

5 Look at the graphs for reactions P, Q and R which all produce a gas.

   a) Which reaction is the fastest?

   b) Which reaction is the slowest?

6 Hydrogen gas is formed when magnesium reacts with sulfuric acid. The table shows how the volume of hydrogen changed with time when some magnesium was reacted with sulfuric acid.

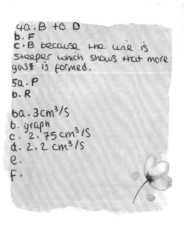

Handwritten notes:
4a. B to D
b. F
c. B because the line is steeper which shows that more gas is formed.
5a. P
b. R
6a. 3cm³/s
b. graph
c. 2.75 cm³/s
d. 2.2 cm³/s
e.
f.

Time in s	0	10	20	30	40	50	60	70	80	90	100
Volume of hydrogen in cm³	0	30	55	75	88	98	102	104	104	104	104

   a) Given that 30 cm³ of hydrogen has formed after 10 seconds, calculate the mean rate of reaction during the first 10 seconds in cm³/s of hydrogen formed.

   b) Plot a graph of the volume of hydrogen against time.

   c) Calculate the rate of reaction at 20 seconds by drawing a tangent to the curve.

   d) Calculate the rate of reaction at 40 seconds by drawing a tangent to the curve.

   e) Write a word equation for this reaction.

   f) Write a balanced equation for this reaction.

e rate of reaction can be
reased by:-
increasing the frequency of
isions.

reasing the energy of
isions

○ Collision theory

For a chemical reaction to take place particles of the reactants must collide with enough energy to react. The minimum amount of energy particles need to react is called the activation energy.

> **KEY TERM**
>
> **Activation energy** The minimum energy particles must have to react.

Successful collisions (i.e. ones which result in a reaction) take place when reactant particles collide with enough energy to react (Figure 6.4). Unsuccessful collisions, ones which do not result in a reaction, take place when reactant particles collide but do not have enough energy to react.

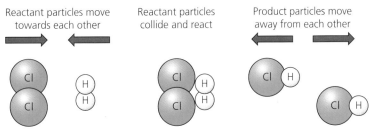

► Figure 6.4 A successful collision.

The rate of a reaction depends on the frequency of successful collisions.

○ Factors affecting the rate of reactions

There are several factors that affect the rate of chemical reactions.

Temperature

The higher the temperature, the faster a reaction. This is because the particles have more energy (they are more energetic) and so more of the particles have enough energy to react when they collide. The particles are also moving faster and so collide more frequently. This means that there are more frequent successful collisions.

An everyday example of the effect of temperature on chemical reactions is the use of fridges (Figure 6.5). When food goes off, chemical reactions are taking place. We can slow down the rate at which these reactions take place and so slow down the rate at which food goes off by cooling food down and putting it in a fridge.

▲ Figure 6.5 Food is kept in a fridge to keep it cool and slow down chemical reactions that make food go off.

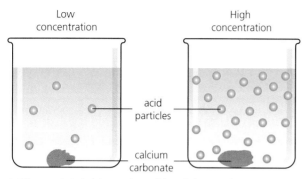

Low concentration   High concentration

acid particles

calcium carbonate

▲ **Figure 6.6** Acid reacting with calcium carbonate.

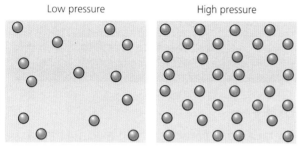

Low pressure   High pressure

▲ **Figure 6.7** Reactions are faster at higher pressure.

More energetic collisions is more useful than more frequent collisions

▶ **Figure 6.8** The effect of changing surface area.

Concentration of reactants in solution

Many chemical reactions involve reactants that are dissolved in solutions. The concentration of a solution is a measure of how much solute is dissolved. The higher the concentration, the more particles of solute that are dissolved (Figure 6.6).

The higher the concentration of reactants in solution, the greater the rate of reaction. This is because there are more reactant particles in the solution and so there are more frequent successful collisions.

Pressure of reacting gases

Some chemical reactions involve reactants that are gases. The greater the pressure of a gas, the closer the reactant particles are together.

The higher the pressure of reactants that are gases, the greater the rate of reaction. This is because the reactant particles are closer together and so there are more frequent successful collisions (Figure 6.7).

Surface area of solid reactants

Some chemical reactions involve reactants that are solids. The surface area of a solid is increased if it is broken up into more pieces (Figure 6.8). When a solid is made into a powder it has a massive surface area. The greater the surface area of a solid reactant, the greater the rate of reaction. This is because there are more particles on the surface that can react and so there are more frequent successful collisions.

Large piece of solid – low surface area

Only reactant particles at the surface can collide with particles of the other reactant

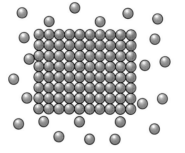

Smaller pieces of solid – bigger surface area

If the solid is broken up into more pieces, there are more reactant particles that can be collided with

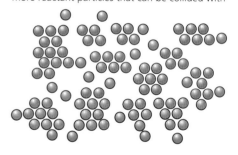

A solid cube with sides of length 2 cm has a total surface area of 24 cm². If it is broken up into eight smaller cubes with sides of length 1 cm then its surface area increases to 48 cm². This means that the surface area has increased while the total volume stays the same (Table 6.2).

The surface area to volume ratio can also be considered to explain this. The single cube with 2 cm sides has a surface area to volume ratio of 3:1. When it is broken up into eight smaller cubes with 1 cm sides, the surface area to volume ratio is greater at 6:1 (Table 6.2).

Concentration- Amount of
particles in a certain volume
of solution

Pressure - Amount of gaseous
particles in a certain volume

more particles in a certain
volume
more frequent collisions
higher chance of successful
collisions
higher rate of reaction

Table 6.2

Cube	One larger cube	Broken up into smaller cubes
	2cm × 2cm × 2cm	1cm × 1cm × 1cm
Surface area	Surface area of each side = $2 \times 2 = 4\,cm^2$ There are six sides, so Total surface area = $6 \times 4 = 24\,cm^2$	Each cube: Surface area of each side = $1 \times 1 = 1\,cm^2$ There are six sides, so Surface area of each cube = $6 \times 1 = 6\,cm^2$ For all eight cubes: Total surface area = $8 \times 6 = 48\,cm^2$
Volume	Volume = $2 \times 2 \times 2 = 8\,cm^3$	Each cube: Volume = $1 \times 1 \times 1 = 1\,cm^3$ For all eight cubes: Total volume = $8 \times 1 = 8\,cm^3$
Surface area : volume ratio	$24:8 = 3:1$	$48:8 = 6:1$

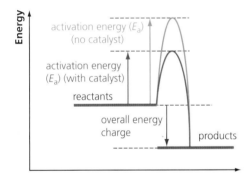

▲ Figure 6.9

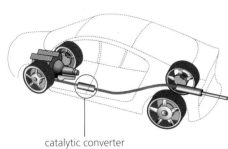

catalytic converter
▲ Figure 6.10

catalysts are usually in the form
of gauzes / honey comb to
increase the surface area

The presence of catalysts

→ A substance that increases the rate of the reaction and is not chemically altered by the end of the reaction

A **catalyst** is a substance that speeds up a chemical reaction but does not get used up. A catalyst works by providing a different route (pathway) for the reaction that has a lower activation energy (Figure 6.9).

Different reactions have different catalysts. Some examples of catalysts include:

- iron in the production of ammonia from reaction of hydrogen with nitrogen (the Haber process)
- nickel in the production of margarine from reaction of vegetable oils with hydrogen
- platinum in catalytic converters to remove some pollutants from the exhaust gases of cars (Figure 6.10).

The catalyst is not used up in a reaction. The amount of catalyst left at the end of the reaction is the same as there was at the start. As the catalyst is not used up, it does not appear in the chemical equation for the reaction. For example, in the reaction to remove nitrogen monoxide (NO) from car exhaust gases, the catalyst platinum (Pt) does not appear in the equation. Sometimes the catalyst is written on top of the arrow in the equation.

$$2NO \rightarrow N_2 + O_2 \quad \text{or} \quad 2NO \xrightarrow{Pt} N_2 + O_2$$

▲ **Figure 6.11** Biological detergents contain enzymes.

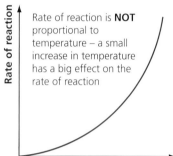

Rate of reaction is **NOT** proportional to temperature – a small increase in temperature has a big effect on the rate of reaction

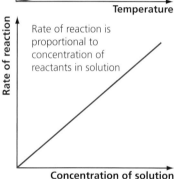

Rate of reaction is proportional to concentration of reactants in solution

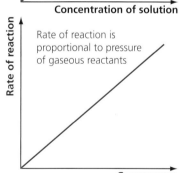

Rate of reaction is proportional to pressure of gaseous reactants

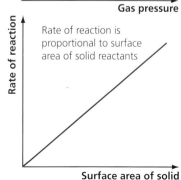

Rate of reaction is proportional to surface area of solid reactants

▲ **Figure 6.13**

Enzymes are molecules that act as catalysts in biological systems. For example:

● amylase is an enzyme that catalyses the breakdown of starch into sugars
● protease is an enzyme that catalyses the breakdown of protein into amino acids.

Enzymes are produced in living organisms and are vital for most life processes such as respiration and digestion. We can also make artificial use of enzymes. For example, biological washing powders contain protease enzymes to break down proteins in dirt on clothes (Figure 6.11).

The relative effect of factors affecting rates

Most factors affecting rates (including the concentration of reactants in solution, the pressure of reactant gases and the surface area of solid reactants) usually have a proportional effect on the rate of reaction. For example, if any of these factors are doubled, the reaction rate doubles. This is because in each case if the factor is doubled, the number of reactant particles available for collisions doubles and so the frequency of successful collisions will double (Figure 6.12).

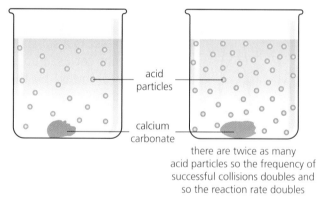

acid particles

calcium carbonate

there are twice as many acid particles so the frequency of successful collisions doubles and so the reaction rate doubles

▲ **Figure 6.12** The effect of doubling the concentration of a reactant in solution.

However, changes in temperature have a much greater effect. A small increase in temperature causes a big increase in rate of reaction (Figure 6.13). This is because a small increase in temperature leads to many more particles having the activation energy.

> **TIP**
>
> In a proportional relationship, any change made to one factor has the same effect on the other factor. For example, if one factor is doubled, the other factor doubles. If one factor is made 10 times bigger, the other factor gets 10 times bigger. If one factor is made 3 times smaller, the other factor becomes 3 times smaller.

Test yourself

7 Hydrogen peroxide decomposes very slowly into water and oxygen. The reaction is much faster if a catalyst such as manganese(IV) oxide is used.

  a) What is a catalyst?

  b) How do catalysts work?

  c) The diagram shows the reaction profile for the decomposition of hydrogen peroxide with and without a catalyst. Which arrow represents the activation energy for the reaction that takes place with a catalyst?

  d) The rate of reaction can also be increased by adding an enzyme. What are enzymes?

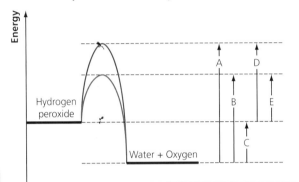

8 Nitrogen gas reacts with hydrogen gas to make ammonia. The reaction is carried out at 200 atmospheres pressure. Explain why the reaction is faster at higher pressures.

9 Sodium thiosulfate solution reacts with hydrochloric acid to form solid sulfur. As the sulfur is formed, the solution becomes cloudy. The reaction can be carried out in a conical flask on top of a piece of paper with a cross drawn on it. The time it takes for the solution to become so cloudy that the cross can no longer be seen can be measured and used to compare reaction rates under different conditions. The table below shows the results of some experiments carried out in this way.

Experiment	Concentration of sodium thiosulfate solution in mol/dm³	Temperature in °C	Time taken to become too cloudy to see cross in s
A	0.10	20	75
B	0.20	20	37
C	0.30	20	25
D	0.20	30	18

  a) Which experiment was fastest?

  b) What is the effect of changing the concentration on the rate of reaction? Which experiments did you use to work this out?

  c) Explain why changing concentration affects the rate of reaction.

  d) What is the effect of changing the temperature on the rate of reaction? Which experiments did you use to work this out?

  e) Explain why changing temperature affects the rate of reaction.

10 Calcium carbonate reacts with hydrochloric acid and produces carbon dioxide gas. An experiment was done to compare the rate of reaction using large pieces, small pieces and powdered calcium carbonate. In each case the same mass of calcium carbonate was used.

Type of calcium carbonate	Time taken to produce 100 cm³ of carbon dioxide in s
Large pieces	75
Small pieces	48
Powder	5

  a) Which type of calcium carbonate had the greatest surface area?

  b) Calculate the mean rate of reaction for each experiment in cm³/s of carbon dioxide produced.

  c) What is the effect of changing the surface area of the calcium carbonate on the rate of reaction?

  d) Explain why changing the surface area affects the rate of reaction.

11 This question compares the surface area to volume ratio of a single cube with 10 cm sides to 1000 smaller cubes with 1 cm sides that can be made from the bigger cube.

  a) Calculate the surface area of a cube with 10 cm sides.

  b) Calculate the surface area to volume ratio of a cube with 10 cm sides.

  c) Calculate the total surface area of 1000 cubes with 1 cm sides.

  d) Calculate the surface area to volume ratio of 1000 cubes with 1 cm sides.

Investigating how changes in concentration affect the rate of reaction by measuring the volume of a gas produced

An experiment was carried out to determine the effect of changing the concentration of hydrochloric acid on the rate of the reaction between magnesium and hydrochloric acid. 1.0g of magnesium turnings was reacted with excess hydrochloric acid of different concentrations and the volume of gas produced recorded every minute. The apparatus used is shown in the figure.

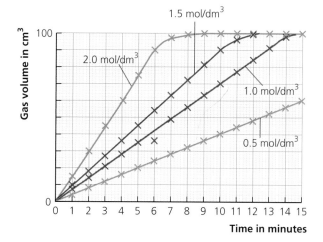

Questions

1 Name the piece of apparatus labelled A.

2 Explain why the bung must be inserted immediately once the magnesium is added to the acid.

3 Name one other piece of apparatus which must be used in this experiment.

4 Write a balanced symbol equation for the reaction between magnesium and hydrochloric acid.

5 What is observed in the flask during the reaction?

6 How would you ensure that this experiment is a fair test?

7 State one source of error in this experiment.

The results from this experiment are plotted on a graph.

8 Use the graph to determine the time at which each reaction ended for acid of concentration 2.0 mol/dm³ and 1.5 mol/dm³.

9 Are there any results which are anomalous? Explain your answer.

10 Which concentration of acid produced the slowest rate of reaction?

11 What is the relationship between the concentration of acid and the rate of reaction?

12 Explain this relationship in terms of collision theory.

Another way of measuring rates of reaction is by measuring the loss of mass of reactants, often by recording the mass of the reacting vessel and its contents over a certain period time, as shown in the example in the practical activity on the next page.

Practical

Investigating the rate of decomposition of hydrogen peroxide solution by measuring the loss in mass

Hydrogen peroxide decomposes in the presence of solid manganese(IV) oxide to produce water and oxygen.

hydrogen peroxide → water + oxygen

The apparatus shown was used to investigate the rate of decomposition of hydrogen peroxide solution. $20\,cm^3$ of hydrogen peroxide solution were added to 1.0 g of solid manganese(IV) oxide at 20°C.

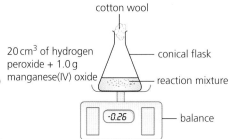

The following results were obtained.

Time in minutes	Mass of oxygen lost in g
1	0.23
2	0.34
3	
4	0.45
5	0.47
6	0.48
7	0.48

Questions

1 a) Write a balanced symbol equation for the decomposition of hydrogen peroxide (H_2O_2).

  b) What is the purpose of the cotton wool plug?

  c) Plot a graph of mass of oxygen lost against time.

  d) Use the graph to state the mass of oxygen lost after 3 minutes.

  e) Suggest an alternative way of measuring the rate of this reaction without measuring the mass of oxygen lost.

  f) Sketch on the same axes the graph you would expect to obtain if the experiment was repeated at 20°C using 1.0 g of manganese(IV) oxide with $20\,cm^3$ of this hydrogen peroxide solution mixed with $20\,cm^3$ of water.

2 At the end of this experiment the manganese(IV) oxide can be recovered.

  a) Draw a labelled diagram of the assembled apparatus which could be used to recover the manganese(IV) oxide at the end of the experiment.

  b) How would you experimentally prove that the manganese(IV) oxide was not used up in this experiment?

  c) Complete this sentence:
In this experiment the manganese dioxide is acting as a _____.

Investigating how changes in concentration affect the rate of the reaction by methods involving a colour change or turbidity

Sodium thiosulfate solution ($Na_2S_2O_3$) reacts with dilute hydrochloric acid according to the equation:

$$Na_2S_2O_3(aq) + 2HCl(aq) \rightarrow 2NaCl(aq) + S(s) + SO_2(g) + H_2O(l)$$

In an experiment 25 cm³ of sodium thiosulfate was placed in a conical flask on top of a piece of paper with a cross drawn on it. Hydrochloric acid was added and the stopwatch started. A precipitate was produced that caused the solution to become cloudy. The stopwatch was stopped when the experimenter could no longer see the cross on the paper through the solution, due to the precipitate formed. The precipitate causes turbidity, which is cloudiness in a fluid caused by large numbers of individual solid particles.

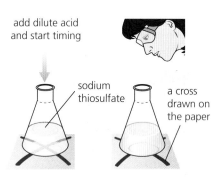

add dilute acid and start timing

sodium thiosulfate

a cross drawn on the paper

KEY TERM

Turbidity The cloudiness of a solution.

Questions

1   Look at the equation and identify the product which causes the solution to become cloudy.

The experiment was repeated using different concentrations of sodium thiosulfate. The results are recorded below.

Experiment	Concentration of sodium thiosulfate in mol/dm³	Time taken for cross to disappear in s	Rate of reaction in s⁻¹ (1/time)
1	0.4	105	0.0095
2	0.8	79	0.0127
3	1.2	54	0.0185
4	1.6	32	

2   Calculate the value for the rate of reaction for experiment 4 using the equation: rate $= \dfrac{1}{time}$. Give your answer to 3 significant figures.

3   Identify three variables that must be kept constant to make this a fair test.

4   From the results of the experiment state the effect of increasing the concentration of sodium thiosulfate solution on the rate of the reaction.

5   Explain in terms of collision theory why increasing the concentration of sodium thiosulfate solution has this effect.

6   State and explain one change which could be made to this experiment to give more accurate results.

Reversible reactions and dynamic equilibrium

A reversible reaction is a reaction where the products can react together to make the original reactants again

○ Reversible reactions

Some chemical reactions are reversible. This means that once the products have been made from the reactants, the products can react to reform the reactants. The ⇌ arrows are used to show the reaction is reversible.

$$\text{reactants} \underset{\text{reverse reaction}}{\overset{\text{forward reaction}}{\rightleftharpoons}} \text{products}$$

(a)

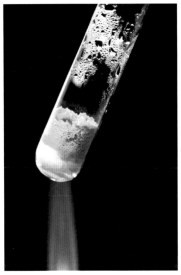

(b)

▲ **Figure 6.14 (a)** Reaction of anhydrous copper sulfate with water; **(b)** heating hydrated copper sulfate.

For example, anhydrous copper sulfate (which is white) reacts with water to form hydrated copper sulfate (which is blue). However, this reaction is reversible as when the blue hydrated copper sulfate is heated it breaks back down into white anhydrous copper sulfate and water (Figure 6.14).

anhydrous copper sulfate	+ water	$\rightleftharpoons$	hydrated copper sulfate
$CuSO_4$	$+ 5H_2O$	$\rightleftharpoons$	$CuSO_4.5H_2O$
white solid	colourless liquid		blue solid

When ammonium chloride (NH_4Cl) is heated it breaks down into ammonia (NH_3) and hydrogen chloride (HCl). However, when cooled the ammonia can react with the hydrogen chloride to reform ammonium chloride (Figure 6.15).

ammonium chloride	$\rightleftharpoons$	ammonia	+	hydrogen chloride
NH_4Cl	$\rightleftharpoons$	NH_3	+	HCl
white solid		colourless gas		colourless gas

Energy changes in reversible reactions

If a reversible reaction is exothermic in one direction, then it will be endothermic in the other (Table 6.3). The amount of energy transferred will be the same in each direction. For example, if the forward reaction is exothermic with an energy change of –92 kJ, then the reverse reaction will be endothermic with an energy change of +92 kJ. This is due to the law of conservation of energy.

▲ **Figure 6.15** Heating ammonium chloride.

Table 6.3

A reversible reaction where the forward reaction is exothermic	A reversible reaction where the forward reaction is endothermic
Exothermic (–92 kJ) →	Endothermic (+58 kJ) →
nitrogen + hydrogen $\rightleftharpoons$ ammonia	dinitrogen tetroxide $\rightleftharpoons$ nitrogen oxide
N_2 + $3H_2$ $\rightleftharpoons$ $2NH_3$	N_2O_4 $\rightleftharpoons$ $2NO_2$
← Endothermic (+92 kJ)	← Exothermic (–58 kJ)

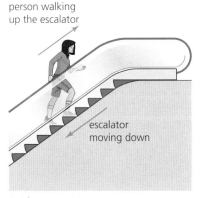

person walking up the escalator

escalator moving down

▲ **Figure 6.16**

Dynamic equilibrium

→ no matter escapes, only energy does

If a reversible reaction takes place in a closed system, that is apparatus where no substances can get in or out, a **dynamic equilibrium** is reached. This happens when both the forward and reverse reactions are taking place simultaneously and at exactly the same rate of reaction.

A good analogy of a system in dynamic equilibrium is an athlete walking up an escalator that is moving down (Figure 6.16). The athlete on the escalator is in a state of dynamic equilibrium if she walks up the escalator at the same rate as the escalator moves down.

namic equilibrium — when the rate of the forward reaction is the same as the rate of e reverse reaction (reversible reaction) (closed system)

*Equilibrium only occurs when the reaction happens in a closed system. the concentration o the reactants and products are constant

6 The rate and extent of chemical change

(H) KEY TERMS

Closed system A system where no substances can get in or out.

Dynamic equilibrium System where both the forward and reverse reactions are taking place simultaneously and at the same rate.

(H) Show you can...

a) Draw a reaction profile for the reaction
$N_2 + 3H_2 \rightleftharpoons 2NH_3$
The energy change for the forward reaction is $-92\,kJ\,mol^{-1}$
Label:
the activation energy for the forward reaction;
the activation energy for the reverse reaction;
the energy change for the forward reaction.
b) Under what condition does this reaction reach dynamic equilibrium?

★ Test yourself

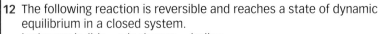

12 The following reaction is reversible and reaches a state of dynamic equilibrium in a closed system.
hydrogen iodide $\rightleftharpoons$ hydrogen + iodine
  a) The forward reaction is endothermic. Will the reverse reaction be endothermic or exothermic?
  b) What is a closed system?
  c) What is happening to the reactions at dynamic equilibrium?
  d) Given the formulae below, write a balanced equation for this reaction. (hydrogen iodide = HI, hydrogen = H_2, iodine = I_2)

13 Ethanol can be made in the following reaction which is reversible and reaches a state of dynamic equilibrium in a closed system. The energy change for the forward reaction is $-42\,kJ$.
ethene + steam $\rightleftharpoons$ ethanol
  a) Is the forward reaction endothermic or exothermic?
  b) Will the reverse reaction be endothermic or exothermic?
  c) What is the energy change for the reverse reaction?

○ Le Châtelier's principle

The position of an equilibrium

The athlete on the escalator can be in state of dynamic equilibrium wherever she is on the escalator, whether she is near the top, middle or bottom (Table 6.4). In a similar way, in a chemical reaction at dynamic equilibrium there could be mainly reactants, mainly products or a similar amount of each. The relative amount of reactants and products can be described by the position of the equilibrium.

Table 6.4

Equilibrium position	Lies to the left	Lies somewhere in the middle	Lies to the right
reactants $\rightleftharpoons$ products	Means there are *more reactants* than products in the mixture of chemicals at equilibrium	Means there is a *similar amount of reactants and products* in the mixture of chemicals at equilibrium	Means there are *more products* than reactants in the mixture of chemicals at equilibrium
Escalator analogy	position lies to the left	position lies somewhere in the middle	position lies to the right

Changing the position of an equilibrium

The relative amounts of reactants and products present at equilibrium (i.e. the position of an equilibrium) depends on the conditions. If the conditions are changed, the position of the equilibrium changes and so the relative amounts of reactants and products changes.

Imagine the athlete on the escalator. If the speed of the escalator moving down was increased then the system would no longer be in

Le chatlier's principle
whenever a change in conditions (concentration, temperature, pressure) is introduced to a system at equilibrium, the position of the equilibrium shifts as to cancel out the change

160

[handwritten, top of page] a system is at equilibrium and a change is made to any of conditions, then the system responds to counteract the change
use this term in the exam

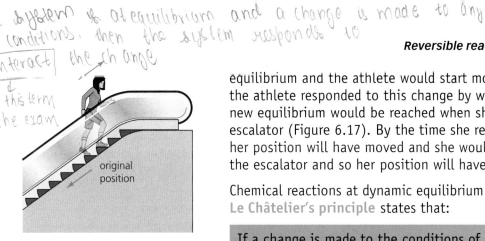

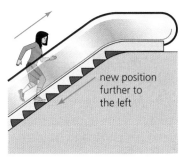

▲ **Figure 6.17** A new equilibrium is established with its position further to the left after the speed of the escalator was increased.

equilibrium and the athlete would start moving down the escalator. If the athlete responded to this change by walking or running faster, a new equilibrium would be reached when she matches the speed of the escalator (Figure 6.17). By the time she reaches this new equilibrium, her position will have moved and she would be nearer the bottom of the escalator and so her position will have moved to the left.

Chemical reactions at dynamic equilibrium respond to changes as well. **Le Châtelier's principle** states that:

> If a change is made to the conditions of a system at equilibrium, then the position of the equilibrium moves to oppose that change in conditions

The effect of changing concentration

If the concentration of a chemical in an equilibrium is changed, the position of the equilibrium will move to oppose that change. If more of a chemical is added, the equilibrium position moves to remove it. If some of a chemical is removed, the equilibrium position moves to make more of it.

Table 6.5 illustrates how changes in concentration affect the position of an equilibrium of the form:

reactants ⇌ products

Table 6.5

	Increase concentration of a reactant	Decrease concentration of a reactant	Increase concentration of a product	Decrease concentration of a product
How the system responds to oppose the change	Equilibrium position moves right to reduce concentration of reactant	Equilibrium position moves left to increase concentration of reactant	Equilibrium position moves left to reduce concentration of product	Equilibrium position moves right to increase concentration of product

Example

Ammonia is made by reacting nitrogen with hydrogen. What would happen to the amount of ammonia formed if the concentration of nitrogen was increased?

Answer

nitrogen + hydrogen ⇌ ammonia

$N_2(g) + 3H_2(g) \rightleftharpoons 2NH_3(g)$

If the concentration of nitrogen is increased:

- The equilibrium position moves right to reduce the concentration of added nitrogen.

- More ammonia is produced because the equilibrium position moves right.

Example

Hydrogen can be made by reacting methane with steam. What would happen to the amount of hydrogen formed if the carbon monoxide was removed as it was formed?

Answer

methane + steam ⇌ hydrogen + carbon monoxide

$CH_4(g) + H_2O(g) \rightleftharpoons 3H_2(g) + CO(g)$

If the carbon monoxide is removed:

- The equilibrium position moves right to increase the concentration of carbon monoxide.

- More hydrogen is produced because the equilibrium position moves right.

In the Haber Process, if the temperature is increased, the yield of the products will decrease

In the Haber Process, if the temperature is decreased, the yield of the products will increase

The effect of changing temperature

If the temperature of an equilibrium is changed, the position of the equilibrium will move to oppose that change. In order to increase the temperature, the equilibrium position moves in the direction of the exothermic reaction. In order to decrease the temperature, the equilibrium position moves in the direction of the endothermic reaction (Table 6.6).

Table 6.6

Equilibrium	Forward reaction is exothermic		Forward reaction is endothermic	
	reactants $\xrightleftharpoons[\text{endothermic}]{\text{exothermic}}$ products		reactants $\xrightleftharpoons[\text{exothermic}]{\text{endothermic}}$ products	
Change	Increase temperature	Decrease temperature	Increase temperature	Decrease temperature
How the system responds to oppose the change	Equilibrium position moves left in endothermic direction to lower the temperature	Equilibrium position moves right in exothermic direction to increase the temperature	Equilibrium position moves right in endothermic direction to lower the temperature	Equilibrium position moves left in exothermic direction to increase the temperature

However, reactions where more product is formed at lower temperatures are not usually carried out at low temperature. This is because they would be far too slow. A compromise temperature is usually used that gives a reasonable yield of product but at a fast rate.

Example

Ammonia is made by reacting nitrogen with hydrogen. The forward reaction is exothermic. What would happen to the amount of ammonia formed if the temperature was increased?

Answer
nitrogen + hydrogen $\rightleftharpoons$ ammonia

$N_2(g) + 3H_2(g) \rightleftharpoons 2NH_3(g)$

If the temperature is increased:

• The equilibrium position moves left in the endothermic direction to reduce the temperature.

• Less ammonia is produced because the equilibrium position moves left.

Example

Hydrogen can be made by reacting methane with steam in an endothermic reaction. What would happen to the amount of hydrogen formed if the temperature was increased?

Answer
methane + steam $\rightleftharpoons$ hydrogen + carbon monoxide

$CH_4(g) + H_2O(g) \rightleftharpoons 3H_2(g) + CO(g)$

If the temperature is increased:

• The equilibrium position moves right in the endothermic direction to reduce the temperature.

• More hydrogen is produced because the equilibrium position moves right.

The effect of changing pressure

The more molecules that are present in a gas, the greater the pressure of the gas (Figure 6.18).

If the pressure of an equilibrium containing gases is changed, the position of the equilibrium will move to oppose that change (if it can). In order to increase the pressure, the equilibrium position moves to the side with the most gas molecules. In order to decrease the pressure, the equilibrium position moves to the side with fewer gas molecules (Table 6.7).

lower pressure   higher pressure

▲ **Figure 6.18** The more gas particles present, the higher the pressure.

Table 6.7

Equilibrium	More gas molecules of reactants than products e.g. $2A(g) \rightleftharpoons B(g)$		Same number of gas molecules of reactants and products e.g. $A(g) \rightleftharpoons B(g)$		More gas molecules of products than reactants e.g. $A(g) \rightleftharpoons 2B(g)$	
Change	Increase pressure	Decrease pressure	Increase pressure	Decrease pressure	Increase pressure	Decrease pressure
How the system responds to oppose the change	Equilibrium position moves right to side with less gas molecules to reduce pressure	Equilibrium position moves left to side with more gas molecules to increase pressure	Equilibrium does not move because there are same number of gas molecules of reactants and products		Equilibrium position moves left to side with less gas molecules to reduce pressure	Equilibrium position moves right to side with more gas molecules to increase pressure

The use of high pressure is very expensive due to the high energy cost of compressing gases and the high cost of pipes to withstand that pressure. In reactions where higher pressure gives more product, the value of the extra product formed is sometimes less than the cost of those higher pressures. This means that sometimes the actual pressure used in a process is not as high as might be predicted because it is not cost effective. There is often a compromise between the amount of product formed and the cost of using higher pressure.

Example

Ammonia is made by reacting nitrogen with hydrogen. What would happen to the amount of ammonia formed if the pressure was increased?

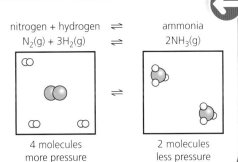

Answer

If the pressure is increased:

- The equilibrium position moves right to the side with fewer gas molecules to decrease the pressure.
- More ammonia is produced because the equilibrium position moves right.

Example

Hydrogen can be made by reaction of methane with steam. What would happen to the amount of hydrogen formed if the pressure was increased?

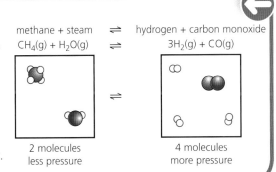

Answer

If the pressure is increased:

- The equilibrium position moves left to the side with fewer gas molecules to decrease the pressure.
- Less hydrogen is produced because the equilibrium position moves left.

Example

Hydrogen gas reacts with iodine gas to form hydrogen iodide gas. What would happen to the amount of hydrogen iodide formed if the pressure was increased?

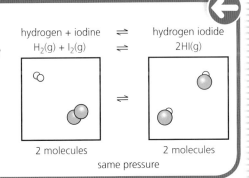

Answer

If the pressure is increased:

- The equilibrium position does not move because there are the same number of gas molecules on both sides of the equation.
- The amount of hydrogen iodide formed remains the same.

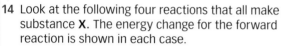

The effect of catalysts

Catalysts have no effect on the position of an equilibrium. However, the catalyst does increase the rate of forward and backward reactions, both by the same amount. This means that the system reaches equilibrium faster and the product is formed faster.

Test yourself

14 Look at the following four reactions that all make substance **X**. The energy change for the forward reaction is shown in each case.

Reaction 1: $A(g) + B(g) \rightleftharpoons X(g)$ exothermic
Reaction 2: $C(g) + 2D(g) \rightleftharpoons X(g)$ endothermic
Reaction 3: $E_2(g) \rightleftharpoons 2X(g)$ exothermic
Reaction 4: $F(g) + G(g) \rightleftharpoons 2X(g)$ exothermic

a) In which of the reactions, if any, would the amount of **X** formed increase as the temperature increases?

b) In which of the reactions, if any, would the amount of **X** formed increase as the pressure increases?

c) In which of the reactions, if any, would the amount of **X** formed remain the same as pressure increases?

15 The colourless gas dinitrogen tetroxide (N_2O_4) forms an equilibrium with the brown gas nitrogen dioxide (NO_2). An equilibrium mixture is a pale brown colour. The energy change for the forward reaction is +58 kJ. What will happen to the colour of the equilibrium mixture if it is heated up? Explain your answer.

$N_2O_4(g) \rightleftharpoons 2NO_2(g)$

16 In aqueous solution, pink cobalt ions (Co^{2+}) react with chloride ions (Cl^-) to form blue cobalt tetrachloride ions ($CoCl_4^{2-}$). This reaction is reversible and forms a dynamic equilibrium.

$Co^{2+} + 4Cl^- \rightleftharpoons CoCl_4^{2-}$
pink   blue

a) State Le Châtelier's principle.

b) An equilibrium mixture was a purple colour as it contained a similar amount of the pink Co^{2+} ions and the blue $CoCl_4^{2-}$ ions. What would happen to the colour if more chloride ions were added? Explain your answer.

c) When this purple mixture is cooled down it goes pink. Is the forward reaction exothermic or endothermic? Explain your answer.

17 Gas R is made when gases P and Q react:
$P(g) + Q(g) \rightleftharpoons R(g)$

a) The table below shows how the percentage of R in the equilibrium mixture varies with temperature. Plot a graph to the show the percentage of R against temperature.

Temperature in °C	100	200	300	400	500
Percentage of R in equilibrium mixture	58	42	30	21	16

b) Is the reaction to form R endothermic or exothermic? Explain your answer.

c) This reaction gives a higher yield of R at lower temperatures. Why might a higher temperature actually be used to produce R in this process?

Show you can...

Chlorodifluoromethane reacts to produce tetrafluoroethene in an endothermic reaction.

chlorodifluoromethane $\rightleftharpoons$ tetrafluoroethene + hydrogen chloride

$2CHClF_2(g)$ $\rightleftharpoons$ $C_2F_4(g)$ + $2HCl(g)$

Explain in terms of reaction rate and equilibrium whether

a) A high or low pressure should be used.
b) A high or low temperature should be used.

Chapter review questions

1 Carbon monoxide reacts with steam to form carbon dioxide and hydrogen in a reversible reaction. This reaches a state of dynamic equilibrium in a closed system. The energy change for the forward reaction is −42 kJ.

carbon monoxide + steam $\rightleftharpoons$ carbon dioxide + hydrogen

a) What is a dynamic equilibrium?

b) Is the forward reaction endothermic or exothermic?

c) What is the energy change for the reverse reaction?

d) This reaction is often done using a catalyst. What is a catalyst?

2 Hydrogen (H_2) can react explosively with oxygen (O_2) to make water. However, a mixture of hydrogen and oxygen does not explode unless a flame or spark is brought near the mixture.

hydrogen + oxygen $\rightarrow$ water

a) Define the term activation energy.

b) Explain why hydrogen and oxygen molecules may not react with each other even when they collide unless a flame or spark is present.

c) Write a balanced equation for this reaction.

3 a) A reaction produced 50 cm^3 of hydrogen gas in 40 seconds. Calculate the mean rate of reaction in cm^3/s of hydrogen produced in these 40 seconds.

b) A reaction produced 0.16 g of oxygen gas in 20 seconds. Calculate the mean rate of reaction in g/s of oxygen produced in these 20 seconds.

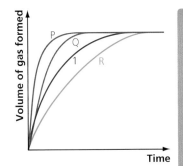

4 Calcium carbonate reacts with hydrochloric acid and forms carbon dioxide gas. A student carried out a series of experiments to investigate how concentration of acid, temperature and surface area of calcium carbonate affected the rate. The table gives details about each reaction (the acid was in excess in each case).

The graphs show how the volume of carbon dioxide gas formed changed with time. Match experiments 2, 3 and 4 with lines P, Q and R. Explain your reasoning.

Experiment	Concentration of acid in mol/dm^3	Temperature in °C	Type of calcium carbonate
1	2.0	20	Small pieces
2	1.0	20	Small pieces
3	2.0	30	Small pieces
4	2.0	20	Powder

5 The metal cobalt reacts slowly with acid. However, it reacts faster if the temperature and the concentration of acid are increased.

a) Explain why increasing the temperature increases the rate of reaction.

b) Explain why increasing the concentration of acid increases the rate of reaction.

c) Which has the biggest relative effect – increasing the temperature or increasing the concentration of acid?

6 Methanol is a useful fuel. Two ways in which it can be made are shown below. Both reactions are exothermic. Method 1 is typically done with a catalyst at 250°C and at 80 atmospheres pressure.

Method 1: $CO(g) + 2H_2(g) \rightleftharpoons CH_3OH(g)$

Method 2: $CO_2(g) + 3H_2(g) \rightleftharpoons CH_3OH(g) + H_2O(g)$

a) Explain why a high pressure is used for method 1.

b) Explain why pressure higher than this is not used for method 1.

c) Explain why the rate of reaction in method 1 increases as the pressure increases.

d) According to Le Châtelier's principle, more methanol would be produced in method 1 at temperatures lower than 250°C. Explain why temperatures lower than 250°C are not used.

e) i) Calculate the atom economy to form methanol in method 1.

   ii) Calculate the atom economy to form methanol in method 2.

   iii) Explain, in terms of atom economy, why method 1 may be preferable to method 2.

7 In aqueous solution, the yellow chromate(VI) ion, CrO_4^{2-}, forms an equilibrium with the orange dichromate(VI) ion, $Cr_2O_7^{2-}$

$$2CrO_4^{2-} + 2H^+ \rightleftharpoons Cr_2O_7^{2-} + H_2O$$
yellow   orange

a) What would happen to the colour of the mixture if sulfuric acid was added? Explain your answer.

b) What would happen to the colour of the mixture if sodium hydroxide was added? Explain your answer.

8 Rhubarb stalks contain some ethanedioic acid. If rhubarb stalks are placed in a purple solution of dilute acidified potassium manganate(VII), the solution goes colourless as the ethanedioic acid reacts with it. The results from two experiments are shown below. In the first experiment one piece of rhubarb stalk was used in the shape of a cuboid measuring $5\,cm \times 1\,cm \times 1\,cm$. In the second experiment, a similar piece of rhubarb stalk was chopped up into five cubes each measuring $1\,cm \times 1\,cm \times 1\,cm$.

	Size of rhubarb stalk used	Time to go colourless in s
Experiment 1	One piece of rhubarb stalk in the shape of a cuboid measuring $5\,cm \times 1\,cm \times 1\,cm$	75
Experiment 2	Five pieces of rhubarb stalk in the shape of cubes each measuring $1\,cm \times 1\,cm \times 1\,cm$	52

a) Show that the same amount of rhubarb was used in each experiment.

b) Calculate the surface area of rhubarb in each experiment.

c) Calculate the surface area to volume ratio of rhubarb in each experiment.

d) What does this experiment show about the effect of surface area on rate of reaction?

e) Explain why changing surface area has the effect that it does.

f) What variables should be controlled between experiments 1 and 2 to ensure that this is a fair test?

9 Carbon dioxide gas is formed when calcium carbonate reacts with hydrochloric acid. The table below shows how the volume of carbon dioxide changed over time when this reaction was done.

Time in s	0	10	20	30	40	50	60	70	80	90	100
Volume of carbon dioxide in cm³	0	22	35	43	48	52	55	57	58	58	58

a) Plot a graph of the volume of carbon dioxide over time.

b) Calculate the rate of reaction at 20 seconds by drawing a tangent to the curve.

c) Calculate the rate of reaction at 60 seconds by drawing a tangent to the curve.

d) Write a word equation for this reaction.

e) Write a balanced equation for this reaction.

Practice questions

1 Which one of the following is used to increase the rate at which ammonia is produced from hydrogen and nitrogen? [1 mark]

  A   catalyst   B   oxidising agent

  C   reducing agent   D   reduced temperature

2 In which one of the following experiments will the rate of reaction be quickest at the start of the reaction? [1 mark]

  A   zinc powder reacting with an excess of $1\,mol/dm^3$ HCl at 20°C

  B   zinc powder reacting with an excess of $1\,mol/dm^3$ HCl at 30°C

  C   zinc powder reacting with an excess of $2\,mol/dm^3$ HCl at 30°C

  D   zinc granules reacting with an excess of $1\,mol/dm^3$ HCl at 30°C

3 In the laboratory preparation of oxygen from hydrogen peroxide using manganese(IV) oxide as catalyst, the mass of manganese(IV) oxide was measured at various times. Which one of the following graphs best shows the experimental results? [1 mark]

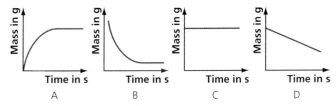

4 In a laboratory experiment 0.5 g of magnesium ribbon was reacted with excess dilute hydrochloric acid at room temperature. The volume of gas produced was noted every 10 seconds. The results were plotted on the axis below as graph C.

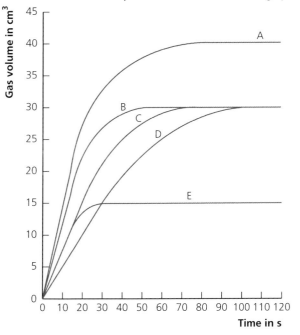

a) At what time did reaction C finish? [1 mark]

b) Calculate the mean rate of reaction of experiment C over the time of the whole reaction. [1 mark]

The experiment was repeated using different conditions and the results obtained given in graphs A, B, D and E.

c) State and explain which of the graphs A, B, D or E would have been obtained if the 0.5 g of magnesium ribbon was replaced by 0.5 g of magnesium powder. [2 marks]

d) State and explain which of the graphs A, B, D or E would have been obtained if the 0.5 g of magnesium ribbon was added to excess dilute hydrochloric acid at a temperature below room temperature. [2 marks]

e) State and explain which of the graphs A, B, D or E would have been obtained if 0.25 g of magnesium ribbon was reacted with excess dilute hydrochloric acid at room temperature. [2 marks]

f) State and explain in terms of collision theory, the effect of increasing the concentration of hydrochloric acid on the rate of the reaction between hydrochloric acid and magnesium. [3 marks]

5 Ethene reacts with steam to form ethanol, in a reversible, exothermic reaction. A catalyst of phosphoric acid is used.

$$C_2H_4(g) + H_2O(g) \rightleftharpoons C_2H_5OH(g)$$

a) What is a reversible reaction? [1 mark]

b) What is an exothermic reaction? [1 mark]

c) State and explain the effect of an increase in pressure on the yield of ethanol. [3 marks]

d) State and explain the effect of an increase in temperature on the yield of ethanol. [3 marks]

e) What is a catalyst? [2 marks]

f) State the effect of the catalyst on the yield of ethanol. [1 mark]

g) Assuming an 80% yield, what mass of ethene is needed to produce 460 kg of ethanol? [3 marks]

6 Nitrogen(IV) oxide (NO_2) is a brown gas which can form from the colourless gas dinitrogen tetroxide N_2O_4 in a dynamic equilibrium. The energy change for the forward reaction is exothermic.

$$2NO_2(g) \rightleftharpoons N_2O_4(g)$$

a) What is meant by the term dynamic equilibrium? [2 marks]

b) Explain what is observed when the pressure on the equilibrium mixture is decreased. [3 marks]

c) Explain what is observed when the temperature of the equilibrium mixture is reduced. [3 marks]

Working scientifically:
Presenting information and data in a scientific way

Recording results

During experimental activities you must often record results in a table. When drawing tables and recording data ensure that:

▶ the table is a ruled box with ruled columns and rows

▶ there are headings for each column or row

▶ there are units for each column and row – usually placed after the heading for example 'Temperature in °C'

▶ there is room for repeat measurements and averages – remember the more repeats you do the more reliable the data.

Times in minutes	Mass of oxygen lost in g	Mass of oxygen lost in g	Average mass of oxygen lost in g
1			
2			
3			
4			
5			

Example of a table to record experimental results.

Questions

1 50 cm³ of hydrogen peroxide and 1.0 g of manganese dioxide were allowed to react at 25°C. The volume of oxygen collected from the reaction at 10 second intervals is:

After 10 seconds, 30 cm³; after 20 seconds, 49 cm³; after 30 seconds, 59 cm³; after 40 seconds, 63 cm³; and after 50 seconds, 63 cm³ – on repeating the experiment the volume of gas obtained at each time interval was 32, 51, 59, 63, 65 cm³ respectively.

Present these results in a suitable table with headings and units. Calculate the mean volume of gas produced and include this in your table.

2 Solubility is measured in g/100 g water. The solubility of potassium sulfate in 100 g water was measured at different temperatures with the following results:

at 90°C, 22.9 g/100 g dissolves; at 50°C, 16.5 g/100 g dissolves; at 70°C, 19.8 g/100 g dissolves; at 10°C, 9.3 g/100 g dissolves; and at 30°C, 13.0 g/100 g dissolves.

a) Present these results in a suitable table with headings.

b) What is the trend shown by the results?

Plotting graphs

When carrying out experiments you must be able to translate data from one form to another. Most often this involves using data from a table to draw a graph. A graph is an illustration of how two variables relate to one another.

When drawing a graph remember that:

▶ The independent variable is placed on the x-axis, while the dependent variable is placed on the y-axis.

▶ Appropriate scales should be devised for the axis, making the most effective use of the space on the graph paper. The data should be critically examined to establish whether it is necessary to start the scale(s) at zero.

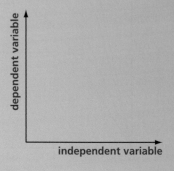

▶ Axes should be labelled with the name of the variable followed by 'in' and the unit of measurement. For example the label may be temperature in °C.

▶ A line of best fit should be drawn. When judging the position of the line there should be approximately the same number of data points on each side of the line. Resist the temptation to simply connect the first and last points. The best fit line could be straight or a curve. Ignore any anomalous results when drawing your line.

TIP

You need to judge the line of best fit with your eyes. Using a see-through plastic ruler or a flexible curve can help you. If drawing a curve by hand, it is easier to get a smoother line if you have your elbow on the inside of the curve as you draw it.

Questions

3 In an experiment some calcium carbonate and acid were placed in a conical flask on a balance and the balance reading recorded every minute. The results were recorded and the graph shown below was drawn.

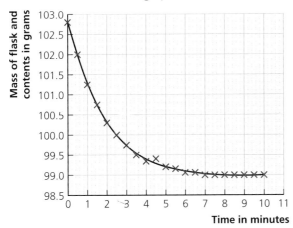

a) Are there any results which you would ignore when drawing a best fit curve?

b) What is the independent variable in this experiment?

c) At what time does the reaction stop?

d) What is the mass of the flask and contents at time 2 minutes?

7 Organic chemistry

Organic chemistry is the study of compounds containing carbon. The number of organic compounds that exist is greater than all other compounds added together. This is because carbon is able to form bonds to other carbon atoms very easily and form chains and rings.

Many medicines, detergents, clothing fibres, solvents and polymers are organic molecules and many of these are made from chemicals that we find in crude oil. Our own bodies and food also contain many organic molecules.

Specification coverage

This chapter covers specification points 4.7.1 to 4.7.3 and is called Organic chemistry.

It covers carbon compounds as fuels and feedstock, reactions of alkenes and alcohols, and synthetic and naturally occurring polymers.

Previously you could have learned:

> Crude oil, coal and natural gas are fossil fuels.
> Fuels are substances that burn in oxygen releasing a lot of thermal energy.
> Non-metal atoms bond to each other by sharing electrons; one covalent bond is made up of two shared electrons.
> Carbon atoms make four covalent bonds in molecules.
> Mixtures of miscible liquids with different boiling points can be separated by fractional distillation.
> Polymers are long chain molecules.

Test yourself on prior knowledge

1 What are fuels?
2 Give three examples of fossil fuels
3 What type of bonds are made when carbon atoms bond with hydrogen atoms?
4 How many covalent bonds do carbon atoms make?
5 What is a polymer?
6 How are mixtures of miscible liquids separated?

Crude oil and alkanes

◯ What is crude oil?

Crude oil is a fossil fuel that is found underground in rocks. It was formed over millions of years from the remains of sea creatures. These creatures were mainly plankton that were buried in mud at the bottom of the oceans (Figure 7.1).

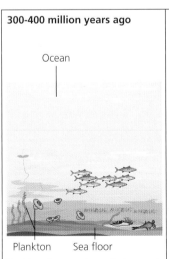
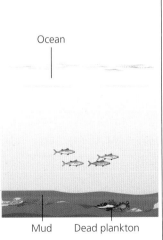
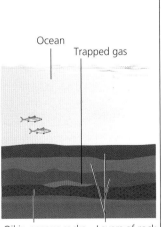
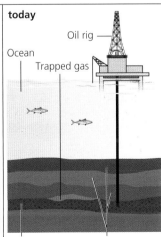

300-400 million years ago — Ocean — Plankton — Sea floor — Plankton (sea creatures that cannot swim against a current, including algae, bacteria, protists and some animals) died and fell onto the sea floor

Ocean — Mud — Dead plankton — The dead plankton were covered in mud

Ocean — Trapped gas — Oil in porous rocks — Layers of rock — Over millions of years, more and more sediment built up. The enormous heat and pressure turned the dead plankton into oil and gas

today — Oil rig — Ocean — Trapped gas — Oil in porous rocks — Layers of rock — Today we drill down through rock to reach the oil and bring it up to the surface

▲ **Figure 7.1** The formation of crude oil.

KEY TERMS

Finite resource A resource that cannot be replaced once it has been used.

Biomass A resource made from living or recently living creatures.

Hydrocarbon A compound containing hydrogen and carbon only.

AlKanes only contain single bonds. AlKanes are saturated because each carbon atom has formed 4 covalent bonds with another atom. This makes alkanes unreactive.

TIP

Carbon atoms always make four covalent bonds. Hydrogen atoms always make one covalent bond. When you draw organic molecules make sure each atom has made the correct number of bonds.

Alkanes are a series of hydrocarbons which have the general formula C_nH_{2n+2}

KEY TERMS

Displayed formula Drawing of a molecule showing all atoms and bonds.

Homologous series A family of compounds with the same general formula, the same functional group and similar chemical properties.

Alkanes A homologous series of saturated hydrocarbons with the general formula C_nH_{2n+2}.

Saturated [in the context of organic chemistry] A molecule that only contains single covalent bonds.

Crude oil is a form of ancient biomass as it was made from the remains of creatures that lived many years ago. Crude oil is a finite resource because we cannot replace it as we use it up.

Crude oil is a mixture of many different compounds. Most of these compounds are hydrocarbons. Hydrocarbons are compounds that contain hydrogen and carbon only. Most of the hydrocarbons in crude oil are alkanes.

○ Alkanes

Alkanes are a family of saturated hydrocarbons. Saturated molecules are ones that only contain single covalent bonds. The structures of the first four alkanes are shown in Table 7.1. The table includes the displayed formula which shows all the atoms and all the bonds in each molecule.

Table 7.1 The first four alkanes.

Alkane	Ball and stick structure	Displayed structure	Structural formula	Molecular formula
Methane			CH_4	CH_4
Ethane			CH_3CH_3	C_2H_6
Propane			$CH_3CH_2CH_3$	C_3H_8
Butane			$CH_3CH_2CH_2CH_3$	C_4H_{10}

All the alkanes have a molecular formula of the form C_nH_{2n+2}. For example, if there are 3 carbon atoms ($n = 3$), then there are 8 hydrogen atoms ($2n + 2 = (2 \times 3) + 2 = 8$). This and other examples are shown in Table 7.2. This means that the general formula of all alkanes is C_nH_{2n+2}.

Table 7.2

	Methane	Ethane	Propane	Butane
Number of C atoms (n)	$n = 1$	$n = 2$	$n = 3$	$n = 4$
Number of H atoms ($2n + 2$)	$2n + 2 = (2 \times 1) + 2 = 4$	$2n + 2 = (2 \times 2) + 2 = 6$	$2n + 2 = (2 \times 3) + 2 = 8$	$2n + 2 = (2 \times 4) + 2 = 10$
Molecular formula	CH_4	C_2H_6	C_3H_8	C_4H_{10}

Homologous series - a sequence of compounds with the same functional group and similar chemical properties

The names of organic compounds are made up of two parts. The first part of the name indicates the number of carbon atoms and the second part indicates which homologous series the molecule belongs to (Figure 7.2).

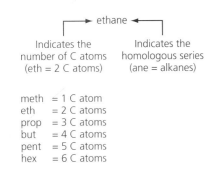

ethane

Indicates the number of C atoms (eth = 2 C atoms)

Indicates the homologous series (ane = alkanes)

meth = 1 C atom
eth = 2 C atoms
prop = 3 C atoms
but = 4 C atoms
pent = 5 C atoms
hex = 6 C atoms

▲ Figure 7.2

KEY TERMS ⭐

Fractional distillation A method used to separate miscible liquids with different boiling points.

Fraction A mixture of molecules with similar boiling points.

The alkanes are an example of a homologous series. A homologous series is a family of compounds with

- the same general formula
- the same functional group
- similar chemical properties.

Alkanes do not contain a functional group. Functional groups will be studied later in the chapter.

◯ Fractional distillation of crude oil

For crude oil to be useful, the hydrocarbons it contains have to be separated. The hydrocarbons have different boiling points and this difference is used to separate them by fractional distillation at an oil refinery. This process separates the hydrocarbons into fractions. A fraction is a mixture of molecules with similar boiling points. In each fraction, the hydrocarbons contain a similar number of carbon atoms.

The crude oil is heated and vaporised. The vaporised crude oil enters the fractionating tower that is hotter at the bottom and cooler at the top. The hydrocarbons cool as they rise up the tower and condense at different heights because they have different boiling points. Hydrocarbons with large molecules are collected as liquids near the bottom of the tower while those with small molecules collect at the top (Figure 7.3).

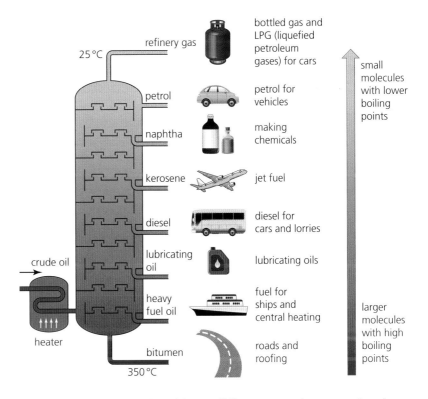

▶ Figure 7.3 Fractional distillation of crude oil.

The fractions from crude oil have different uses because they have different properties. Many are used as fuels, such as liquefied petroleum gases (cars and gas heaters), petrol (cars), kerosene (aeroplanes), diesel (cars and lorries) and heavy fuel oil (ships and heating). Other

Complete combustion

Fuel + oxygen → carbon dioxide + water

Incomplete combustion

Fuel + oxygen → carbon dioxide + water

CO → carbon monoxide produced

fractions are used as feedstock for processes to make useful substances such as medicines, detergents, lubricants and polymers.

The petrochemical industry is a huge industry that deals with the fractional distillation of crude oil and provides fuels and other substances made from crude oil. Modern life would be very different without these fuels and other substances produced from crude oil.

The use of alkanes as fuels

The main use of alkanes from crude oil is as fuels. Alkanes are good fuels because they release a lot of energy when they burn.

When alkanes burn, they react with oxygen. **Complete combustion** takes place if there is a good supply of oxygen from the air. The carbon atoms in the alkane are oxidised, combining with the oxygen to form carbon dioxide. The hydrogen atoms in the alkane are also oxidised, combining with the oxygen to form water.

Complete combustion: alkane + oxygen → carbon dioxide + water

If there is a poor supply of oxygen when alkanes burn, **incomplete combustion** can happen which forms water along with carbon monoxide and/or carbon (in the form of soot). The carbon monoxide formed is toxic and the soot causes a smoky flame.

Incomplete combustion: alkane + oxygen → carbon monoxide + water

Incomplete combustion: alkane + oxygen → carbon + water

When writing balanced equations for the complete combustion of alkanes

- Balance the C atoms: for each C atom in the alkane there will be one CO_2 molecule formed.
- Balance the H atoms: for every two H atoms in the alkane there will be one H_2O molecule formed.
- Count the number of O atoms in the CO_2 and H_2O: the number of O_2 molecules will be half this number.
- If the number of O_2 molecules has a half in it, double all the balancing numbers to get rid of the half.

KEY TERMS

Complete combustion When a substance burns with a good supply of oxygen.

Incomplete combustion When a substance burns with a poor supply of oxygen.

Complete combustion	Incomplete combustion
- Produces CO_2 and H_2O	- Produces H_2O, CO and carbon
- occurs in a plentiful supply of oxygen	- occurs when there is limited supply of oxygen
- Releases more energy to the surroundings	- Releases less energy to the surroundings
	- Produces toxic gas

Carbon monoxide is a toxic, odourless, colourless gas

Example

Write a balanced equation for the complete combustion of methane, CH_4.

Answer

Word equation: methane + oxygen → carbon dioxide + water

Unbalanced equation: $CH_4 + O_2 \rightarrow CO_2 + H_2O$

CO_2: there is 1 C atom in CH_4 so there will be 1 CO_2 formed

H_2O: there are 4 H atoms in CH_4 so there will be 2 H_2O formed

O_2: $CO_2 + 2H_2O$ contains 4 O atoms and so 2 O_2 needed

Balanced equation: $CH_4 + 2O_2 \rightarrow CO_2 + 2H_2O$

Example

Write a balanced equation for the complete combustion of butane, C_4H_{10}.

Answer

Word equation: butane + oxygen → carbon dioxide + water

Unbalanced equation: $C_4H_{10} + O_2 \rightarrow CO_2 + H_2O$

CO_2:   there are 4 C atoms in C_4H_{10} so there will be 4 CO_2 formed

H_2O: there are 10 H atoms in C_4H_{10} so there will be 5 H_2O formed

O_2: $4CO_2 + 5H_2O$ contains 13 O atoms and so 6½ O_2 needed

As there is a half in the equation we will double all the values

Balanced equation: $2C_4H_{10} + 13O_2 \rightarrow 8CO_2 + 10H_2O$

KEY TERMS

Flammability How easily a substance catches fire; the more flammable the more easily it catches fire.

Viscosity How easily a liquid flows; the higher the viscosity the less easily it flows.

The properties of alkanes depend on the size of the molecules and this affects their use as fuels (Figure 7.4). The flammability of a fuel is how easily it catches fire. A flammable fuel catches fire easily. The viscosity of a liquid is how easily it flows. A runny liquid has a low viscosity while a thick, slow moving liquid (e.g. syrup) has a high viscosity.

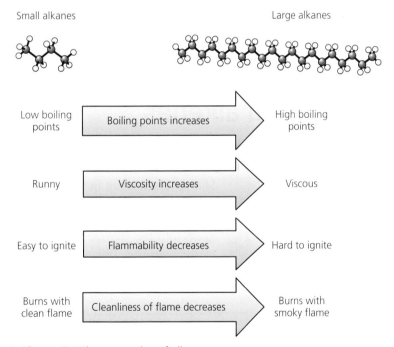

▲ **Figure 7.4** The properties of alkanes.

This means that shorter alkanes are more in demand as fuels because they flow more easily, are more flammable and burn with a cleaner flame. Although larger alkanes are less easy to use, some are used as fuels. For example, fuel oil is used in ships and in some central heating systems.

Test yourself

1. a) Describe how crude oil was formed.
   b) Explain why crude oil can be described as an ancient source of biomass.
   c) Crude oil is a finite resource. Explain what this means.
2. Crude oil is a mixture of hydrocarbons. They are separated by fractional distillation at an oil refinery.
   a) What is a hydrocarbon?
   b) What property of the hydrocarbons allows them to be separated in this way?
   c) Describe how fractional distillation separates the hydrocarbons.
3. Hexane is an alkane containing six carbon atoms.
   a) Alkanes are saturated hydrocarbons. What does saturated mean in this context?
   b) Give the molecular formula of hexane.
   c) Draw the displayed formula of hexane.
4. Pentane and decane are both alkanes. Pentane has the formula C_5H_{12}. Decane has the formula $C_{10}H_{22}$.
   a) Which one has the highest boiling point?
   b) Which one is the most flammable?
   c) Which one is most viscous?
5. Write a balanced equation for the complete combustion of the following alkanes:
   a) pentane, C_5H_{12}
   b) ethane, C_2H_6

Show you can...

A hydrocarbon P (C_xH_y) burns in excess air as shown in the equation below:

$$C_xH_y + 5O_2 \rightarrow 3CO_2 + 4H_2O$$

a) Explain why C_xH_y is a hydrocarbon.
b) Determine the values of x and y using the equation given above.
c) Name hydrocarbon P and draw its displayed formula.

Cracking and alkenes

Cracking is the breaking down of larger hydrocarbons into a smaller alkane and an alkene

○ Cracking

KEY TERM

Cracking The thermal decomposition of long alkanes into shorter alkanes and alkenes.

Shorter alkanes are in very high demand as fuels but longer alkanes are in less demand. This means that there is a surplus of the longer alkanes from the fractional distillation of crude oil. These longer alkanes can be broken down into the shorter alkanes that are in higher demand by a process called cracking. This process also produces unsaturated hydrocarbons called alkenes which can be used as a starting material to make many other substances such as polymers and medicines (Figure 7.5).

$C_{10}H_{22}$ $\longrightarrow$ C_8H_{18} + C_2H_4

longer alkanes

shorter alkanes (used as fuels)

+ alkenes (used to make many useful substances including polymers)

▲ **Figure 7.5** An example of a cracking reaction.

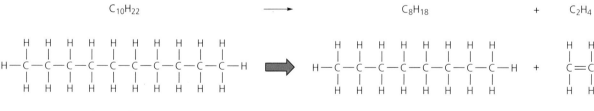

long alkane ――Heat, catalyst―→ Short alkane + Alkene

$C_{16}H_{34} \rightarrow C_{14}H_{30} + C_2H_4$
$C_{36}H_{74} \rightarrow C_{34}H_{70} + C_2H_4$
$C_9H_{20} \rightarrow C_3H_6 + C_6H_{14}$

conditions for catalytic cracking
High temperature
Add a catalyst

thermal decomposition reaction

Cracking can be done in a number of ways. Two methods of cracking are:

- **Catalytic cracking**: heat the alkanes to vaporise them and then pass them over a hot catalyst.
- **Steam cracking**: heat the alkanes to vaporise them, mix them with steam and then heat them to very high temperature.

Cracking is a thermal decomposition reaction because the alkanes are broken down into smaller molecules using heat.

○ Alkenes

→ hydrocarbons which contain a carbon to carbon double bond

The alkenes are a homologous series of unsaturated alkenes. They are unsaturated because they contain a carbon-carbon double bond ($C=C$). The first four alkenes are shown in the table. Their names all end in –ene to show that they are alkenes.

There is no alkene with one carbon atom because at least two carbon atoms are needed in alkenes as they all contain a $C=C$ double bond (Table 7.3).

Alkenes are able to take part in chemical reactions by opening the double bond and bonding each carbon to another atom.

Table 7.3 The first four alkenes.

Alkene	Ball and stick structure	Displayed structure	Structural formula	Molecular formula
Ethene	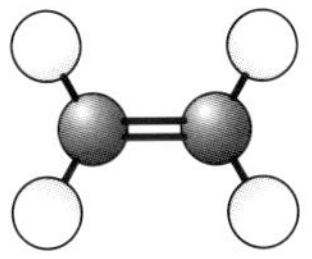	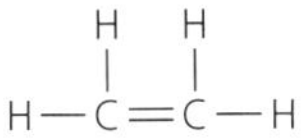 H—C=C—H (with H H above each C)	$CH_2=CH_2$	C_2H_4
Propene	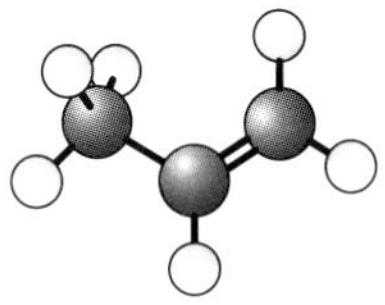	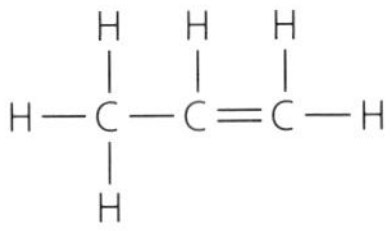 H—C—C=C—H (with H H H above)	$CH_3CH=CH_2$	C_3H_6
Butene	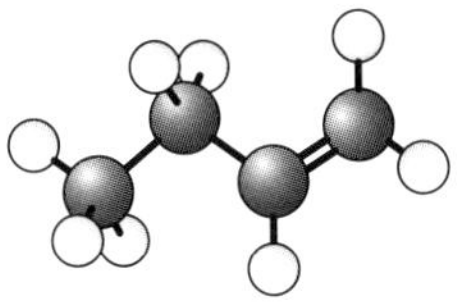	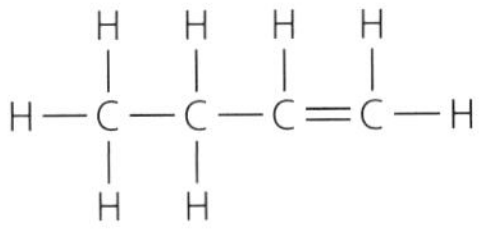 H—C—C—C=C—H	$CH_3CH_2CH=CH_2$	C_4H_8
Pentene	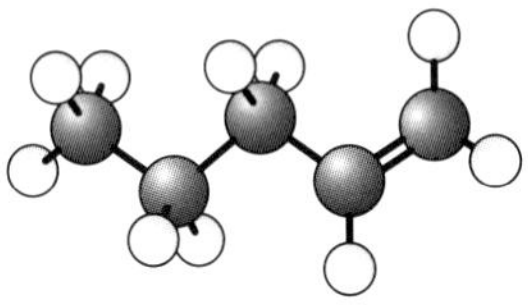	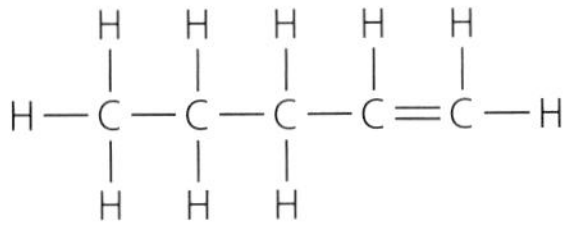 H—C—C—C—C=C—H	$CH_3CH_2CH_2CH=CH_2$	C_5H_{10}

When drawing organic molecules, it does not usually matter what angle you draw the atoms at or which way round you draw the molecules. Molecules are three dimensional and we are representing them in two dimensions on paper and so no diagram will represent them as they actually are. For example, each of the four structures in Figure 7.6 is a correct displayed formula of propene (C_3H_6).

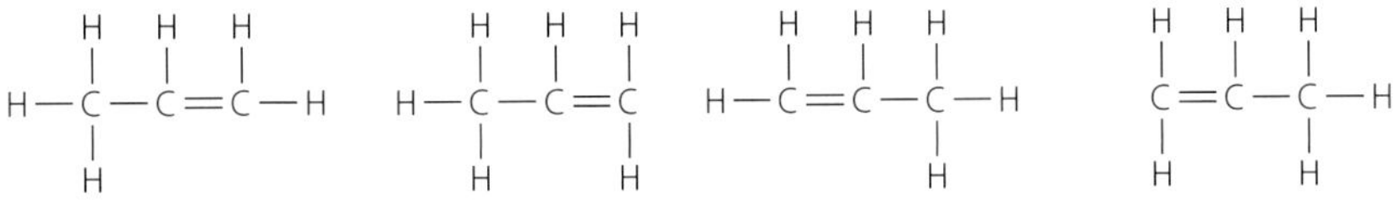

▶ **Figure 7.6** These are all ways of drawing the displayed structure of propene (C_3H_6).

Alkenes contain two fewer hydrogen atoms than alkanes with the same number of carbon atoms. This is due to the double bond between two carbon atoms. The general formula of alkenes therefore is C_nH_{2n}. For example, propene has 3 carbon atoms ($n = 3$) and so must contain 6 hydrogen atoms ($2n = 2 \times 3 = 6$) and so has the formula C_3H_6.

The functional group in alkenes is the C=C double bond. A functional group is an atom or group of atoms that is responsible for most of the chemical reactions of a compound.

Reactions of alkenes

Alkenes are much more reactive than alkanes because they contain a C=C double bond. Most of the reactions of alkenes are addition reactions. In these reactions the C=C double bond breaks open and atoms add onto the two carbon atoms. The C=C double bond becomes a C—C single bond so a saturated molecule is produced (Figure 7.7).

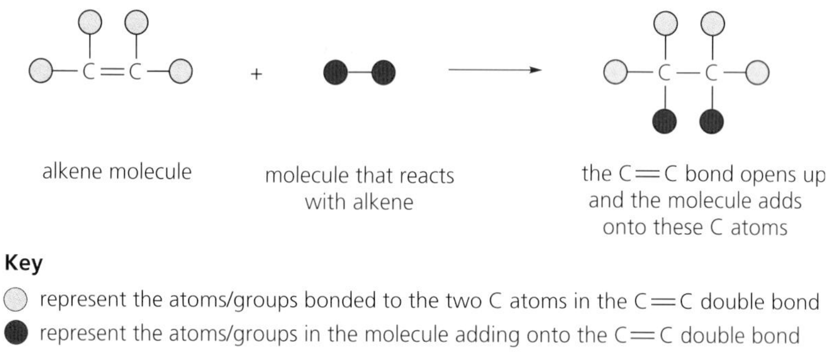

alkene molecule

molecule that reacts with alkene

the C=C bond opens up and the molecule adds onto these C atoms

Key

⊙ represent the atoms/groups bonded to the two C atoms in the C=C double bond
● represent the atoms/groups in the molecule adding onto the C=C double bond

▶ **Figure 7.7** Molecules add onto alkenes.

Reaction of alkenes with halogens

The halogens chlorine (Cl_2), bromine (Br_2) and iodine (I_2) all react readily with alkenes. In each case, the C=C double bond opens up and one halogen atom adds onto each of the carbon atoms in the double bond. Some examples are shown in Figure 7.8.

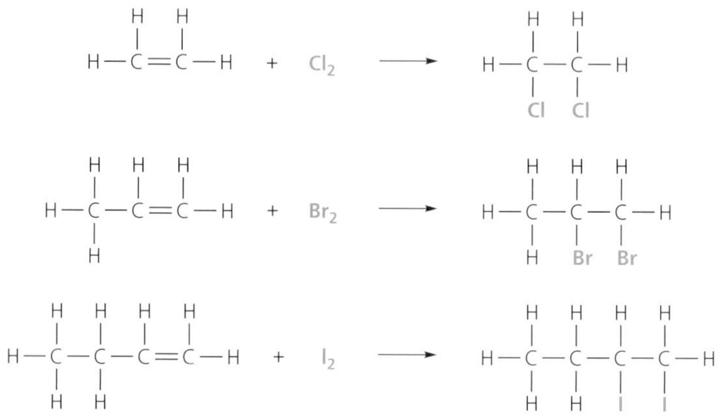

▶ **Figure 7.8** Reactions of some alkenes with halogens.

The reaction with bromine is used to test for the presence of C=C double bonds in compounds. Bromine water is a solution of bromine in

Alkenes are more reactive than alkanes
an alkene will react with bromine water whereas an alkane will
t.
When an alkene reacts with
rine water, it turns crystal clear
here as an alkane will have no
reaction

water and has an orange colour due to the dissolved Br_2 molecules. When the bromine water reacts with the C=C double bond, the Br_2 molecules are used up and so the reaction mixture goes colourless (Table 7.4).

Table 7.4

Type of compound	Saturated compound	Unsaturated compound
C=C double bonds	No C=C double bonds	Contains C=C double bond(s)
Reaction with bromine water	Stays orange 	Orange → colourless 

TIP ✓

When bromine water reacts with alkenes, it goes colourless. It is wrong to say that it goes clear. Clear does not mean the same thing as colourless. Colourless means there is no colour but clear means see-through. Bromine water itself is an example of a clear, coloured solution. It is a clear, orange solution.

Reaction of alkenes with hydrogen

Alkenes react with hydrogen at 150°C in the presence of a nickel catalyst to form alkanes. In this reaction, hydrogen atoms add onto the carbon atoms in the C=C double bond. This converts an unsaturated alkene into a saturated alkane (Figure 7.9).

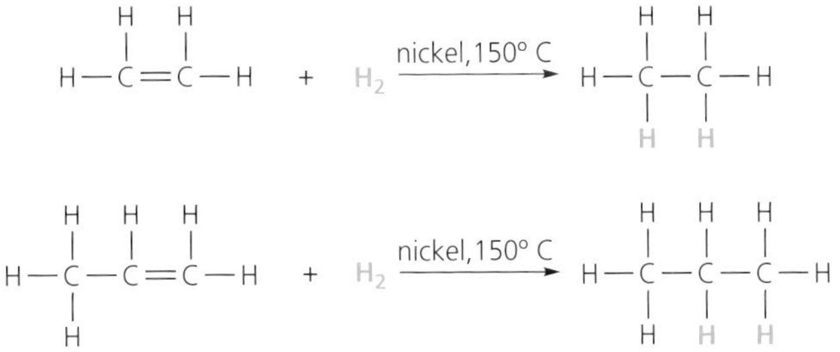

▶ **Figure 7.9** Reactions of some alkenes with hydrogen to make alkanes.

The reaction is called hydrogenation because hydrogen is being added across the C=C double bond. The metal nickel is often used as the catalyst.

KEY TERMS ★

Hydrogenation Addition of hydrogen.

Hydration Addition of water.

▶ **Figure 7.10** Reactions of some alkenes with steam to make alcohols.

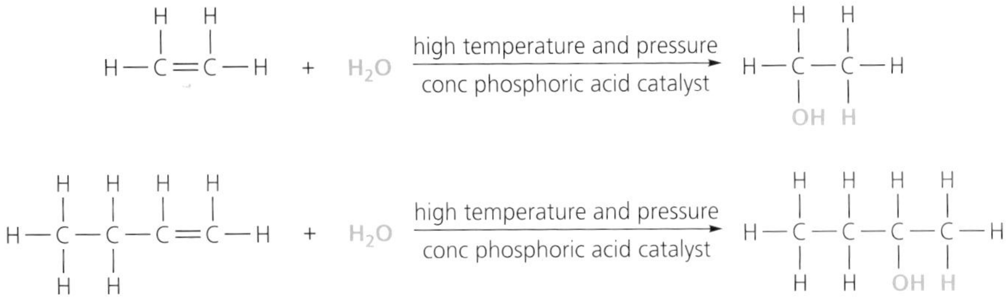

Reaction of alkenes with steam

Alkenes react with steam at high temperature and high pressure in the presence of concentrated phosphoric acid as catalyst to form alcohols. In this reaction, the atoms of H_2O add onto the carbon atoms in the C=C double bond (Figure 7.10).

TIP ✓

Take care not to mix up the terms hydrogenation and hydration. *Hydrogenation* is the addition of hydrogen. *Hydration* is the addition of water.

The reaction is called hydration because water is being added across the C=C double bond. Concentrated phosphoric acid is often used as the catalyst.

Reaction of alkenes with oxygen

Alkenes will also burn in oxygen like alkanes do. However, we do not usually burn alkenes for two reasons:

1 Alkenes tend to burn in air with smoky flames due to incomplete combustion.
2 Alkenes are very valuable as they can be used to make polymers or as a starting material for many other chemicals.

Test yourself

6 Some long alkanes are cracked to produce shorter alkanes and alkenes.
   a) Give three reasons why longer alkanes are in less demand as fuels than shorter alkanes.
   b) Outline two ways in which long alkanes are cracked.
   c) Cracking is a thermal decomposition reaction. What does the term thermal decomposition mean?
   d) Give some examples of the uses of the shorter alkanes and alkenes produced in cracking.

7 Balance the following equations for reactions that can take place when alkanes are cracked.
   a) $C_{18}H_{38} \rightarrow C_{12}H_{26} + C_3H_6$
   b) $C_{20}H_{42} \rightarrow C_{12}H_{26} + C_2H_4$
   c) $C_{18}H_{38} \rightarrow C_{10}H_{22} + C_4H_8 + C_2H_4$
   d) $C_{25}H_{52} \rightarrow C_{11}H_{24} + C_4H_8 + C_3H_6$

8 Alkenes are unsaturated hydrocarbons.
   a) What does the term unsaturated mean?
   b) What does the term hydrocarbon mean?
   c) Explain why alkenes are much more reactive than alkanes.
   d) Alkenes can be burned but they tend to burn with smoky flames. Why do they burn with smoky flames?

9 Hexene is an alkene with six carbon atoms.
   a) Give the molecular formula of hexene.
   b) Draw the displayed formula of hexene.
   c) Describe what would be seen if some bromine water was added to some hexene.
   d) Write an equation for this reaction, showing the displayed formula of the organic molecules.

10 a) Alkenes are very reactive and do addition reactions. What is an addition reaction?
   b) Write equations for the following addition reactions of alkenes. Show the displayed formula of each organic molecule.
      i) butene + H_2
      ii) propene + H_2O
      iii) ethene + Br_2
      iv) propene + I_2
      v) pentene + Cl_2
   c) In which of the reactions in part (b) is/are saturated molecules produced?
   d) Which of the reactions in part (b) is/are hydration reactions?
   e) Which of the reactions in part (b) is/are hydrogenation reactions?

Show you can...

The diagram shows some of the reactions of ethene.

a) Write the molecular formula of the products (A, B, C, D) of the reactions of ethene shown.
b) What is observed in the reaction to produce B?
c) Which product (A, B, C or D) is produced in the hydration of ethene?
d) Which product (A, B, C or D) is produced in a hydrogenation reaction?

A ← combustion — ethene — + Br_2 → B
C ← + H_2 — ethene — + H_2O → D

Alcohols, carboxylic acids and esters

○ Alcohols

Alcohols are a homologous series of compounds containing the functional group –OH. The first four alcohols are shown in Table 7.5. The names of alcohols end in –ol.

Table 7.5 The first four alcohols.

Alcohol	Ball and stick structure	Displayed structure	Structural formula	Molecular formula
Methanol		H—C—O—H with H above and H below	CH_3OH	CH_4O
Ethanol		H—C—C—O—H with H's	CH_3CH_2OH	C_2H_6O
Propanol		H—C—C—C—O—H with H's	$CH_3CH_2CH_2OH$	C_3H_8O
Butanol		H—C—C—C—C—O—H with H's	$CH_3CH_2CH_2CH_2OH$	$C_4H_{10}O$

Making ethanol

Ethanol is made by the fermentation of sugars in these conditions:

- the sugars are dissolved in water
- yeast is added
- the mixture is kept in a warm place at about 30°C
- air is kept out of the mixture.

The reaction takes a few days and the ethanol formed has to be separated from the other substances in the flask (Figure 7.11).

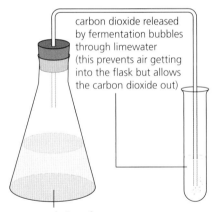

carbon dioxide released by fermentation bubbles through limewater (this prevents air getting into the flask but allows the carbon dioxide out)

aqueous solution of sugar with yeast in conical flask

▲ **Figure 7.11**

Reactions of alcohols

Combustion reactions

Alcohols burn well in oxygen. In a good supply of oxygen, complete combustion takes place to form carbon dioxide and water. For example:

ethanol + oxygen → carbon dioxide + water

$$C_2H_5OH + 3O_2 \rightarrow 2CO_2 + 3H_2O$$

propanol + oxygen → carbon dioxide + water

$$2C_3H_7OH + 9O_2 \rightarrow 6CO_2 + 8H_2O$$

Mild oxidation of alcohols

When alcohols are burned they form carbon dioxide and water if combustion is complete. If dilute solutions of alcohols in water are left standing in air, the alcohols will only be partially oxidised and form carboxylic acids (Figure 7.12).

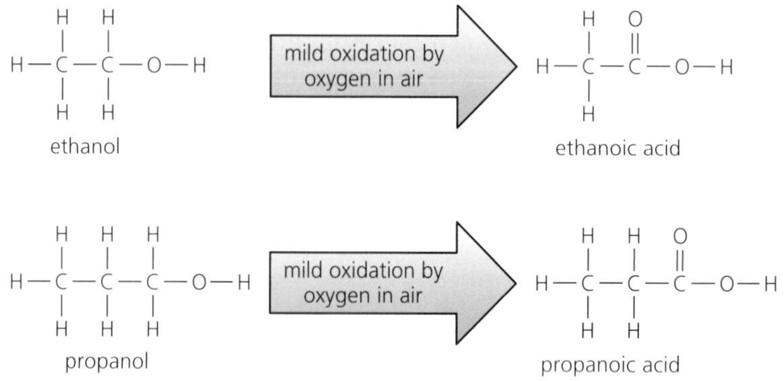

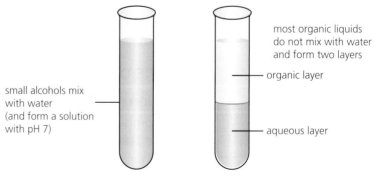

ethanol → mild oxidation by oxygen in air → ethanoic acid

propanol → mild oxidation by oxygen in air → propanoic acid

▲ **Figure 7.12**

Wine contains ethanol. If wine is left standing in air, the ethanol will be oxidised to ethanoic acid by the oxygen and the wine is turned into wine vinegar (Figure 7.13). Vinegar is a solution of ethanoic acid.

Reaction of alcohols with carboxylic acids

Alcohols react with carboxylic acids in the presence of a catalyst to form esters. This is studied on page 184.

Reaction of alcohols with sodium

Alcohols react with sodium. In this reaction hydrogen gas is produced (Figure 7.14).

Reaction of alcohols with water

Alcohols do not react with water but small alcohol molecules do dissolve in water and form a neutral solution of pH 7. Most other organic compounds do not mix with water and instead form two layers which are made up of a layer of the organic compound that is usually on top of a layer of water (Figure 7.15).

▲ **Figure 7.13** Wine vinegar is made from the oxidation of wine by the oxygen in air.

▲ **Figure 7.14** Hydrogen gas is formed when sodium reacts with alcohols.

small alcohols mix with water (and form a solution with pH 7)

most organic liquids do not mix with water and form two layers

organic layer

aqueous layer

▲ **Figure 7.15** Reactions of alcohols with water.

Uses of alcohols

Alcohols are very useful substances. Some of these uses are shown below.

- **As fuels**: Alcohols burn very well and can be used as fuels. Camping gas stoves often use alcohols as fuels. In every litre of petrol sold in the UK, 50 cm³ (5%) of the fuel is actually ethanol. In Brazil, many cars run on ethanol as a fuel instead of petrol.
- **As solvents**: Alcohols are very good solvents. For example, many solutions of medicines and perfumes are made with alcohols because these substances do not dissolve in water.
- **In alcoholic drinks**: Ethanol is the main alcohol in alcoholic drinks. Alcoholic drinks are made by fermentation of crops such as grapes (to make wine), apples (to make cider) or malted barley (to make beer). In each case, ethanol is produced by yeast in fermentation.

> **KEY TERM**
>
> Carboxylic acids A homologous series of compounds containing the functional group —COOH.

○ Carboxylic acids

Carboxylic acids are a homologous series of compounds containing the functional group –COOH. The first four carboxylic acids are shown in Table 7.6. The names of carboxylic acids end in –oic acid.

Table 7.6 The first four carboxylic acids.

Carboxylic acid	Ball and stick structure	Displayed structure	Structural formula	Molecular formula
Methanoic acid		$H-\overset{\overset{O}{\|\|}}{C}-O-H$	HCOOH	CH_2O_2
Ethanoic acid		$H-\overset{\overset{H}{\|}}{\underset{H}{C}}-\overset{\overset{O}{\|\|}}{C}-O-H$	CH_3COOH	$C_2H_4O_2$
Propanoic acid		$H-\overset{\overset{H}{\|}}{\underset{H}{C}}-\overset{\overset{H}{\|}}{\underset{H}{C}}-\overset{\overset{O}{\|\|}}{C}-O-H$	CH_3CH_2COOH	$C_3H_6O_2$
Butanoic acid		$H-\overset{\overset{H}{\|}}{\underset{H}{C}}-\overset{\overset{H}{\|}}{\underset{H}{C}}-\overset{\overset{H}{\|}}{\underset{H}{C}}-\overset{\overset{O}{\|\|}}{C}-O-H$	$CH_3CH_2CH_2COOH$	$C_4H_8O_2$

Many everyday substances also contain the carboxylic acid functional group (Figure 7.16).

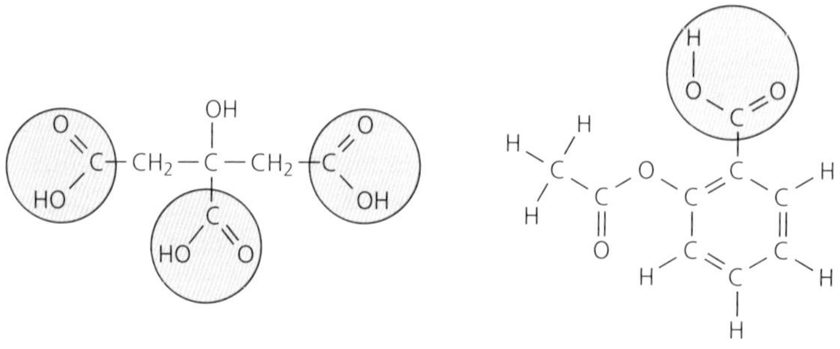

▶ **Figure 7.16** Some molecules that contain the –COOH functional group.

citric acid (in citrus fruits, e.g. oranges, lemons)　　　　aspirin

Properties of carboxylic acids

Weak acids

Carboxylic acids dissolve in water to produce acidic solutions (Figure 7.17). Acids are substances that produce H^+ ions when added to water.

CH_3COOH 　　　　　　　　　　CH_3COO^- ＋ H^+

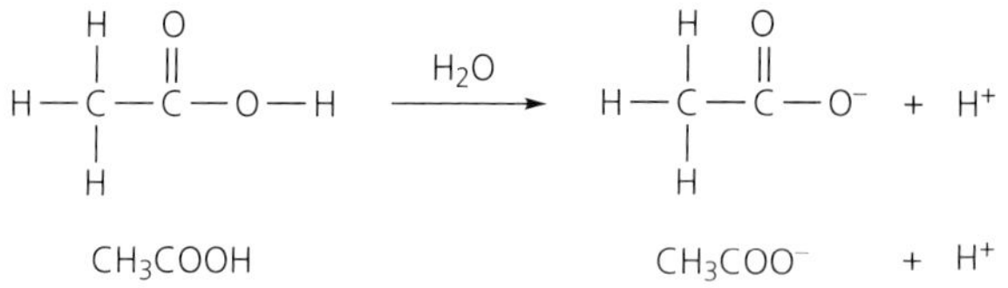

▲ **Figure 7.17**

Carboxylic acids are weak acids, which means only a small fraction of the molecules break down into ions when added to water (see page 109). As they are only weak acids, carboxylic acids do not cause us harm when we eat or drink them in small amounts.

Reaction with carbonates

Acids react with metal carbonates to form a salt, carbon dioxide and water. This means that there is fizzing when metal carbonates are added to aqueous solutions of carboxylic acids. However, the reaction is slow as carboxylic acids are weak acids.

Reaction with alcohols

Carboxylic acids react with alcohols in the presence of an acid catalyst to form compounds called esters. For example, ethanoic acid reacts with ethanol to form the ester ethyl ethanoate which is used as the solvent in nail varnish (Figures 7.18 and 7.19).

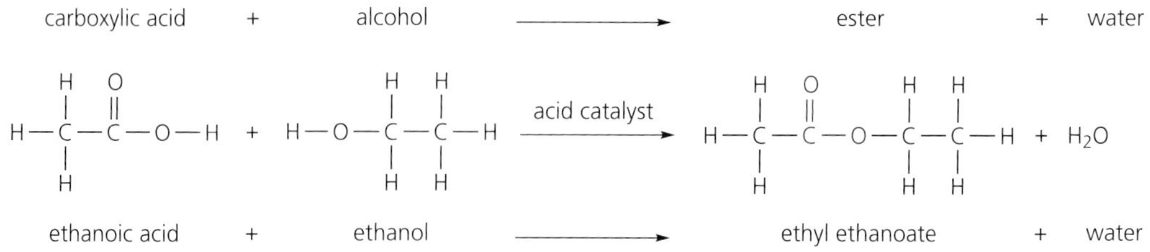

▲ **Figure 7.18** The solvent in nail varnish is the ester ethyl ethanoate.

carboxylic acid　　＋　　alcohol　　⟶　　ester　　＋　water

ethanoic acid　　＋　　ethanol　　⟶　　ethyl ethanoate　　＋　water

▶ **Figure 7.19**

Esters all contain the functional group –COO–. Some examples of esters found naturally in fruits are shown in Figure 7.20. These esters are largely responsible for the smell and taste of fruit. We can use these esters, either extracted from fruits or made artificially, as flavourings for food and drink or to make perfumes.

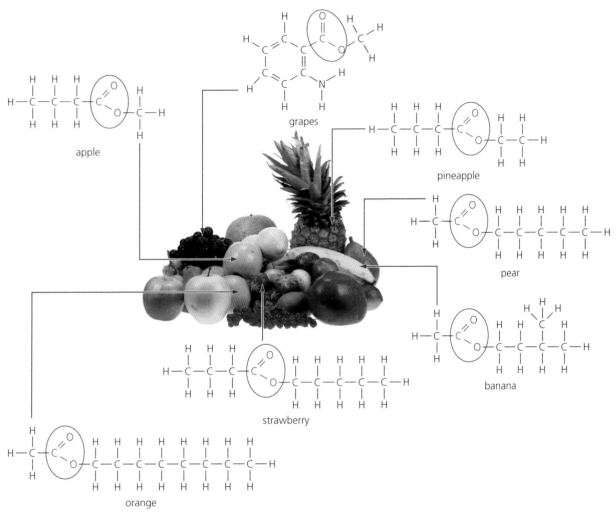

▲ **Figure 7.20** Some of the esters in fruits (the ester functional group is circled in each case).

Test yourself

11 Look at the eight molecules A to H.

$$
\begin{array}{c}
\ \ \ \ H\ \ \ \ H\ \ \ \ O \\
\ \ \ \ |\ \ \ \ \ \ |\ \ \ \ \ \ || \\
H-C-C-C-O-H \\
\ \ \ \ |\ \ \ \ \ \ | \\
\ \ \ \ H\ \ \ \ H
\end{array}
$$

molecule A

CH_3OH

molecule B

$CH_2{=}CHCH_3$

molecule C

$$
\begin{array}{c}
\ \ \ \ H\ \ \ \ H\ \ \ \ H \\
\ \ \ \ |\ \ \ \ \ \ |\ \ \ \ \ \ | \\
H-C-C-C-H \\
\ \ \ \ |\ \ \ \ \ \ |\ \ \ \ \ \ | \\
\ \ \ \ H\ \ \ \ O\ \ \ \ H \\
\ \ \ \ \ \ \ \ \ \ \ | \\
\ \ \ \ \ \ \ \ \ \ \ H
\end{array}
$$

molecule D

$CH_3CH_2COOCH_2CH_3$

molecule E

$$
\begin{array}{c}
\ \ \ \ H\ \ \ \ H\ \ \ \ H\ \ \ \ H \\
\ \ \ \ |\ \ \ \ \ \ |\ \ \ \ \ \ |\ \ \ \ \ \ | \\
H-C-C-C-C-O-H \\
\ \ \ \ |\ \ \ \ \ \ |\ \ \ \ \ \ |\ \ \ \ \ \ | \\
\ \ \ \ H\ \ \ \ H\ \ \ \ H\ \ \ \ H
\end{array}
$$

molecule F

$$
\begin{array}{c}
\ \ \ \ H\ \ \ \ O\ \ \ \ \ \ \ \ H \\
\ \ \ \ |\ \ \ \ \ \ ||\ \ \ \ \ \ \ \ | \\
H-C-C-O-C-H \\
\ \ \ \ |\ \ \ \ \ \ \ \ \ \ \ \ \ \ | \\
\ \ \ \ H\ \ \ \ \ \ \ \ \ \ \ H
\end{array}
$$

molecule G

CH_3COOH

molecule H

a) Which of these molecules is/are alcohols?

b) Which of these molecules is/are carboxylic acids?

c) Which of these molecules is/are esters?

12 a) Substance P dissolves in water to form a solution that turns universal indicator solution orange. It fizzes when it reacts with potassium carbonate. What functional group does organic substance P contain?

b) Substance Q dissolves in water to form a solution that turns universal indicator solution green. It fizzes when it reacts with sodium. What functional group does organic substance Q contain?

13 Ethanol is an alcohol.

a) Draw the displayed formula of ethanol.

b) What is the functional group in ethanol?

c) Give three uses for ethanol.

d) What would happen if sodium was added to ethanol?

e) Wine contains ethanol. What happens to the ethanol in wine if a glass of wine is left to stand in air for a few days?

14 Alcohols are very good fuels. Write a balanced equation for the complete combustion of:

a) propanol, C_3H_7OH

b) butanol, C_4H_9OH

15 Propanoic acid is a carboxylic acid.

a) Draw the displayed formula of propanoic acid.

b) What is the functional group in propanoic acid?

c) What would happen if sodium carbonate was added to propanoic acid?

16 Ethanol reacts with ethanoic acid to form an ester.

a) Name the ester produced in this reaction.

b) A catalyst is needed for this reaction. Identify a suitable catalyst.

Show you can...

Look at the following reaction sequence and answer the questions that follow.

a) Write the name of the ester formed when ethanol and ethanoic acid react in reaction C.

b) What condition is needed for reaction C?

c) What is added to ethanol in reaction B?

d) Write a balanced symbol equation for reaction A.

e) Name a solid which is added to ethanoic acid in reaction D.

carbon dioxide — A → ethanol — B → hydrogen

D

C

ethanoic acid — C → ester

Polymers

▲ **Figure 7.21** Joining paper-clips together to make a long chain.

○ Addition polymers

Paper-clips can be joined together to make a long chain (Figure 7.21).

In a similar way, molecules containing a C=C double bond can react with each other in addition reactions. They join onto each other to create a long chain molecule called a polymer.

The C=C double bonds open up and the molecules join onto each other to make a long chain molecule. The exact number of molecules that join together varies, but is likely to be several hundred. For example, lots of ethene molecules join together to make the polymer poly(ethene), better known as polythene (Figure 7.22).

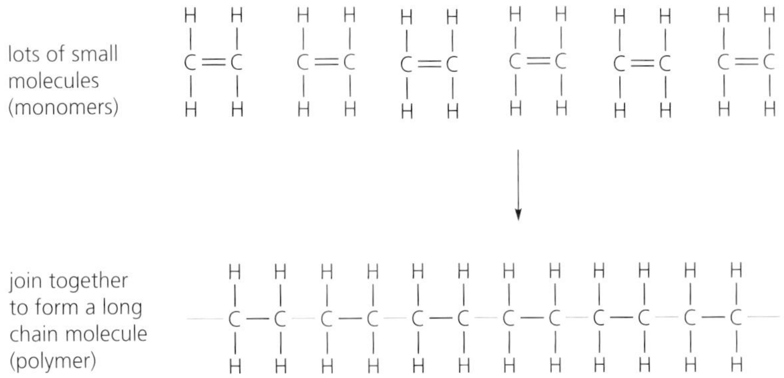

▶ **Figure 7.22** Making poly(ethene).

The chemical equation for this reaction is shown in Figure 7.23.

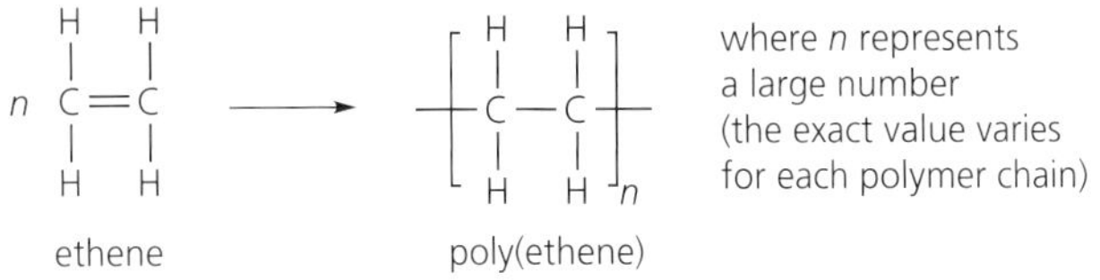

▶ **Figure 7.23**

The polymer is made up of a unit that repeats many times – this is known as the repeating unit. We can draw a single repeating unit or show the full structure of the polymer (Figure 7.24).

repeating unit of poly(ethene)   structure of poly(ethene)

▲ **Figure 7.24** Poly(ethene).

The small molecules that we join together to make a polymer are called monomers. In addition polymerisation, the monomers all contain a C=C double bond. Some more examples of monomers and the polymers that are formed are shown in Table 7.7. The name of a polymer is the word *poly* followed by the name of the monomer in brackets.

Table 7.7 Some polymers and their uses.

Monomer	Repeating unit of polymer	Structure of polymer	Uses of polymer
H H | | C=C | | H H Ethene	H H | | —C—C— | | H H	[H H] [| |] [C—C] [| |] [H H]$_n$ Poly(ethene) (often called polythene)	Bags Crates/boxes
H Cl | | C=C | | H H Chloroethene	H Cl | | —C—C— | | H H	[H Cl] [| |] [C—C] [| |] [H H]$_n$ Poly(chloroethene) (often called PVC)	Water pipes Coating for window frames
H CH$_3$ | | C=C | | H H Propene	H CH$_3$ | | —C—C— | | H H	[H CH$_3$] [| |] [C—C] [| |] [H H]$_n$ Poly(propene) (often called polypropylene)	Carpets Crates/boxes
F F | | C=C | | F F Tetrafluoroethene	F F | | —C—C— | | F F	[F F] [| |] [C—C] [| |] [F F]$_n$ Poly(tetrafluoroethene) (often called PTFE or Teflon)	PTFE tape (used by plumbers) Non-stick coating in frying pans

When monomers with C=C double bonds join together in addition polymerisation, no other product is formed.

Show you can...

Part of the structure of a polymer is shown here:

H F H F H F H F
\| \| \| \| \| \| \| \|
—C—C—C—C—C—C—C—C—
\| \| \| \| \| \| \| \|
Cl CH$_3$ Cl CH$_3$ Cl CH$_3$ Cl CH$_3$

a) Draw the repeating unit of this polymer.
b) Draw the structure of the monomer used to make this polymer.
c) Is this polymer a hydrocarbon? Explain your answer.

Test yourself

17 a) What is a polymer?
  b) What is a monomer?
18 a) What functional group do monomers that form addition polymers contain?
  b) What else is produced, if anything, when an addition polymer is formed from monomer molecules?
19 a) Name the addition polymer formed from styrene.
  b) Name the monomer used to make poly(bromoethene).
20 The structure of dichloroethene is shown. Write a balanced equation for the polymerisation of this monomer.
21 Copy and complete the table.

Structure of monomer	Repeating unit of polymer	Structure of polymer
CH$_3$ H | | C=C | | CH$_3$ H		

◯ Condensation polymers

Condensation polymerisation involves monomers with two functional groups joining together to make a polymer. As the monomers join together, small molecules such as water are released which is why this is called condensation polymerisation.

There are many examples of condensation polymers. Some are synthetic ones including polyesters such as terylene, and polyamides such as nylon and Kevlar. There are also many naturally occurring condensation polymers such as starch, cellulose and proteins and these are studied on page 193.

Polyesters

Polyesters are made when molecules with two carboxylic acid functional groups react with molecules with two alcohol functional groups. The carboxylic acid groups react with the alcohol groups to form ester linkages (–COO–) and give off water (Figure 7.25).

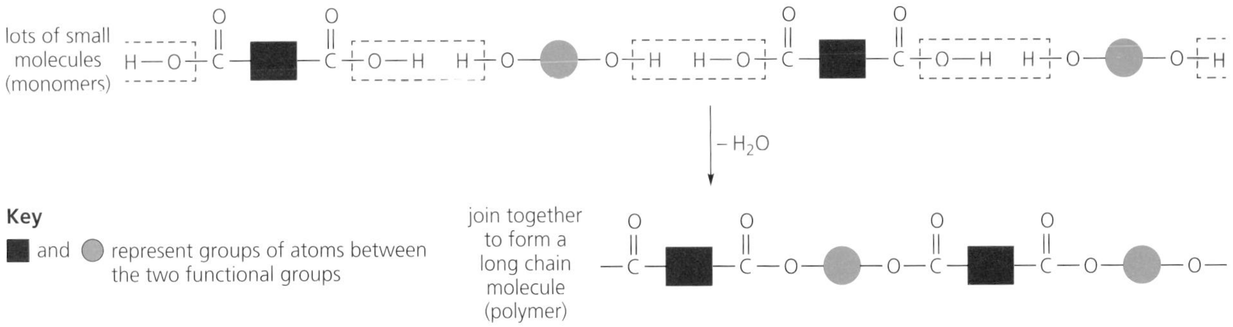

Key

■ and ● represent groups of atoms between the two functional groups

▲ **Figure 7.25** Making polyesters from two monomers.

The overall equation for this reaction can be written as in Figure 7.26.

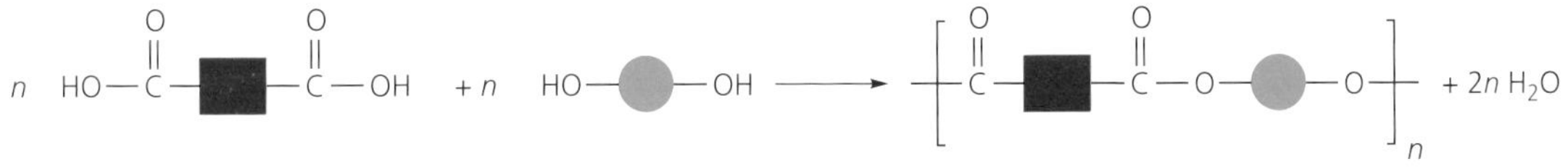

▲ Figure 7.26

The structure of the repeating unit and of the polymer can represented as in Figure 7.27.

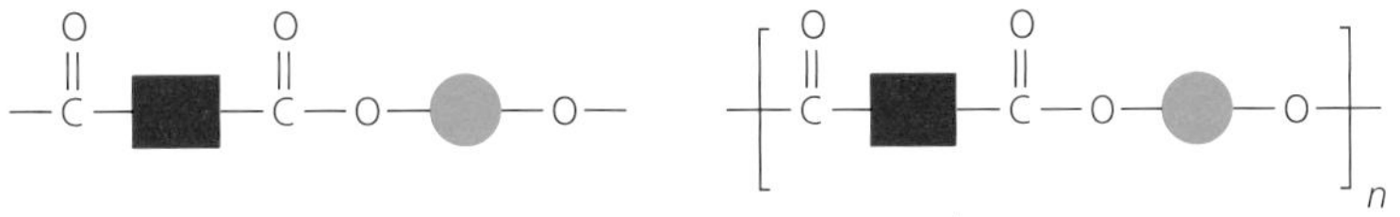

repeating unit structure   polymer structure

▲ Figure 7.27

Polyesters can also be formed from monomers that each contain one carboxylic and one alcohol group (Figure 7.28).

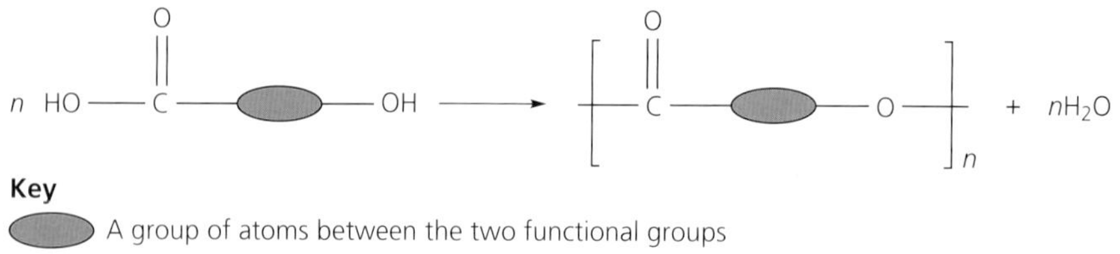

$$n \ \text{HO} - \overset{\overset{\displaystyle O}{\|}}{C} - \bigoplus - \text{OH} \longrightarrow \left[\overset{\overset{\displaystyle O}{\|}}{C} - \bigoplus - O \right]_n + \ n\text{H}_2\text{O}$$

Key

⬭ A group of atoms between the two functional groups

▲ **Figure 7.28** Making polyesters from one monomer.

Polyamides

Polyamides are made when molecules with two carboxylic acid functional groups react with molecules with two amine functional groups. The amine functional group is $-\text{NH}_2$. The carboxylic acid groups react with the amine groups to form amide linkages ($-\text{CONH}-$) and give off water (Figure 7.29).

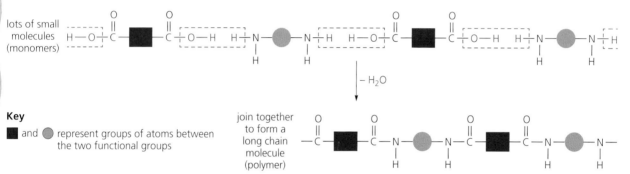

lots of small molecules (monomers)

$-\text{H}_2\text{O}$

Key

■ and ● represent groups of atoms between the two functional groups

join together to form a long chain molecule (polymer)

▲ **Figure 7.29** Making polyamides from two monomers.

The overall equation for this reaction can be written as in Figure 7.30.

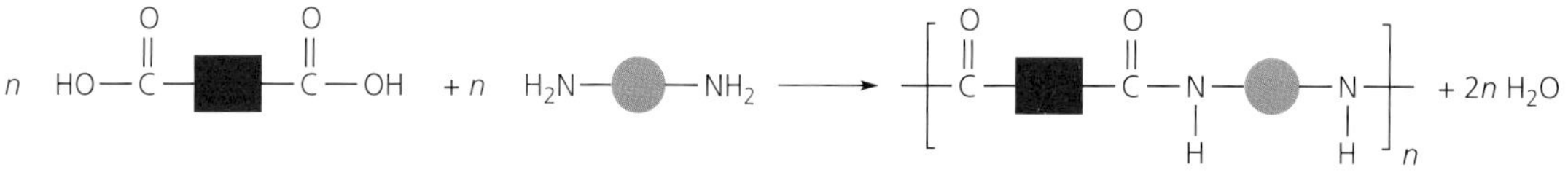

$$n \ \text{HO} - \overset{\overset{\displaystyle O}{\|}}{C} - \blacksquare - \overset{\overset{\displaystyle O}{\|}}{C} - \text{OH} + n \ \text{H}_2\text{N} - \bigcirc - \text{NH}_2 \longrightarrow \left[\overset{\overset{\displaystyle O}{\|}}{C} - \blacksquare - \overset{\overset{\displaystyle O}{\|}}{C} - \underset{\overset{|}{\text{H}}}{N} - \bigcirc - \underset{\overset{|}{\text{H}}}{N} \right]_n + 2n \ \text{H}_2\text{O}$$

▲ **Figure 7.30**

Polyamides can also be formed from monomers that each contain one carboxylic acid group and one amine group (Figure 7.31).

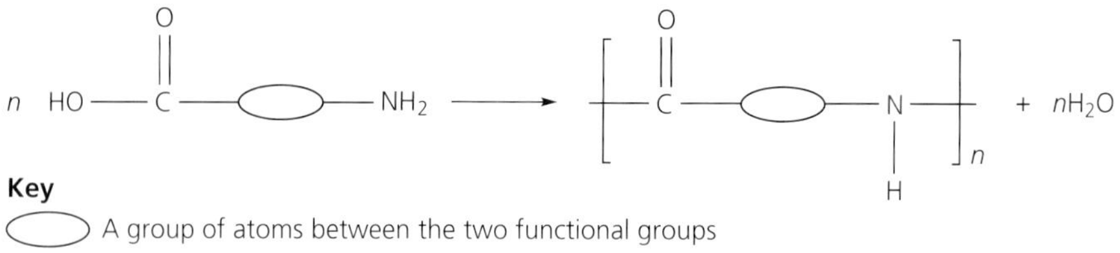

$$n \ \text{HO} - \overset{\overset{\displaystyle O}{\|}}{C} - \bigoval - \text{NH}_2 \longrightarrow \left[\overset{\overset{\displaystyle O}{\|}}{C} - \bigoval - \underset{\overset{|}{\text{H}}}{N} \right]_n + \ n\text{H}_2\text{O}$$

Key

⬭ A group of atoms between the two functional groups

▲ **Figure 7.31** Making polyamides from one monomer.

Table 7.8 gives some specific examples of condensation polymers formed in this way.

Table 7.8 Some condensation polymers and their monomers.

Monomers	Polymer	
HO—C(=O)—CH₂—CH₂—C(=O)—OH and HO—CH₂—OH	Repeating unit of polymer	—C(=O)—CH₂—CH₂—C(=O)—O—CH₂—O—
	Structure of polymer	[—C(=O)—CH₂—CH₂—C(=O)—O—CH₂—O—]ₙ
HO—C(=O)—CH₂—CH₂—CH₂—CH₂—C(=O)—OH and HO—CH₂—CH₂—OH	Repeating unit of polymer	—C(=O)—CH₂—CH₂—CH₂—CH₂—C(=O)—O—CH₂—CH₂—O—
	Structure of polymer	[—C(=O)—CH₂—CH₂—CH₂—CH₂—C(=O)—O—CH₂—CH₂—O—]ₙ
HO—C(=O)—CH₂—OH	Repeating unit of polymer	—C(=O)—CH₂—O—
	Structure of polymer	[—C(=O)—CH₂—O—]ₙ
HO—C(=O)—CH₂—CH₂—CH₂—C(=O)—OH and H₂N—CH₂—CH₂—NH₂	Repeating unit of polymer	—C(=O)—CH₂—CH₂—CH₂—C(=O)—N(H)—CH₂—CH₂—N(H)—
	Structure of polymer	[—C(=O)—CH₂—CH₂—CH₂—C(=O)—N(H)—CH₂—CH₂—N(H)—]ₙ
HO—C(=O)—CH₂—CH₂—CH₂—CH₂—C(=O)—OH and H₂N—CH₂—NH₂	Repeating unit of polymer	—C(=O)—CH₂—CH₂—CH₂—CH₂—C(=O)—N(H)—CH₂—N(H)—
	Structure of polymer	[—C(=O)—CH₂—CH₂—CH₂—CH₂—C(=O)—N(H)—CH₂—N(H)—]ₙ
HO—C(=O)—CH₂—CH₂—NH₂	Repeating unit of polymer	—C(=O)—CH₂—CH₂—N(H)—
	Structure of polymer	[—C(=O)—CH₂—CH₂—N(H)—]ₙ

Test yourself

22 a) What does the word *condensation* refer to in the term condensation polymerisation?

b) How do the monomers in condensation polymerisation compare to those in addition polymerisation?

23 a) Give three examples of synthetic condensation polymers.

b) Give three examples of naturally occurring condensation polymers.

24 The structure of a condensation polymer is shown. Draw the structure of the two monomers that could be used to make this polymer.

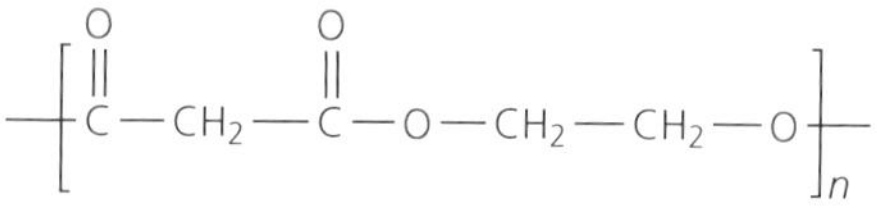

$$\left[\begin{array}{c} O \\ \| \\ C - CH_2 - \overset{\displaystyle O}{\overset{\|}{C}} - O - CH_2 - CH_2 - O \end{array} \right]_n$$

25 The two molecules below can be used to make a condensation polymer.

$$HO - \overset{\displaystyle O}{\overset{\|}{C}} - CH_2 - CH_2 - CH_2 - \overset{\displaystyle O}{\overset{\|}{C}} - OH \quad \text{and} \quad H_2N - CH_2 - NH_2$$

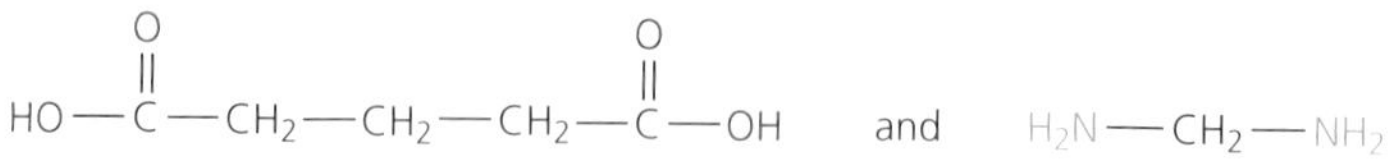

a) Draw the repeating unit of the condensation polymer.

b) Draw the structure of the condensation polymer.

Show you can...

The repeating unit of a polyester is shown here:

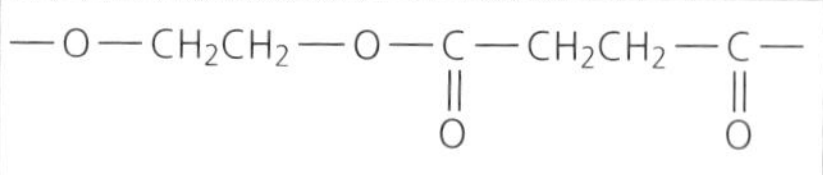

$$- O - CH_2CH_2 - O - \overset{\displaystyle}{\underset{\displaystyle O}{\overset{\|}{C}}} - CH_2CH_2 - \overset{\displaystyle}{\underset{\displaystyle O}{\overset{\|}{C}}} -$$

a) What is the empirical formula of the repeating unit?

b) Draw the structure of the acid which could be used in the preparation of this polyester.

Biochemistry

KEY TERM

Amino acids Molecules containing both a carboxylic acid and an amine functional group.

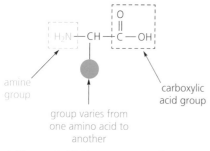

▲ **Figure 7.32** The structure of amino acids.

○ Amino acids

Amino acids have the general structure shown in Figure 7.32. They contain two functional groups:

- carboxylic acid (–COOH), and
- amine (–NH₂).

There are 20 amino acids that are essential to life. Three examples are shown in Figure 7.33.

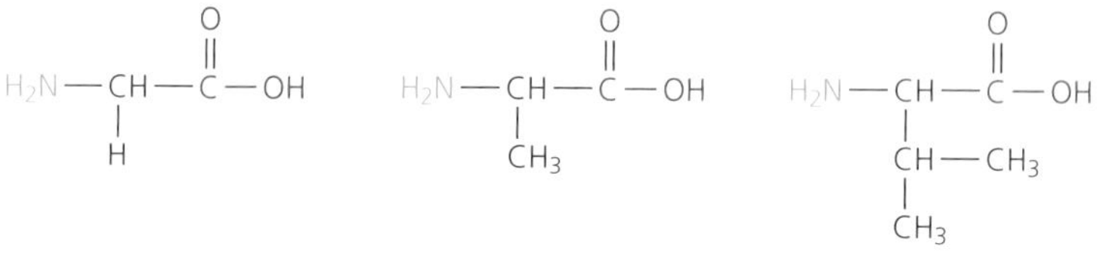

$$H_2N - \overset{\displaystyle O}{\underset{\displaystyle H}{\overset{\|}{\underset{|}{CH}}}} - C - OH \qquad H_2N - \overset{\displaystyle O}{\underset{\displaystyle CH_3}{\overset{\|}{\underset{|}{CH}}}} - C - OH \qquad H_2N - \overset{\displaystyle O}{\underset{\displaystyle CH - CH_3}{\overset{\|}{\underset{|}{CH}}}} - C - OH$$

glycine   alanine   valine

▲ **Figure 7.33** Some amino acids.

Amino acids join to each other in condensation polymerisation reactions to form polypeptides. The equation in Figure 7.34 shows the polymerisation of glycine.

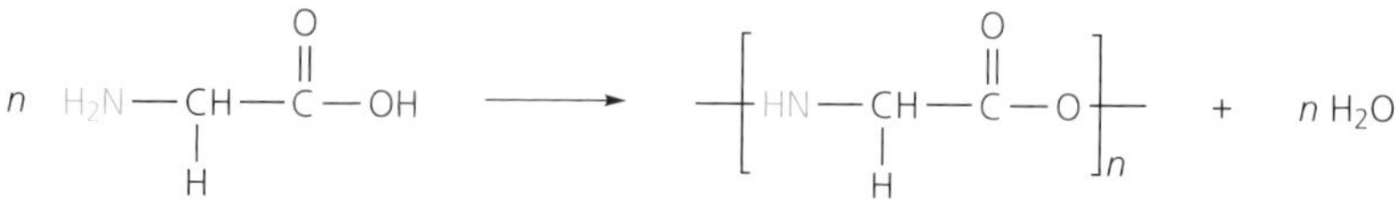

▶ **Figure 7.34** The polymerisation of the amino acid glycine to form a polypeptide.

$$n \; H_2N - \overset{\displaystyle O}{\underset{\displaystyle H}{\overset{\|}{\underset{|}{CH}}}} - C - OH \longrightarrow \left[HN - \overset{\displaystyle O}{\underset{\displaystyle H}{\overset{\|}{\underset{|}{CH}}}} - C - O \right]_n + \; n \, H_2O$$

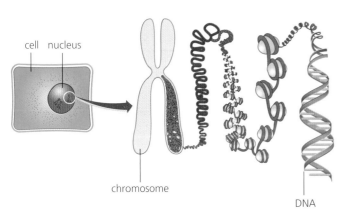

Table 7.9 shows the structure of this and some other polypeptides formed from amino acids.

Table 7.9 Some amino acids and the condensation polymers that they form.

Structure of monomer	Repeating unit of polymer	Structure of polymer
$H_2N-CH-C(=O)-OH$, with H below	$-HN-CH-C(=O)-O-$, with H below	$[HN-CH-C(=O)-O]_n$, with H below
$H_2N-CH-C(=O)-OH$, with CH_3 below	$-HN-CH-C(=O)-O-$, with CH_3 below	$[HN-CH-C(=O)-O]_n$, with CH_3 below
$H_2N-CH-C(=O)-OH$, with $CH-CH_3$ and CH_3 below	$-HN-CH-C(=O)-O-$, with $CH-CH_3$ and CH_3 below	$[HN-CH-C(=O)-O]_n$, with $CH-CH_3$ and CH_3 below

Different amino acids can be combined in the same chain to form proteins.

○ Naturally occurring polymers

Proteins

Proteins are polymer molecules made from lots of different amino acids joined together in a long chain. There are 20 amino acids that make proteins. In each protein, these different amino acids are joined together in a different but very specific sequence.

Starch and cellulose

Starch and cellulose are both polymer molecules made from the sugar glucose. They differ in the way the glucose molecules join together. Starch, cellulose and glucose are all carbohydrates. Carbohydrates are biological molecules containing carbon, hydrogen and oxygen.

DNA

The nucleus of a cell contains chromosomes which contain genetic information. This information is needed for the development and function of all living organisms and viruses. These chromosomes are made of DNA (deoxyribonucleic acid) (Figure 7.35).

DNA consists of two long polymer chains that are held together in the form of a double helix. Each polymer chain in DNA is made up from four different monomers called nucleotides. These nucleotides can bond together in very many different sequences. In humans more than 99% of this sequence is the same but there is some variation from one person to another.

cell nucleus
chromosome
DNA

▲ **Figure 7.35** DNA in cells.

Groups on the side of the nucleotides hold the two strands in the helix together (Figure 7.36). There are four different nucleotides. They can each attract one of the other nucleotides and so work in pairs to hold the two polymer strands together in the helix. These groups are sometimes abbreviated to A, T, C and G.

A
T
C
G

▲ **Figure 7.36** DNA consists of two polymer strands held together in a double helix.

Show you can...

Classify each of the following as monomer or polymer:

DNA, polyester, nucleotide, ethanediol, glucose, starch, cellulose, amino acids, protein, chloroethene, poly(propene), polypeptide, glycine, hexanedioic acid.

Test yourself

26 Proteins, starch and cellulose are polymers. What are the monomers in:
   a) proteins?   **b)** starch?   **c)** cellulose?

27 **a)** What are carbohydrates?
   b) Give three examples of carbohydrates.

28 DNA is made of two long polymer strands held together in a double helix.
   a) What does DNA stand for?
   b) **i)** What is the general name for the molecules that are joined together in the polymer strands in DNA?
       ii) How many different types of these molecules are there in the DNA chains?

29 **a)** What are amino acids?
   b) What are proteins?

30 The structure of the amino acid leucine is shown.
   a) Draw the repeating unit of the polypeptide that can be formed from leucine.
   b) Write an equation for the formation of this polypeptide from leucine.

$$H_2N-CH-\overset{\displaystyle O}{\underset{\displaystyle \underset{\displaystyle CH_3}{\underset{\displaystyle |}{CH-CH_3}}}{\underset{\displaystyle |}{\underset{\displaystyle CH_2}{|}}}}C-OH$$

Identification of functional groups

A series of experiments were carried out on an organic substance X which has one functional group. The observations are recorded in the table below.

Experiment	Observation
1 Place 2 cm³ of X on a watch glass and ignite with a splint.	Smoky orange flame
2 In a fume cupboard add 1 cm³ of X to a test tube, add 1 cm³ of bromine water and mix well.	Orange solution changes to colourless
3 Add a spatula of sodium carbonate to 1 cm³ of X in a test tube.	No bubbles

Questions
1 What type of combustion do you think is occurring in experiment 1?

2 Based on the experiments above, suggest a functional group which may be present in X.

3 Based on the experiments above, suggest a functional group which may be absent from X.

In another experiment a series of tests was carried out on ethanoic acid.

Experiment	Observation
4 Place 2 cm³ of ethanoic acid in a test tube and add 2 cm³ of water. Shake well.	
5 Place 2 cm³ of ethanoic acid in a test tube and add 1 spatula of sodium carbonate.	
6 Place 2 cm³ of ethanoic acid in a test tube and add 2 cm³ of ethanol and 1 cm³ of acid catalyst. Warm gently in a water bath.	sweet smell

4 What would be the observation in reaction 4?

5 What would be the observation in reaction 5?

6 Name the organic substance produced in experiment 6 above.

7 What is the functional group present in ethanoic acid?

Chapter review questions

1 The structure of five molecules are shown.

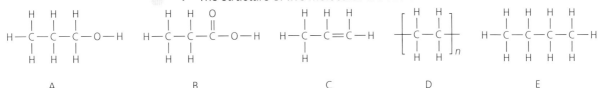

A   B   C   D   E

a) Which of these molecules is:

   i) an alkane

   ii) an alkene

   iii) an alcohol

   iv) a carboxylic acid

   v) a polymer?

b) Name each of these molecules.

2 Hexane is a *saturated hydrocarbon*. Hexene is an *unsaturated* hydrocarbon.

a) Define the three words in italics.

b) These two compounds can be distinguished using bromine water. What would happen to orange bromine water if it was added to:

   i) hexane?

   ii) hexene?

3 Identify the functional group in each of the following molecules.

a) There is fizzing when sodium carbonate is added to molecule **F**.

b) Molecule **G** mixes with water and forms a solution with pH 7.

c) Molecule **H** turns bromine water from orange to colourless.

d) Molecule **I** forms a solution with pH 4 when added to water.

e) There is fizzing when sodium is added to molecule **J**. Molecule **J** also burns very well.

4 Draw the displayed formula of each of the following molecules.

a) methane

b) ethanol

d) butanoic acid

c) propene

5 Write a balanced equation for the complete combustion of each of the following compounds.

a) pentane, C_5H_{12}

b) pentanol, $C_5H_{12}O$

6 **a)** Crude oil is split into fractions by fractional distillation. Explain how this process separates crude oil into fractions.

b) Fuel oil is one of the fractions produced from crude oil. It is made of long alkane molecules.

i) Give two reasons why long alkanes are in less demand as fuels as short alkanes.

ii) Name the process used to produce short alkanes from long alkanes.

iii) Describe one way in which this process is done.

iv) What other type of compound is produced in this process besides short alkanes?

v) Balance the following equation for this reaction that takes place in this process.

$$C_{18}H_{38} \rightarrow C_{10}H_{22} + C_3H_6 + C_2H_4$$

7 The molecule shown here is an alkene that can be made into an addition polymer.

a) What group does this molecule possess that allows it to form an addition polymer?

b) What else, if anything, is formed when this molecule reacts to form an addition polymer?

c) Draw the repeating unit of the polymer formed.

d) Write a balanced equation for the formation of this polymer.

8 Terylene is a condensation polymer. It is the main polyester used in clothes. It can be made from these two monomers.

a) Draw the repeating unit of the polymer formed.

b) Write a balanced equation for the formation of this polymer.

c) Why is this called a condensation polymer?

9 Isoleucine is an amino acid. A polypeptide can be formed when isoleucine molecules join together in a condensation polymerisation reaction.

a) Draw out the molecule and draw a circle around the amine functional group.

b) Draw a square around the carboxylic acid functional group.

c) Draw the repeating unit of the polypeptide formed.

d) Write a balanced equation for the formation of this polypeptide.

10 The molecule shown here is an alkene with the molecular formula C_4H_8. Write a balanced equation for the reaction of this alkene with each of the substances shown. Draw a displayed formula for each of the organic molecules.

a) bromine

b) steam

c) hydrogen

Practice questions

1 What type of reaction takes place when C_2H_5OH is converted into CH_3COOH? [1 mark]

  A combustion

  B oxidation

  C reduction

  D polymerisation

2 One molecule of an alkene with relative formula mass 42 reacts with one molecule of bromine to produce a product X. What is the relative formula mass of X? [1 mark]

  A 42

  B 122

  C 160

  D 202

3 Which one of the following is not a hydrocarbon? [1 mark]

  A butane

  B ethane

  C ethanol

  D methane

4 Which one of the following molecular formulae represents a compound which is a member of the same homologous series as C_2H_6? [1 mark]

  A C_2H_4

  B C_3H_6

  C C_4H_8

  D C_4H_{10}

5 Crude oil is the source of hydrocarbons such as alkanes.

  a) i) What is crude oil? [1 mark]

      ii) How are alkanes obtained from crude oil? [1 mark]

      iii) Name two fuels which are obtained from crude oil. [2 marks]

  b) Octane, C_8H_{18}, is an alkane which is a constituent of petrol.

      i) Octane is a saturated hydrocarbon. What is meant by the terms **saturated** and **hydrocarbon**? [2 marks]

      ii) What is the general formula for the alkanes? [1 mark]

  c) Alkenes, such as propene, can be obtained from larger alkanes such as octane.

      i) What name is given to the process of forming alkenes from large alkanes? [1 mark]

      ii) Write an equation for the formation of propene and an alkane, from octane. [2 marks]

      iii) Describe a chemical test for an unsaturated hydrocarbon such as propene. [2 marks]

      iv) Write an equation for the reaction involved in the test described in part (iii). [1 mark]

6 Fractional distillation of crude oil is used to produce hydrocarbon fuels. The diesel fraction contains heptadecane molecules which have the formula $C_{17}H_{36}$. Incomplete combustion of diesel produces carbon monoxide.

  a) Write an equation for the incomplete combustion of heptadecane to form carbon monoxide. [2 marks]

  b) State which has the higher boiling point, heptadecane or decane ($C_{10}H_{22}$)? [1 mark]

  c) Cracking of long chain hydrocarbons produces shorter chain hydrocarbons.

      i) State two different conditions which could be used for cracking. [2 marks]

      ii) Write an equation for the thermal cracking of heptadecane in which ethene and propene are produced in a 2:1 ratio with only one other product. [2 marks]

  d) Ethene and propene can be converted into poly(ethene) and poly(propene) respectively.

      i) Name the type of reaction involved in the production of poly(ethene) and poly(propene). [1 mark]

      ii) Write the equation for the conversion of propene into poly(propene). [2 marks]

7 a) The structure of tetrafluoroethene is shown.

F   F
| |
C == C
| |
F   F

      i) Name the polymer formed from the polymerisation of tetrafluoroethene. [1 mark]

      ii) Write an equation for the polymerisation of tetrafluoroethene. [2 marks]

  b) Amino acids can be converted into polypeptides. The structure of the amino acid glycine is shown.

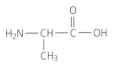

i) Name the type of reaction involved in the conversion of glycine to a polypeptide. [1 mark]

ii) Write an equation for this reaction. [2 marks]

8 Perfume is a mixture of essential oils dissolved in a solvent.

a) Myrcene, one of the essential oils used in making perfume is shown here.

i) Explain why myrcene is a hydrocarbon. [1 mark]

ii) What is the functional group present in myrcene? [1 mark]

iii) Is myrcene is a saturated or unsaturated compound? [1 mark]

b) Linalool is another essential oil used in perfume making. It has a sweet lavender like smell.

i) Identify the two functional groups in linalool. [2 marks]

H ii) Like all alcohols, linalool ($C_{10}H_{17}OH$) undergoes complete combustion in excess air. Complete and balance the symbol equation below for this reaction. [1 mark]

$$C_{10}H_{17}OH + O_2 \rightarrow \rule{1cm}{0.4pt} + \rule{1cm}{0.4pt}$$

iii) Compare the reactions of myrcene and linalool with bromine water and with sodium. In your answer decide if the organic compound reacts, and if so list any observations seen during the reaction. [5 marks]

c) Ethanol is an alcohol which is often used as a solvent in perfumes.

i) Draw the displayed formula of ethanol. [1 mark]

ii) Ethene can be used to manufacture the ethanol used in perfumes. Write a balanced symbol equation for the reaction of ethene to produce ethanol and state the conditions necessary. [3 marks]

iii) Draw the displayed formula of ethene. [1 mark]

iv) What type of reactions do alkenes undergo? [1 mark]

d) Ethanoic acid is a carboxylic acid which can be used to make other solvents. These solvents are also used in perfumes.

i) Draw the displayed formula of ethanoic acid. [1 mark]

ii) State two observations you would make when a spatula of sodium carbonate reacts with ethanoic acid. [2 marks]

Whenever experiments and investigations are carried out in the laboratory, you need to decide if the experiment is safe by carrying out a risk assessment. A risk assessment is a judgement of how likely it is that someone might come to harm if a planned action is carried out and how these risks could be reduced.

A good risk assessment includes:

1 A list of all the hazards in the experiment.

2 A list of the risks that the hazards could cause.

3 Suitable control measures you could take which will reduce or prevent the risk.

For chemicals there should be a COSHH hazard warning sign on the container.

Dangerous to the environment	Toxic	Gas under pressure
Corrosive	Explosive	Flammable
Caution – used for less serious health hazards like skin irritation	Oxidising	Longer term health hazards such as carcinogenicity

> **KEY TERMS**
>
> **Hazard** Something that could cause harm.
>
> **Risk** An action involving a hazard that might result in danger.

> **TIP**
>
> An oxidising substance does not burn itself. It should be kept away from flammable substances though as it will provide the oxygen allowing the flammable substance to burn fiercely.

The hazards for each chemical can be found by looking up the CLEAPSS student safety sheets or hazcards. These should be recorded in your risk assessment; for example, 'when using pure ethanol it should be labelled **HIGHLY FLAMMABLE (Hazcard 60).**' Hazcards will tell you about control measures and list when a fume cupboard needs to be used, when to wear gloves and the suitable concentrations and volumes for a safe experiment.

Some examples of hazards, risks and control measures.

Hazard	Risk	Control measure
Concentrated sulfuric acid	Corrosive	Handle with gloves; use small volumes; wear eye protection
Ethanol	Flammable	Keep away from flames / no naked flames to be used / use water bath to heat
Bromine	Toxic	Handle with gloves; use dilute solutions; wear eye protection; use in fume cupboard
Cracked glassware	Could cause cuts	Check for cracks before use
Hot apparatus	Could cause burns	Allow apparatus to cool before touching
Bags and stools	Could be tripping hazard	Tuck stools under benches; leave bags in bag store
Chemicals being heated in test tubes	Chemicals could spit out of test tubes	Wear eye protection; point tubes away from people
Beaker being heated on a tripod and gauze	Could fall over spilling hot liquid	Keep apparatus away from edge of bench; work standing up; wear eye protection
Long hair	Could catch fire	Tie back long hair

Questions

1 The diagram shows the apparatus used to heat some hydrated copper sulfate in a boiling tube. Copy and complete the table to give a risk assessment for this experiment.

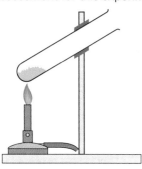

Hazard	Risk	Control measure

2 To prepare a sample of an ester, 5 cm³ of ethanol and 5 cm³ of ethanoic acid were mixed in a test tube with 5 drops of concentrated sulfuric acid and warmed. Write a risk assessment for this experiment.

3 The instructions to react sodium with ethanol were:
'Place a piece of sodium in a test tube of ethanol and observe the reaction.' Rewrite these experimental instructions in view of health and safety precautions.

4 What hazards and risks are associated with carrying out the electrolysis of a solution in the laboratory?

8 Chemical analysis

Blood is a mixture of many substances. It is very important that medics can identify which substances are in blood and how much of these substances are present. Analysis is a key area of chemistry and there are many tests and techniques that can be used to identify, measure and test the purity of substances.

Specification coverage

This chapter covers specification points 4.8.1 to 4.8.3 and is called Chemical analysis.

It covers purity, formulations, chromatography, identification of common gases, and identification of ions by chemical and spectroscopic means.

There is related work about writing ionic equations in the Appendix.

Purity, formulations and chromatography

◯ Pure substances

What is a pure substance?

In everyday language, a pure substance is regarded as a natural substance that has had nothing added to it. For example, 'pure orange juice' is considered to be the juice taken from oranges with nothing else, such as colourings or sweeteners, added; 'pure soap' is considered to be soap with nothing else, such as perfumes, added; milk may be regarded as a pure substance because it is taken straight from the cow and nothing else is added. However, a scientist uses the word 'pure' in a different way and would not consider the orange juice, soap or milk to be pure substances.

A **pure substance** is a single element or compound. For example:

● Diamond (C) is a pure substance because it contains only carbon atoms.
● Oxygen (O_2) is a pure substance because it contains only oxygen molecules.
● Glucose ($C_6H_{12}O_6$) is a pure substance because it contains only glucose molecules.

KEY TERM

Pure substance A single element or compound that is not mixed with any other substance.

▲ **Figure 8.1** A scientist would not consider this orange juice to be 'pure'.

A mixture contains more than one substance. For example,

● Orange juice is a mixture of water molecules, citric acid molecules, vitamin C molecules, glucose molecules, etc. (Figure 8.1).
● Soap is a mixture of several salts made from different fatty acids.
● Milk is a mixture of several substances including water, animal fats, emulsifiers, minerals, etc.
● Mineral water is a mixture because it contains water molecules, calcium ions, magnesium ions, nitrate ions and many other ions.
● Air is a mixture because it contains nitrogen molecules, oxygen molecules, argon atoms, etc.

Melting and boiling points of pure substances and mixtures

Pure substances melt and boil at specific temperatures. For example, water has a boiling point of 100°C and a melting point of 0°C. While a pure substance changes state, the temperature remains constant at these values. For example, while pure water is boiling, the temperature stays at 100°C. When pure water is freezing, the temperature stays at 0°C.

However, mixtures change state over a range of temperatures. Some everyday examples of this are shown in Figure 8.2.

The car radiator contains a mixture of antifreeze and water so that the mixture freezes below 0°C.

Salt is put on roads to create a mixture with water that freezes below 0°C.

Petrol is a mixture of hydrocarbons that boil over a range of temperatures from about 60 to 100°C.

▶ **Figure 8.2** Some very useful mixtures.

● **Salt water** freezes at a range of temperatures below 0°C, for example between −5°C and −10°C depending how much salt is dissolved. This is used to stop ice forming on roads in winter. Grit, which contains salt, is put on roads to prevent water freezing if the temperature drops below 0°C. Salt water boils at temperatures above 100°C, for example between 101 and 103°C depending how much salt is dissolved.

- **Antifreeze** is mixed with water in car radiators to stop the water freezing in cold weather. The melting point of antifreeze is –13°C and that of water is 0°C. The mixture typically has a melting point range between –30°C to –40°C depending how much antifreeze is used.
- **Petrol** is a mixture of hydrocarbons and boils over a range of temperatures from about 60°C to 100°C. Each individual substance in petrol has its own specific boiling point, but the mixture boils over a range of temperatures.

Formulations

A formulation is a mixture that has been designed as a useful product. It is made by mixing together several different substances in carefully measured quantities to ensure the product has the required properties. Some examples are listed in Table 8.1.

Table 8.1

Product	Comments
Alloys	• Alloys are specific mixtures of metals with other elements (e.g. steel is a mixture of iron and carbon; brass is a mixture of copper and zinc) • Alloys are harder than pure metals • There are many different alloys • Each alloy is designed to have the specific properties required for its use
Fertilisers	• Most fertilisers contain specific mixtures of different substances (for example, many fertilisers contain ammonium nitrate, phosphorus oxide and potassium oxide) • There are many different types of fertiliser with different amounts of different substances in them that are suitable for different plants and/or different soil types
Fuels	• Many fuels are specific mixtures (e.g. petrol and diesel are complex and carefully controlled mixtures of hydrocarbons designed to burn well and power a car engine)
Medicines	• Many medicines are specific mixtures of substances (e.g. aspirin tablets contain several other substances besides aspirin, including corn starch which is there to bind the tablet together; Calpol contains paracetamol in malitol liquid so that the medicine can be taken off a spoon)
Cleaning agents	• Cleaning agents are specific mixtures of many different substances (e.g. some dishwasher tablets contain detergents, alkalis, bleaches, rinse aid, etc.; some toilet cleaners contain bleaches, alkalis, detergents, etc.; some oven cleaner sprays contain sodium hydroxide to react with dirt, butane as a propellant, etc.)
Foods	• Many foods are very specific mixtures (e.g. margarine is a mixture of vegetable oils, water, emulsifiers, salt, etc.; tomato ketchup is a mixture of tomatoes, vinegar, sweeteners, spices, salt, etc.; vegetable soup is a mixture of water, vegetables, spices, etc.)
Paints	• Paints are mixtures whose contents include a solvent (water for emulsion, hydrocarbons for gloss), pigments (for the colour) and binder (to hold the pigments in place when the paint dries)

▲ **Figure 8.3** Many everyday products are formulations.

Test yourself

1 Some people may say that a bottle of mineral water is pure water.
   a) Explain why some people may regard this as being pure.
   b) Explain why a scientist would not say that it is pure.
2 A sample of water was found to freeze between –2°C and –4°C. Was this water pure? Explain your answer.
3 Some students made samples of aspirin in the laboratory. The melting point of aspirin is 136°C. Which student(s) made pure aspirin? Explain your answer.

Student	A	B	C	D	E	F
Melting point in °C	137–138	130–132	136	139–144	136–137	131–136

4 a) What is a formulation?
   b) Give four examples of formulations.

Show you can...

The table shows the melting points and boiling points of some metallic elements and alloys named by letters A to C.

	Melting point in °C	Boiling point in °C
A	−34	356
B	420	913
C	1425–1540	2530–2545

a) What state does substance A exist in at room temperature (20°C) and pressure?

b) What state does substance B exist in at room temperature (20°C) and pressure?

c) What state does substance C exist in at room temperature (20°C) and pressure?

d) What is an alloy?

e) Classify the substances in the table as elements or alloys and explain your answer.

○ Chromatography

What happens in paper chromatography?

Chromatography is a very useful technique that can be used to separate and analyse mixtures. There are several types of chromatography including paper, thin layer, column and gas chromatography.

Paper chromatography is often used to analyse coloured substances. In paper chromatography:

1 A pencil line is drawn on the chromatography paper near the bottom. Pencil is used as it will not dissolve in the solvent.

2 Small amounts of the substances being analysed are placed in spots on the pencil line.

3 The paper is hung in a beaker of the solvent. The pencil line and spots must be above the level of the solvent so that the spots do not dissolve into the solvent in the beaker.

4 Over the next few minutes, the solvent soaks up the paper.

5 When the solvent is near the top, the paper is taken out of the solvent and the level that the solvent reached is marked. This is known as the solvent front.

6 The paper is left to dry.

A pure substance can only ever produce one spot in chromatography, whatever solvent is used. Mixtures will usually produce more than one spot, one for each substance in the mixture (Figure 8.4). It is possible two substances in a mixture will move the same distance and appear as a single spot in some solvents.

In the example in Figure 8.4, Y is a mixture of two substances as it produces two spots. We can see by comparing the spots to A, B and C, that Y is a mixture of substances A and C.

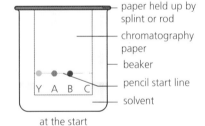

at the start

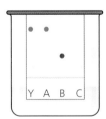

after the solvent has soaked up the paper

▲ **Figure 8.4** Paper chromatography.

How chromatography works

In each type of chromatography there is a mobile phase and a stationary phase. In paper chromatography, the stationary phase is the piece of chromatography paper and the mobile phase is a solvent.

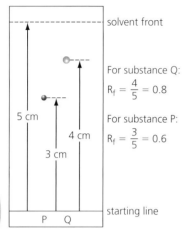

For substance Q:
$R_f = \frac{4}{5} = 0.8$

For substance P:
$R_f = \frac{3}{5} = 0.6$

▲ **Figure 8.5** Finding R_f values.

How far each substance moves depends on its relative attraction to the paper and the solvent. Substances that have a stronger attraction to the solvent move quickly and travel a long way up the paper. Substances that have a stronger attraction to the paper move slowly and only travel a short distance up the paper. In Figure 8.4 it can be seen that substance Q moves the greatest distance and so has a stronger attraction to the solvent than the paper while substance P moves the shortest distance and so has a stronger attraction to the paper than Q.

R_f values

The ratio of the distance a substance moves to the distance moved by the solvent is called the R_f value (Figure 8.5). The distance is measured to the centre of the spot.

$$R_f = \frac{\text{distance moved by substance}}{\text{distance moved by solvent}}$$

The R_f value for a substance is always the same in the same solvent. However, substances will have different R_f values in different solvents. R_f values can be used to identify substances.

Test yourself

5 Food colouring S was analysed by paper chromatography and compared to substances 1–6. The samples were placed on a pencil line on a piece of chromatography paper which was hung in a solvent.

   a) Explain why the starting line was drawn in pencil.

   b) Explain why the level of the solvent must be below the level of the pencil line.

   c) Which of the substances 1 to 6 appear to be pure substances? How can you tell?

   d) How many substances are in colouring S?

   e) Which substances are in colouring S?

   f) Calculate the R_f values for each spot in colouring S. Give your answer to 2 significant figures.

   g) Which colour spot moved slowest during the experiment?

6 A substance was analysed by chromatography in two different solvents.

   a) Calculate the R_f value in each solvent. Give your answer to 2 significant figures.

   b) Explain why the substance moved further in solvent 2 than solvent 1.

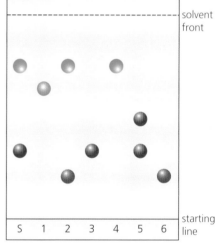

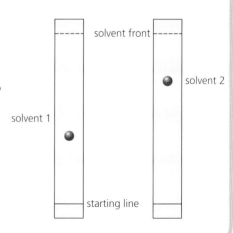

Identification of common gases

The gases oxygen, hydrogen, carbon dioxide and chlorine are common gases. There is a simple chemical test to identify each one (Table 8.2).

Table 8.2

Gas	Test		Result
Oxygen, O_2	Insert a glowing splint (one that has just been blown out) into a tube of the gas		The splint relights
Hydrogen, H_2	Insert a burning splint into a tube of the gas		There is a squeaky pop sound
Carbon dioxide, CO_2	The gas is shaken with or bubbled through limewater (a solution of calcium hydroxide in water)		Limewater goes cloudy
Chlorine, Cl_2	Insert damp (red or blue) litmus paper into the gas		Litmus paper is bleached (it loses its colour and turns white)

Show you can...

A student carried out three tests on a gas X and recorded the results in the table.

Test	Observations
Test 1: damp red litmus paper	Stays red
Test 2: bubble into limewater	Stays colourless
Test 3: lighted splint	Squeaky pop
Test 4: bubble into bromine water	Stays orange

a) What does the result of test 1 tell you about gas X?

b) What does the result of test 2 tell you about gas X?

c) Use the results of tests 3 and 4 to decide if gas X is hydrogen or ethene.

Test yourself

7 Magnesium reacts with hydrochloric acid to produce a gas. This gas was tested and found to give a squeaky pop with a burning splint.

   a) Identify the gas produced.

   b) Write a word equation for the reaction between magnesium and hydrochloric acid.

   c) Write a balanced equation for the reaction between magnesium and hydrochloric acid.

8 Copper carbonate reacts with nitric acid to produce a gas. This gas was tested and found to turn a solution of limewater cloudy.

   a) Identify the gas produced.

   b) What is limewater?

   c) Write a word equation for the reaction between copper carbonate and nitric acid.

   d) Write a balanced equation for the reaction between copper carbonate and nitric acid.

9 The electrolysis of a solution of sodium chloride was carried out. A gas was produced at the positive electrode that turned damp blue litmus paper white. Another gas was produced at the negative electrode which gave a squeaky pop with a burning splint.

   a) Identify the gas produced at the positive electrode.

   b) Identify the gas produced at the negative electrode.

   c) Write a half equation for the process at the positive electrode and state whether this is reduction or oxidation.

   d) Write a half equation for the process at the negative electrode and state whether this is reduction or oxidation.

10 The electrolysis of a solution of magnesium sulfate was carried out. A gas was produced at the positive electrode that was found to re-light a glowing splint. Another gas was produced at the negative electrode which gave a squeaky pop with a burning splint.

   a) Identify the gas produced at the positive electrode.

   b) Identify the gas produced at the negative electrode.

   c) Write a half equation for the process at the positive electrode and state whether this is reduction or oxidation.

   d) Write a half equation for the process at the negative electrode and state whether this is reduction or oxidation.

Identification of ions by chemical and spectroscopic means

○ Chemical analysis of ions

Compounds made from a combination of metals and non-metals are made of ions. These compounds all contain positive ions and negative ions. For example: sodium chloride contains sodium ions (Na^+) and chloride ions (Cl^-).); magnesium sulfate contains magnesium ions (Mg^{2+}) and sulfate ions (SO_4^{2-}). Positive ions are called cations. Negative ions are called anions.

We can test ionic compounds to identify which ions they contain. The tests for some ions are given in this chapter.

KEY TERMS

Cation Positive ion.

Anion Negative ion.

Tests for positive ions (cations)

Flame tests

Some positive ions give distinctive colours in flame tests. A simple way to do a flame test is to dip a damp splint into the compound and then put the splint into a roaring Bunsen flame. Table 8.3 shows the colours produced by some common ions.

Table 8.3

Ion	Lithium, Li^+	Sodium, Na^+	Potassium, K^+	Calcium, Ca^{2+}	Copper, Cu^{2+}
Result	Crimson-red flame	Yellow-orange flame	Lilac flame	Red-orange flame	Green flame

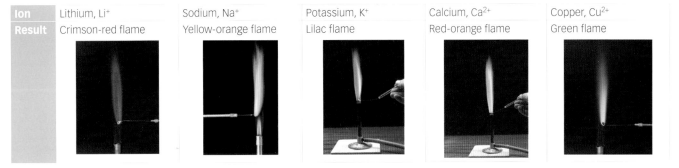

If a flame test is carried out on a mixture containing more than one type of positive ion, these colours will be mixed and it may not be possible to identify the ions this way.

Test with sodium hydroxide solution

Sodium hydroxide solution can be used to identify some positive ions. When sodium hydroxide solution is added to a solution of the substance being analysed, some positive ions produce a precipitate. A precipitate is a solid formed when two solutions are mixed.

KEY TERM

Precipitate A solid formed when two solutions are mixed.

For example, when solutions of copper(II) sulfate and sodium hydroxide are mixed, a blue precipitate of copper(II) hydroxide is formed as the copper ions from the copper(II) sulfate react with the hydroxide ions from the sodium hydroxide.

- **Word equation:** copper(II) sulfate (aq) + sodium hydroxide (aq) → copper(II) hydroxide (s) + sodium sulfate (aq)
- **Balanced equation:** $CuSO_4(aq) + 2NaOH(aq) \rightarrow Cu(OH)_2(s) + Na_2SO_4(aq)$
 - **Ionic equation:** $Cu^{2+}(aq) + 2OH^-(aq) \rightarrow Cu(OH)_2(s)$

Table 8.4

Ion	Copper(II), Cu^{2+}	Iron(II), Fe^{2+}	Iron(III), Fe^{3+}
Result and ionic equation for reaction	Blue precipitate	Green precipitate	Brown precipitate

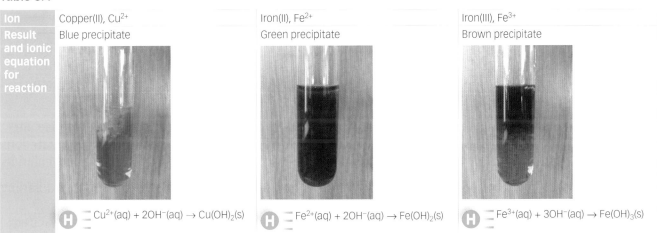

$Cu^{2+}(aq) + 2OH^-(aq) \rightarrow Cu(OH)_2(s)$ $Fe^{2+}(aq) + 2OH^-(aq) \rightarrow Fe(OH)_2(s)$ $Fe^{3+}(aq) + 3OH^-(aq) \rightarrow Fe(OH)_3(s)$

Table 8.5

Ion	Magnesium, Mg^{2+}	Calcium, Ca^{2+}	Aluminium, Al^{3+}
Result and ionic equation for reaction	White precipitate 	White precipitate 	White precipitate 
	$Mg^{2+}(aq) + 2OH^-(aq) \rightarrow Mg(OH)_2(s)$	$Ca^{2+}(aq) + 2OH^-(aq) \rightarrow Ca(OH)_2(s)$	$Al^{3+}(aq) + 3OH^-(aq) \rightarrow Al(OH)_3(s)$
Addition of excess NaOH(aq)	White precipitate remains	White precipitate remains	White precipitate dissolves

Solutions containing magnesium ions, calcium ions and aluminium ions all produce a white precipitate when sodium hydroxide solution is added. Aluminium ions can be distinguished because the white precipitate of aluminium hydroxide dissolves when excess sodium hydroxide solution is added, but this does not happen for calcium ions or magnesium ions. These can be distinguished with a flame test as calcium ions give a red-orange flame while magnesium ions do not give any colour at all.

Tests for negative ions (anions)

Test for carbonate ions

There is a simple test for carbonate ions in a compound. When dilute acid is added to a compound containing carbonate ions, there is fizzing due to the formation of carbon dioxide gas. The identity of the carbon dioxide gas can be confirmed by testing with limewater. Carbon dioxide turns limewater cloudy (Figure 8.6).

Most carbonate compounds are insoluble in water. However, there are some carbonate compounds including sodium carbonate and potassium carbonate that do dissolve in water and produce solutions containing carbonate ions.

▲ **Figure 8.6** Dilute acids react with carbonate ions to form carbon dioxide gas that turns limewater cloudy.

Test for sulfate ions

There is a simple test for compounds containing sulfate ions. Dilute hydrochloric acid followed by barium chloride solution is added to a solution of the compound being tested. If a white precipitate forms, the compound contains sulfate ions (Figure 8.7).

For example, when solutions of sodium sulfate and barium chloride are mixed, a white precipitate of barium sulfate is formed as the sulfate ions from the sodium sulfate react with the barium ions from barium chloride.

▲ **Figure 8.7** A white precipitate of barium sulfate showing the presence of sulfate ions.

- **Word equation:** barium chloride (aq) + sodium sulfate (aq) $\rightarrow$ barium sulfate (s) + sodium chloride (aq)
- **Balanced equation:** $BaCl_2(aq) + Na_2SO_4(aq) \rightarrow BaSO_4(s) + 2NaCl(aq)$
- **Ionic equation:** $Ba^{2+}(aq) + SO_4^{2-}(aq) \rightarrow BaSO_4(s)$

Table 8.6

Ion	Chloride, Cl⁻	Bromide, Br⁻	Iodide, I⁻
Result and ionic equation for reaction	White precipitate  $Ag^+(aq) + Cl^-(aq) \rightarrow AgCl(s)$	Cream precipitate  $Ag^+(aq) + Br^-(aq) \rightarrow AgBr(s)$	Yellow precipitate  $Ag^+(aq) + I^-(aq) \rightarrow AgI(s)$

Test for halide ions

Halide ions include chloride ions (Cl^-), bromide ions (Br^-) and iodide ions (I^-). There is a simple test for compounds containing halide ions. Dilute nitric acid followed by silver nitrate solution is added to a solution of the compound being tested. Table 8.6 shows the results for different halide ions.

For example, when solutions of potassium iodide and silver nitrate are mixed, a yellow precipitate of silver iodide is formed as the iodide ions from the potassium iodide react with the silver ions from the silver nitrate.

- **Word equation:** silver nitrate (aq) + potassium iodide (aq) → silver iodide (s) + potassium nitrate (aq)
- **Balanced equation:** $AgNO_3(aq) + KI(aq) \rightarrow AgI(s) + KNO_3(aq)$
- **Ionic equation:** $Ag^+(aq) + I^-(aq) \rightarrow AgI(s)$

Using the results of tests to identify an ionic compound

Example

Compound **X** was analysed. It gave a lilac flame in a flame test. When hydrochloric acid followed by barium chloride solution was added to a solution of **X**, a white precipitate was produced.

1 Identify the positive ion in compound **X**.

2 Identify the negative ion in compound **X**.

3 Write the name and formula of compound **X**.

4 Write a balanced equation for the formation of the white precipitate.

5 Write an ionic equation for the formation of the white precipitate.

Answers

1 Potassium ions (K^+) – due to the lilac flame in the flame test.

2 Sulfate ions (SO_4^{2-}) – due to the white precipitate when hydrochloric acid followed by barium chloride solution was added

3 Potassium sulfate; contains K^+ ions and SO_4^{2-} ions, therefore formula = K_2SO_4

4 Barium chloride (aq) + potassium sulfate (aq) → barium sulfate (s) + potassium chloride (aq)

   $BaCl_2(aq) + K_2SO_4(aq) \rightarrow BaSO_4(s) + 2KCl(aq)$

5 $Ba^{2+}(aq) + SO_4^{2-}(aq) \rightarrow BaSO_4(s)$

TIP

Ionic compounds contain both positive and negative ions. To identify an ionic compound, both the positive and negative ions must be identified.

Test yourself

11 Compound **A** was analysed. It produced a yellow flame in a flame test. When dilute acid was added to a sample of **A**, a gas was given off that turned limewater cloudy.

   a) Identify the positive ion in compound **A**.
   b) Identify the negative ion in compound **A**.
   c) Write the name and formula of compound **A**.
   d) Name the gas that turns limewater cloudy.

12 Compound **B** was analysed. It gave a green flame in a flame test. When nitric acid followed by silver nitrate solution was added to a solution of **B**, a white precipitate was produced.

   a) Identify the positive ion in compound **B**.
   b) Identify the negative ion in compound **B**.
   c) Write the name and formula of compound **B**.
   d) Write a balanced equation for the formation of the white precipitate.
   e) Write an ionic equation for the formation of the white precipitate.

13 Compound **C** was analysed. A green precipitate was produced when sodium hydroxide solution was added to a solution of **C**. A yellow precipitate was produced when nitric acid followed by silver nitrate solution was added to a separate solution of **C**.

   a) Identify the positive ion in compound **C**.
   b) Identify the negative ion in compound **C**.
   c) Write the name and formula of compound **C**.
   d) Write a balanced equation for the formation of the green precipitate.
   e) Write a balanced equation for the formation of the yellow precipitate.
   f) Write an ionic equation for the formation of the green precipitate.
   g) Write an ionic equation for the formation of the yellow precipitate.

14 Compound **D** was analysed. It gave a crimson flame in a flame test. When dilute acid was added to a sample of **D**, a gas was given off that turned limewater cloudy.

   a) Identify the positive ion in compound **D**.
   b) Identify the negative ion in compound **D**.
   c) Write the name and formula of compound **D**.

15 Compound **E** was analysed. It gave a red-orange flame in a flame test. It produced a white precipitate when sodium hydroxide solution was added to a solution of **E**. This white precipitate did not dissolve when more sodium hydroxide solution was added. A cream precipitate was produced when nitric acid followed by silver nitrate solution was added to a separate solution of **E**.

   a) Identify the positive ion in compound **E**.
   b) Identify the negative ion in compound **E**.
   c) Write the name and formula of compound **E**.
   d) Write a balanced equation for the formation of the white precipitate.
   e) Write a balanced equation for the formation of the cream precipitate.
   f) Write an ionic equation for the formation of the white precipitate.
   g) Write an ionic equation for the formation of the cream precipitate.

16 Compound **F** was analysed. A white precipitate was produced when sodium hydroxide solution was added to a solution of **F**. This white precipitate dissolved when more sodium hydroxide solution was added. When hydrochloric acid followed by barium chloride solution was added to a separate solution of **F**, a white precipitate was produced.

   a) Identify the positive ion in compound **F**.
   b) Identify the negative ion in compound **F**.
   c) Write the name and formula of compound **F**.
   d) Write a balanced equation for the formation of the white precipitate when sodium hydroxide is added.
   e) Write a balanced equation for the formation of the white precipitate when barium chloride is added.
   f) Write an ionic equation for the formation of the white precipitate when sodium hydroxide is added.
   g) Write an ionic equation for the formation of the white precipitate when barium chloride is added.

Show you can ...

In an experiment solutions containing each of barium, cadmium and silver ions were added separately to solutions containing carbonate, chloride or sulfate ions. The table shows which anions formed precipitates with the three different cations. A tick (✓) indicates that a precipitate formed and a cross (✗) indicates that no precipitate formed.

Metal cations	Anions		
	Carbonate	Chloride	Sulfate
Barium	✓	✗	✓
Cadmium	✓	✗	✗
Silver	✓	✓	✓

A student has a solution containing a mixture of barium, cadmium and silver ions. He wishes to make a solid precipitate containing each metal ion from the mixture, and separate it by filtering. State the order in which he must add the anion solutions in order to filter out a compound of each metal in turn.

Investigating how paper chromatography can be used to tell the difference between different coloured solutions

When freezing water for ice rinks, it is important that the water does not contain too many dissolved metal ions as this makes poor quality ice for skating. To determine if a sample of water contained some dissolved metal ions a chromatography experiment was carried out using the water sample (A) and known metal ion solutions B (containing copper(II) ions), C (containing iron(II) ions) and D (containing iron(III) ions). The method used is shown below.

Method

1 Draw a base line on the chromatography paper 1.5 cm from the bottom using a pencil.

2 Place a concentrated drop of each solution to be tested on the base line.

3 Place the chromatography paper into a chromatography tank containing water at a depth of 1 cm.

4 After the water soaks up the paper, dry the paper and spray with sodium hydroxide solution.

The spots which appeared on the paper are shown.

Questions

1 Why is it necessary to draw a pencil line as a base line 1.5 cm from the bottom?

2 Why is the water solvent at a depth of only 1 cm?

3 How is a concentrated drop of each solution added to the chromatography paper?

4 The chromatogram obtained is shown here. Name two metal ions which are present in the water sample A.

5 Write the formula of the compound formed when spot C reacts with sodium hydroxide.

6 Write an equation which is used to calculate R_f value and calculate the R_f value for spot B.

7 If the experiment was repeated using a different solvent would the R_f value be the same?

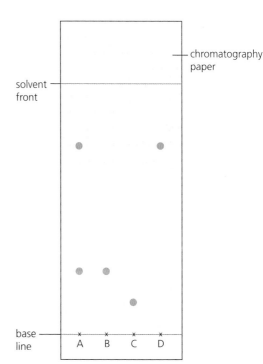

○ **Instrumental analysis**

In addition to carrying out chemical tests in the laboratory to identify elements and compounds, there are also instrumental methods of analysis. These use a wide range of sophisticated machines for analysis and have developed a great deal with recent advances in computing and electronics.

Instrumental methods of analysis have many uses, for example:

- to analyse blood, urine and tissue samples in hospitals (Figure 8.8)
- to analyse the contents of foods
- to monitor the quality of water and air
- to monitor the purity of medicines.

There are many advantages to using these modern, instrumental methods of analysis compared to chemical tests. The advantages include that they are:

- sensitive – they only use a tiny amount of a substance
- informative – they can provide a lot of valuable information about a substance
- rapid – the information can be found very quickly and
- accurate.

▲ **Figure 8.8** Hospital laboratories use instrumental methods of analysis.

Flame emission spectroscopy

When some metal ions are heated in a flame, they give off a distinctive colour. For example, sodium ions gives off yellow-orange light, lithium ions give off crimson-red light and calcium ions give off red-orange light. In flame tests we judge the colour by eye. Sometimes it can be difficult to judge which ion is giving the colour. For example, some students can mix up the colour of lithium ions and calcium ions.

Flame emission spectroscopy is an example of an instrumental method of analysis. It is used to:

- identify metal ions in solution and
- measure the concentration of metal ions in solution.

Flame emission spectroscopy is effectively a sophisticated way to do flame tests. It identifies metal ions with more certainty than flame tests and can measure the concentration of metal ions in a solution which flame tests cannot.

In flame emission spectroscopy, the actual wavelength of the light given off is measured by a spectroscope in the machine. Each metal ion gives off light at its own specific wavelength producing a line spectrum. This allows the identity of ions to be found with certainty. The emission spectra of some ions are shown in Figure 8.9, as well as the full spectrum of visible light. It can be seen, for example, that the emission spectrum of lithium ions is different from that for calcium ions even though their flames have similar colours (Figure 8.9).

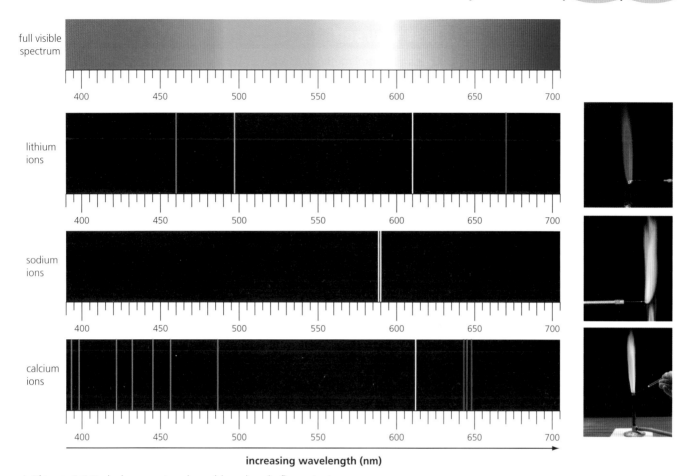

full visible spectrum

lithium ions

sodium ions

calcium ions

increasing wavelength (nm)

▲ **Figure 8.9** Emission spectra alongside colour in flame tests.

The concentration of the ions in the solution can also be found by measuring the intensity of a wavelength of light given out. The more intense the light it gives out, the higher the concentration of that ion. Graphs are used that show the relationship between the intensity of the light and the concentration of the ion (Figure 8.10). The intensity of the light from the sample can be measured and the graph used to find the concentration of the ion, although this is usually done directly by a computer controlling the instrument.

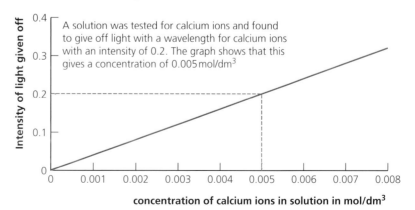

A solution was tested for calcium ions and found to give off light with a wavelength for calcium ions with an intensity of 0.2. The graph shows that this gives a concentration of 0.005 mol/dm³

Intensity of light given off

concentration of calcium ions in solution in mol/dm³

▶ **Figure 8.10** Finding the concentration of ions in solution by flame emission spectroscopy.

Show you can...

a) Copy and complete the table to give the flame colour of some metal ions.

Ion present	Flame colour
Barium	Green
Calcium	
Copper	
Lithium	
Sodium	

To analyse a compound and identify the metal ion present, a flame test was carried out. A green flame colour was observed.

b) Does this result prove that the compound contains barium ions? Explain your answer.

c) How would a line emission spectrum help identify the metal ion present?

Test yourself

17 Instrumental methods of analysis have developed rapidly in modern times. Give four advantages of instrumental methods over chemical analysis.

18 The wavelengths of the main lines in the emission spectra for lithium, sodium and potassium ions are shown in the table.

Metal ion	Li^+	Na^+	Ca^{2+}
Wavelength of main lines in emission spectrum/nm	610, 670	589, 590	612, 643, 645, 648

A compound was analysed with a flame test. It gave a flame with a red colour. Analysis by flame emission spectroscopy gave a spectrum with four lines in the red-orange region at 648 nm, 645 nm, 643 nm and 612 nm.

a) Identify the metal ion.

b) Give two advantages of flame emission spectroscopy over a flame test.

19 The table below shows the intensity of the emission of light with wavelength 670 nm in the flame emission spectroscopy of lithium ions at different concentrations.

Concentration/ mmol in dm³	0.0	0.1	0.2	0.3	0.4	0.5	0.6	0.7	0.8	0.9	1.0
Intensity	0.00	0.06	0.13	0.20	0.26	0.32	0.39	0.45	0.52	0.58	0.65

a) Plot a graph to show how the intensity of the light varies with concentration.

b) Use your graph to find the concentration of lithium ions in a solution that emits light with an intensity of 0.40 at 670 nm.

Use of chemical tests to identify the ions in unknown single ionic compounds

A student carried out a series of tests on an ionic compound labelled A.

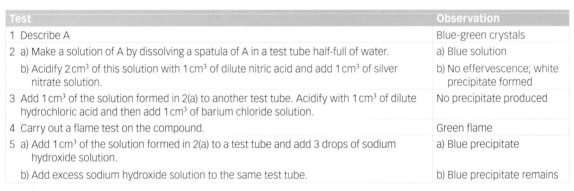

Test	Observation
1 Describe A	Blue-green crystals
2 a) Make a solution of A by dissolving a spatula of A in a test tube half-full of water.	a) Blue solution
b) Acidify 2 cm³ of this solution with 1 cm³ of dilute nitric acid and add 1 cm³ of silver nitrate solution.	b) No effervescence; white precipitate formed
3 Add 1 cm³ of the solution formed in 2(a) to another test tube. Acidify with 1 cm³ of dilute hydrochloric acid and then add 1 cm³ of barium chloride solution.	No precipitate produced
4 Carry out a flame test on the compound.	Green flame
5 a) Add 1 cm³ of the solution formed in 2(a) to a test tube and add 3 drops of sodium hydroxide solution.	a) Blue precipitate
b) Add excess sodium hydroxide solution to the same test tube.	b) Blue precipitate remains

Questions

1 Give an appropriate deduction for tests 2, 3, 4 and 5.

2 Name compound A.

3 Give the formula of compound A.

4 Give the formula of the white precipitate formed in test 2(b).

5 Give the formula of the blue precipitate formed in test 5(a).

Chapter review questions

1 The diagram shows some particles in air.

a) Is air a pure substance? Explain your answer.

b) Give the formula of all the elements shown in the diagram.

c) Give the formula of all the compounds shown in the diagram.

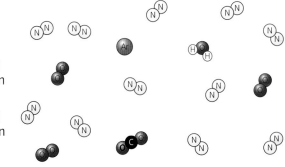

2 Identify gases **A**, **B**, **C** and **D** using the results in the table.

Test	Gas A	Gas B	Gas C	Gas D
Effect on damp red litmus paper	No effect	Paper goes white	No effect	No effect
Effect on burning splint	Flame continues to burn	Flame goes out	Squeaky pop	Flame goes out
Effect on limewater	No effect	No effect	No effect	Goes cloudy
Effect on glowing splint	Flame relights	No effect	No effect	No effect

3 A food colouring **K** was analysed by paper chromatography. A spot of **K** was placed on a pencil line on a piece of chromatography paper along with spots of dyes **W**, **X**, **Y** and **Z**. The paper was placed in a beaker of water and left for a few minutes. The paper was removed from the beaker, the level of the solvent marked and the paper left to dry. The results are shown below.

a) How many substances are in food colouring **K**?

b) Which of the substances **W**, **X**, **Y** and **Z** are in food colouring **K**?

c) Which of the substances **W**, **X**, **Y** and **Z** are likely to be:

  i)   pure substances

  ii) mixtures?

d) Why was the line drawn in pencil?

e) Calculate the R_f value for the spots in food colouring **K**. Give your answer to 2 significant figures.

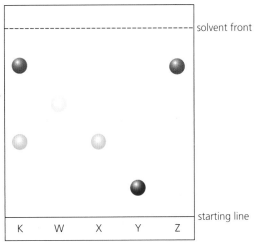

f) What would happen to the R_f value of each spot if a different solvent was used?

4 Sun cream is used by many people to protect themselves from harmful UV rays from the Sun. Sun cream is a formulation. What is meant by the term formulation?

5   Some ice was taken out of a freezer and allowed to warm up. The table shows the temperature over the next 15 minutes as the ice melted.

Time in min	0	1	2	3	4	5	6	7	8	9	10	11	12	13	14	15
Temperature in °C	−20	−15	−10	−5	−4	−3	−2	3	8	13	18	20	20	20	20	20

a) Plot a graph to show the temperature changed over time.

b) Was the ice pure? Explain your answer.

c) Why do you think the temperature stopped rising when it reached 20°C?

6   The table below shows the intensity of the emission of light with wavelength 589 nm in the flame emission spectroscopy of sodium ions at different concentrations.

Concentration in mol/dm³	0.00	0.01	0.02	0.03	0.04	0.05	0.06	0.07	0.08	0.09	0.10
Intensity	0.00	0.09	0.18	0.28	0.37	0.48	0.57	0.66	0.76	0.86	0.96

a) Plot a graph to show how the intensity of the light varies with concentration.

b) Use your graph to find the concentration of sodium ions in a solution that emits light at 589 nm with an intensity of 0.30.

c) Give four advantages of using instrumental analysis over methods of chemical analysis like flame tests.

7   Compound **R** was analysed. It produced a crimson-red flame in a flame test. A yellow precipitate was produced when nitric acid followed by silver nitrate solution was added to a solution of **R**.

a) Identify the positive ion in compound **R**.

b) Identify the negative ion in compound **R**.

c) Write the name and formula of compound **R**.

8   Compound **T** was analysed. A white precipitate was produced when sodium hydroxide solution was added to a solution of **T**. This white precipitate did not dissolve when more sodium hydroxide solution was added. When dilute acid was added to a sample of **T**, a gas was given off that turned limewater cloudy.

a) Write the name and formula of two possible compounds that **T** could be.

b) What further chemical test could be done to identify **T**? Give the possible results of this test for each compound you identified in (a).

9   Compound **S** was analysed. A brown precipitate was produced when sodium hydroxide solution was added to a solution of **S**. When hydrochloric acid followed by barium chloride solution was added to a separate solution of **S**, a white precipitate was produced.

a) Identify the positive ion in compound **S**.

b) Identify the negative ion in compound **S**.

c) Write the name and formula of compound **S**.

d) Write a balanced equation for the formation of the brown precipitate when sodium hydroxide is added.

e) Write a balanced equation for the formation of the white precipitate when barium chloride is added.

f) Write an ionic equation for the formation of the brown precipitate when sodium hydroxide is added.

g) Write an ionic equation for the formation of the white precipitate when barium chloride is added.

Practice questions

1 A solid is thought to be pure benzoic acid. Which of the following is the best way to test its purity? [1 mark]

  A determine the density

  B determine the pH

  C determine the melting point

  D determine the flame colour

2 Which one of the following solutions will give a coloured precipitate when a few drops of sodium hydroxide solution are added? [1 mark]

  A aluminium sulfate   B calcium chloride

  C copper sulfate   D magnesium chloride

3 The gas produced when a solid X reacts with dilute hydrochloric acid causes colourless limewater to turn cloudy. Which of the following could be solid X? [1 mark]

  A calcium carbonate   B calcium chloride

  C calcium hydroxide   D calcium oxide

4 Test tube A contains dilute nitric acid and test tube B contains dilute hydrochloric acid. Which one of the following substances can be added to each test tube and produce different observations in each? [1 mark]

  A barium chloride solution   B copper carbonate

  C silver nitrate solution   D sodium hydroxide solution

5 On 14 April 2010 the volcano Eyjafjallajökull erupted in Iceland, creating an ash cloud which was dangerous for aircraft and led to the closure of many airports for about ten days. A large number of gases were released into the atmosphere from the volcano. These volcanic gases included carbon dioxide, hydrogen and hydrogen chloride.

  a) Copy and complete the table below to describe the tests used to positively identify carbon dioxide and hydrogen. [4 marks]

Gas	Test	Result of positive test
Carbon dioxide		
Hydrogen		

  b) Hydrogen chloride gas is soluble and dissolves in water to form hydrochloric acid. Describe how you would prove that a sample of hydrochloric acid was acidic and contained chloride ions. [4 marks]

  c) Volcanic ash contains many different minerals, including iron compounds and can be used in cosmetics including skin scrubs. The iron compounds may contain the iron(II) ion or the iron(III) ion.

The presence of these ions in solution may be detected by adding sodium hydroxide solution.

    i) State what would be observed when sodium hydroxide solution is added separately to a solution of iron(II) ions and to a solution of iron(III) ions. [2 marks]

    ii) Write an ionic equation, including state symbols for the reaction of iron(II) ions with sodium hydroxide solution. [2 marks]

6 a) In an experiment to determine which Group 7 ion was present in each of three different acids, a few drops of silver nitrate solution were added to a sample of the acid solution.

    i) Copy and complete the table below to show the results of these tests. [3 marks]

Acid	Observation on addition of a few drops of aqueous silver nitrate
Hydrobromic acid	
Hydrochloric acid	
Hydroiodic acid	

    ii) Write a balanced **ionic** equation for the reaction of hydrochloric acid with silver nitrate solution. [1 mark]

  b) Hydrochloric acid reacts with bases to form salts such as sodium chloride and magnesium chloride. An antiseptic mouthwash is thought to contain both of these salts.

    i) Describe how you would confirm that the mouthwash contained sodium ions. [2 marks]

    ii) Describe how you would experimentally confirm that the mouthwash contained magnesium ions. In your answer, refer to the validity of your test. [6 marks]

7 A **mixture** of two ionic compounds was analysed to determine the ions present in the mixture. The two ionic compounds have the **same anion**. The results of the tests are given in the table below. Use the information in the table to answer the questions which follow.

Description of test	Observations
Test 1: A flame test was carried out on a solid sample of the mixture	Lilac flame
Test 2: A sample of the mixture was dissolved in deionised water and sodium hydroxide solution was added	White precipitate which redissolves in excess sodium hydroxide solution
Test 3: A sample of the mixture was dissolved in deionised water and hydrochloric acid followed by barium chloride solution was added	White precipitate

a) Using the evidence from Test 1 only, name a cation which is present in the mixture. [1 mark]

b) Using the evidence from Test 2 only, name a cation which may be present in the mixture. [2 marks]

c) Using the evidence from Test 3, write the formula of the anion which is present in the mixture. [1 mark]

d) Suggest the names of **two** compounds which may be present in the mixture. [2 marks]

e) Write an ionic equation for the formation of the white precipitate in Test 3. [1 mark]

8 a) The presence of anions in salts can be determined by mixing two solutions and observing the formation of a precipitate and its colour.

   i) What is an anion? [1 mark]

   ii) What is meant by the term precipitate? [1 mark]

   iii) Name the solution which is used to test for the presence of sulfate ions. [1 mark]

   iv) Potassium iodide solution was mixed with silver nitrate solution and a precipitate formed. State the colour of the precipitate. [1 mark]

b) i) The presence of cations in salts can be determined using a flame test, or by adding sodium hydroxide solution. Copy and complete the table. [4 marks]

	Copper ion	Calcium ion
Flame test result		
Result on adding some sodium hydroxide solution to the metal ion solution		

   ii) Name an instrumental method which can be used to identify metal ions in solution. [1 mark]

   iii) State two advantages of using instrumental methods. [2 marks]

c) A sports drink contains many different salts and some food colouring. In a laboratory a sports drink was analysed using paper chromatography. Describe how this experiment was carried out. [6 marks]

Working scientifically:
Recording observations

Making and recording observations is an important skill in chemistry. Qualitative observations are what we see and smell during reactions. Important types of observations in chemistry and notes on how to record these are shown in the table.

Type of observation	Notes on recording observations	Examples	
Colour change	Always state the colour of the solution before the reaction and after	Bubbling an alkene into bromine water – the colour change is orange solution to colourless solution	
Bubbles produced	If a gas is produced, then bubbles are often observed in the liquid	When sodium carbonate reacts with an acid, the observation is bubbles	
Temperature change	Often in a reaction the temperature changes – in an exothermic reaction it increases and in an endothermic reaction it decreases	When acids react with alkalis, the temperature increases. When acids react with sodium hydrogencarbonate, the temperature decreases. When water reacts with anhydrous copper sulfate, the temperature increases	
Precipitate produced	When two solutions mix, an insoluble precipitate may form. Ensure you use the word precipitate in your observation. Also state the colour of the precipitate and the colour of the solution before adding the reagent	When barium chloride solution is added to a solution containing sulfate ions a white precipitate is formed from the colourless solution	
Solubility of solids	When a spatula of a solid is added to water, the solid may dissolve. If it dissolves, the observation is often that the solid dissolves to form a solution. Make sure you state the colour of the solution formed. If it does not dissolve, state that it is insoluble	Copper(ii) sulfate crystals dissolve in water to produce a blue solution	
Solubility of liquids	When a liquid is added to water always record if it is miscible or immiscible with water	Ethanol and water are miscible (left-hand tube). The liquids in the right-hand test tube are immiscible.	

TIP ✓
Remember that clear means see-through and does not mean colourless.

TIP ✓
Effervescence is another term to describe the production of bubbles in a reaction.

TIP ✓
Note that writing that carbon dioxide is formed is not an observation.

Questions

1 In a laboratory experiment a student added some magnesium metal to some copper sulfate solution and recorded the observation that 'the magnesium became covered in copper and magnesium sulfate was formed.'

   a) Write a word equation for the reaction occurring.

   b) State and explain if the observations given by the student are correct.

2 Some hydrochloric acid was placed in a conical flask and a spatula of calcium carbonate powder was added.

   a) Write a balanced symbol equation for this reaction.

   b) State the observations which occurred.

3 Copy and complete the table.

Reaction	Observations
ethanoic acid + sodium carbonate	
potassium iodide solution + silver nitrate solution	
bromine water + alkene	
hydrochloric acid + magnesium	
acidified barium chloride + sulfuric acid	

9 Chemistry of the atmosphere

There is air all around us even though we cannot see it. This air contains oxygen which is vital for life, but there was no oxygen in the atmosphere when the Earth was young. This chapter looks at what air is and how the atmosphere has changed over time. It also looks at how the oxygen in the air is used to burn fuels, and how this can produce gases that pollute our atmosphere.

Specification coverage

This chapter covers specification points 4.9.1 to 4.9.3 and is called Chemistry of the atmosphere.

It covers the composition and evolution of the Earth's atmosphere, carbon dioxide and methane as greenhouse gases, and common atmospheric pollutants and their sources.

Previously you could have learned:

> The Earth's atmosphere contains mainly nitrogen and oxygen.
> Plants absorb carbon dioxide and water and use it in photosynthesis to produce oxygen and glucose.
> The Earth is getting warmer – this is called global warming.
> Global warming is changing the Earth's climate.
> Coal, oil and natural gas are fossil fuels.
> When substances burn they react with oxygen to produce oxides.

Test yourself on prior knowledge

1 What are the main two gases in the Earth's atmosphere?
2 Write a word equation for photosynthesis.
3 What is global warming?
4 Give three examples of fossil fuels.
5 Write word equations for reactions of the following substances with oxygen.
   a) hydrogen, H_2
   b) carbon, C
   c) sulfur, S
   d) methane, CH_4
   e) hydrogen sulfide, H_2S

The composition and evolution of the Earth's atmosphere

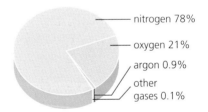

nitrogen 78%
oxygen 21%
argon 0.9%
other gases 0.1%

▲ **Figure 9.1** The gases in dry air.

○ The atmosphere today

The Earth is about 4.6 (4600 million) billion years old. For the last 200 million years or so, the composition of the air has been much the same. Most of the air is nitrogen (about four-fifths) and oxygen (about one-fifth), with small amounts of noble gases (mainly argon) and carbon dioxide (Figure 9.1). There is also a small amount of water vapour but this does vary with weather conditions.

○ The early atmosphere of the Earth

Scientists believe that the Earth's atmosphere was very different when it was young. For example, there is evidence that there was little or no oxygen in the atmosphere. However, there is much uncertainty about the Earth when it was young and there are many theories about what the atmosphere was like and how and when it changed.

Venus and Mars have very similar atmospheres to each other. The atmospheres on both planets are mainly carbon dioxide with some nitrogen but with little or no oxygen (Figure 9.2). Given that the Earth is between Venus and Mars, it may have been that the early atmosphere of the Earth was like this. It is thought that the evolution of life on Earth changed our atmosphere, but this did not happen on Venus or Mars.

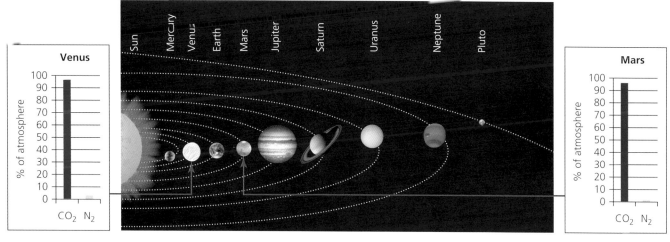

▲ Figure 9.2

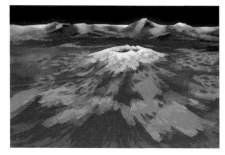

▲ **Figure 9.3** There is intense volcanic activity on Venus today.

▲ **Figure 9.4** The oceans may have been formed as water vapour released by volcanoes cooled and condensed.

One theory about the Earth is that it was very hot when it was formed and there was intense volcanic activity on the planet. Volcanoes release gases from the inside of the planet and the gases in the early atmosphere could have been released from volcanoes in this way (Figure 9.3).

The volcanoes on Earth may have released:

- carbon dioxide (CO_2) – which is likely to have been the main gas in the early atmosphere
- nitrogen (N_2) – this may have gradually built up in the atmosphere over time
- methane (CH_4) – probably in small amounts only
- ammonia (NH_3) – probably in small amounts only
- water vapour (H_2O) – this could have condensed and formed the oceans as the Earth cooled down (Figure 9.4).

○ How the Earth's atmosphere changed

Where did the oxygen come from?

Life evolved on Earth and it is thought that the oxygen in the atmosphere was formed by photosynthesis in living creatures. During photosynthesis, carbon dioxide reacts with water to form glucose and oxygen.

carbon dioxide + water → glucose + oxygen

$$6CO_2 + 6H_2O \rightarrow C_6H_{12}O_6 + 6O_2$$

▲ **Figure 9.5** Cyanobacteria.

Cyanobacteria were one of the earliest forms of life on Earth (Figure 9.5). They are a type of algae and the first known organism that photosynthesised. Current fossil evidence is that the oldest cyanobacteria are 2.7 billion years old and it was soon after this that oxygen started to appear in the atmosphere.

Over the next billion years, more complex life forms including plants evolved. As the number of creatures that photosynthesised increased on Earth, the amount of oxygen in the air gradually increased and reached a point where animals could evolve.

Where did the carbon dioxide go?

There were two main ways in which carbon dioxide was removed from the atmosphere (Figure 9.6). Much of the carbon dioxide was:

- dissolved in the oceans or
- used in photosynthesis.

Some of the carbon dioxide that dissolved in the oceans reacted to form insoluble carbonate compounds, such as calcium carbonate, that became sediment. Some produced compounds in the oceans that became part of the shells and skeletons of sea creatures. When these creatures died, their shells and skeletons fell into sediment on the sea floor. Over millions of years, all this sediment became sedimentary rocks, such as limestone which is mainly calcium carbonate.

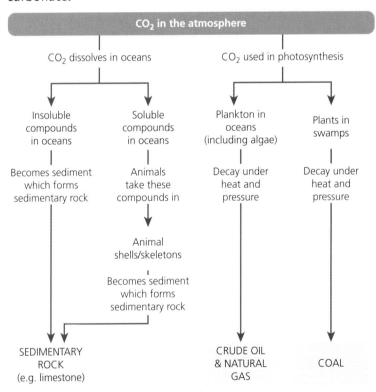

▶ **Figure 9.6** Carbon dioxide has been removed from the atmosphere in many ways.

A lot of carbon dioxide was absorbed by algae and plants for photosynthesis. In the oceans, as algae and other plankton died, their remains were buried in the mud on the sea floor and compressed. Over millions of years this formed crude oil and natural gas that was trapped under rocks. In swamps, the remains of plants were buried and compressed and formed coal, which is a sedimentary rock, over millions of years.

Show you can...

The table shows suggested percentages of different gases found in the atmosphere from early times to present day.

	% in the early atmosphere	% in today's atmosphere
Carbon dioxide	85	0.04
Nitrogen	Very small	78.08
Oxygen	0	20.95

Compare the percentage of each of the following gases in the early atmosphere with the percentage in today's atmosphere. Explain the changes in the percentages of each gas

a) carbon dioxide
b) nitrogen
c) oxygen

Test yourself

1 a) Draw a bar chart to show the main two gases that make up about 99% of the present atmosphere on Earth.
   b) What gases make up the other 1% of the present atmosphere?
2 a) Where may the gases that formed the Earth's early atmosphere have come from?
   b) How may the oceans have been formed?
3 a) What process produced the oxygen in the atmosphere?
   b) Write a word equation for this process.
   c) Write a balanced equation for this process.
4 A lot of the carbon from the carbon dioxide in the Earth's early atmosphere has ended up in the fossil fuels coal, natural gas and crude oil.
   a) Outline how coal was formed.
   b) Outline how natural gas and crude oil were formed.
5 Limestone is a sedimentary rock that is mainly calcium carbonate. Give two sources of the sediment that forms limestone.

Greenhouse gases

◯ What are greenhouse gases?

The Sun gives off radiation. Some of this reaches the Earth and we call this sunlight. This sunlight contains electromagnetic radiation in the ultraviolet, visible and infrared regions. Much of this radiation passes through the gases in the atmosphere and reaches the surface of the Earth. This provides energy and warmth to the planet.

The Earth also gives off radiation. As the Earth is cooler than the Sun, the radiation given off by the Earth has a longer wavelength than the Sun's radiation. The radiation given off by the Earth is in the infrared region.

Some of the gases in the atmosphere absorb infrared radiation given off by the Earth but do not absorb the radiation from the Sun. These are known as greenhouse gases and are important for keeping the Earth warm (Figure 9.7). Many occur naturally and include water vapour (H_2O), carbon dioxide (CO_2) and methane (CH_4) (Figure 9.8).

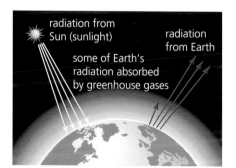

radiation from Sun (sunlight)
radiation from Earth
some of Earth's radiation absorbed by greenhouse gases

▲ Figure 9.7

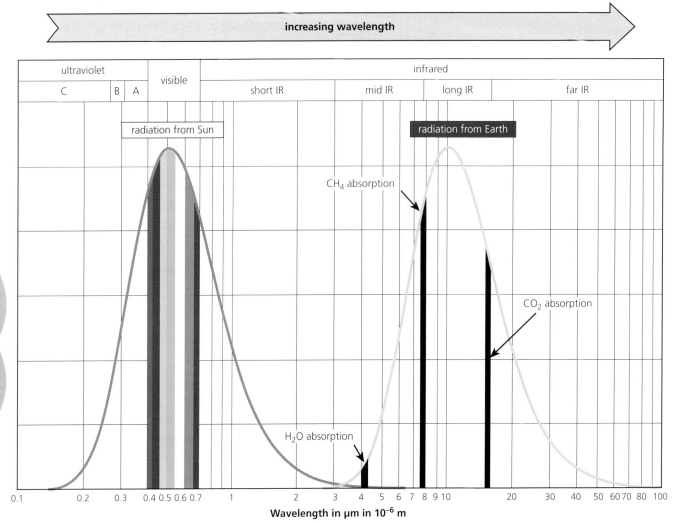

▲ Figure 9.8

⬭ The increase in the amount of greenhouse gases in the atmosphere

The amount of water vapour in the atmosphere varies with weather conditions, and human activities do not have much impact. However, human activities are leading to an increased amount of other greenhouse gases, such as carbon dioxide and methane, in the atmosphere.

The amount of carbon dioxide in the atmosphere

There is a lot of accurate data about the amount of carbon dioxide in the air. It shows clearly that the amount is steadily increasing (Figure 9.9).

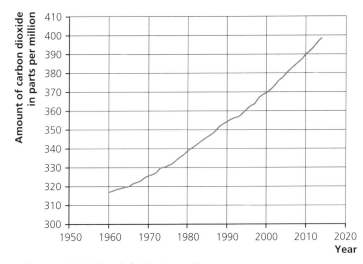

▲ **Figure 9.9** CO_2 levels in the atmosphere.

One reason why the amount of carbon dioxide is increasing is that very large quantities of fossil fuels are being burned. Carbon dioxide is produced when fossil fuels are burned. Figure 9.10 shows how the amount of fossil fuels being burned has increased in recent years. This increase matches the increase in the amount of carbon dioxide in the atmosphere.

▶ **Figure 9.10** Global carbon emissions from burning fossil fuels.

A second reason for the increase in carbon dioxide is deforestation (Figure 9.11). Trees remove carbon dioxide from the air for photosynthesis. In recent years, there has been significant deforestation on the planet as many forests have been cut down but not replaced.

The amount of methane in the atmosphere

There is also data that shows that the amount of methane in the air is increasing (Figure 9.12).

▲ **Figure 9.11** There has been large scale deforestation in the last century.

Animal farming produces a lot of methane. Methane is a product of normal digestion in animals. Most of this methane comes from cattle due to the nature of their digestive system (Figure 9.13). It is also produced when manure decomposes. The amount of animal farming has increased in recent years and this is responsible for some of the increase in methane in the atmosphere.

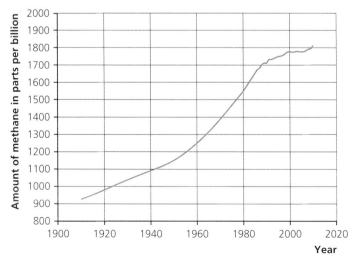

▲ **Figure 9.12** Methane (CH_4) levels in the atmosphere.

▲ **Figure 9.13** Cattle produce a lot of methane.

229

▲ **Figure 9.14** Methane gas is produced by rubbish decomposing in landfill sites.

Show you can...

Look again at Figure 9.9. Which of the following statements can be made using the information from this graph alone?

a) **The concentration of carbon dioxide has risen steadily since 1960.**

b) **There has been no decrease in carbon dioxide concentration over the past 50 years.**

c) **The concentration of carbon dioxide has increased very rapidly over the last 20 years.**

d) **The carbon dioxide concentration increases as the amount of fossil fuels burned increases.**

KEY TERMS

Peer-reviewed When scientific research is studied and commented on by experts in the same area of science to check that it is scientifically valid.

Global warming An increase in the temperature at the Earth's surface

▲ **Figure 9.15** Scientists publish their work in scientific journals.

A large amount of waste is buried in the ground in landfill sites (Figure 9.14). Underground the waste decomposes and this also produces a lot of methane gas.

Test yourself

6 What is different, in terms of wavelength, about the radiation the Earth receives from the Sun and the radiation the Earth gives off?

7 a) What does a greenhouse gas do?

b) Name three important greenhouse gases in the Earth's atmosphere.

8 a) What is happening to the amount of carbon dioxide released into the atmosphere?

b) Give two reasons for this change.

9 a) What is happening to the amount of methane released into the atmosphere?

b) Give two reasons for this change.

◯ Global warming and climate change

Scientific opinions on global warming

Scientists publish their research in scientific journals for other scientists and the general public to see (Figure 9.15). Before it can be published, this work is **peer-reviewed**. This means that it is examined by other scientists who are experts in the same area of science to check that it is scientifically valid.

Based on published peer-reviewed evidence, many scientists believe that the Earth will become warmer due to the human activities that are increasing the amount of greenhouse gases in the atmosphere. They believe that an increase in the temperature at the Earth's surface, known as **global warming**, will result in global climate change. However, models that predict the climate are simplifications as there are so many factors that affect the climate and some of these factors are not fully understood. This means that it is very difficult to model the climate fully and there is some uncertainty in this area. For example, there are some scientists who do not believe that the increasing levels of greenhouse gases will make the Earth warmer.

Due to the uncertainty of climate models and significant public interest, there is much speculation and a wide range of opinions are presented in the media on climate change. Some of this may be biased, for example being put forward by industries that use fossil fuels or that promote renewable energy sources. Also, some may be based on incomplete evidence, perhaps using evidence from one scientific study on its own that does not agree with the majority of published studies.

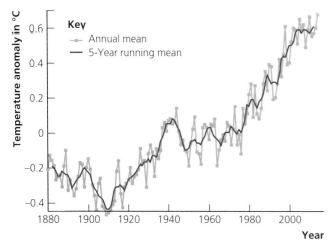

▲ **Figure 9.16** Global mean surface temperature change.

Global warming

There is a great deal of evidence from many sources that the temperature at the Earth's surface has increased in recent years, possibly by about 0.5°C in the last 30 years (Figure 9.16). Most scientists believe that this is largely due to the increased amount of greenhouse gases in the atmosphere.

Climate change and its effects

It is very difficult to predict the effects of the increasing surface temperature of the Earth. Some possible effects are given below.

Sea level rise

As the Earth becomes warmer, sea levels are expected to rise. This may be due to polar ice caps and glaciers melting (Figure 9.17), leading to increased volume of water in the oceans. Another significant factor is that as the water in the oceans becomes warmer, it expands and so has a greater volume. It is difficult to predict by how much sea levels may rise, but there are estimates it could rise by as much as a metre over this century.

(a)

(b)

▲ **Figure 9.17** Glaciers (ice rivers) are in retreat as they are melting (Mer de Glace, Chamonix, France). Photographs taken in (a) 1875 and (b) 2013.

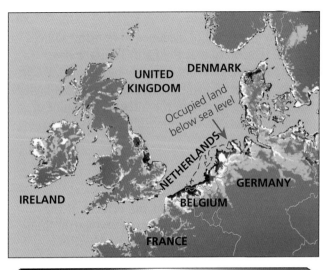

▲ Figure 9.18 Some areas at sea level are at risk of flooding.

As sea levels rise, some parts of the world are at risk of flooding (Figure 9.18). For example, there are some Pacific islands that could be completely submerged. Large areas of countries such as Vietnam, Bangladesh and the Netherlands could be flooded. Some major cities such as London and New York could be at risk in tidal surges if sea levels are higher.

Increased sea levels will also increase coastal erosion. This is where the sea wears away the rock along the coast. Figure 9.19 show the effect of coastal erosion in Sussex.

Storms

As the Earth warms and the climate changes, there may be more frequent and severe storms (Figure 9.20). However, this is uncertain and there is mixed evidence that this is happening. For example, while there is evidence that there is more rain falling in the most severe storms, there is mixed evidence as to whether the number of hurricanes is increasing or not.

(a)

(b)

▲ Figure 9.19 Coastal erosion (Birling Gap, Sussex). Photographs taken in (a) 1906 and (b) 2012.

▲ Figure 9.20 Hurricane Katrina led to massive flooding in New Orleans, USA in 2005.

KEY TERM

Water stress A shortage of fresh water.

▲ Figure 9.21 Water stress has impact on humans, other animals and crops.

Rainfall

It is expected that global warming will affect the amount, timing and distribution of rainfall. This could increase rainfall in some parts of the world but decrease it in others. For example there has been more rainfall in northern Europe in recent years but there has been less rainfall in regions near the equator. The number of heavy rainfall events has increased in recent years, but so has the number of droughts, especially near the equator.

Temperature and water stress

As the temperature and climate changes, it is expected that this will have an impact on the availability of fresh water. This causes water stress which is a shortage of fresh water and affects all living creatures (Figure 9.21). There have been more droughts in areas near the equator in recent years and this leads to water shortages for humans but also for plants and animals affecting food chains.

▲ **Figure 9.22** Climate change could reduce the population of polar bears.

Wildlife

Changes in the climate affect wildlife in many ways. For example, plants may flower earlier, birds may lay eggs earlier and animals may come out of hibernation earlier. Some species may migrate further north as temperatures rise. For example, there are species of dragonfly now being seen in the south of the UK that had only been seen in warmer countries previously. The number of polar bears is expected to fall significantly as the temperature increases as there are fewer suitable places to live (Figure 9.22).

Food production

The weather has a very significant impact on crop production. Changes in the climate may affect the capacity of some regions to produce food due to changes in rainfall patterns, drought, flooding, higher temperatures and the type and number of pests in the region.

Test yourself

10 Before a scientist can publish their research it is peer-reviewed. What is peer review in this context?
11 **a)** What is global warming?
 b) What do most scientists believe is causing global warming?
12 One impact of global warming could be rising sea levels.
 a) Give two reasons why global warming would cause sea levels to rise.
 b) Why are rising sea levels a problem?
 c) Rising sea levels could increase coastal erosion. What is coastal erosion?
13 **a)** What is water stress?
 b) Why might global warming cause water stress?
14 **a)** In what ways could global warming affect global climate?
 b) Why could global warming affect food production?
15 In what ways could global warming affect wildlife?

○ Carbon footprint

What is a carbon footprint?

The amount of carbon dioxide and other greenhouse gases in the air is increasing and likely to be causing global warming and climate change. Therefore it is important that we monitor the amount of greenhouse gases released by different activities. A carbon footprint is defined as the amount of carbon dioxide and other greenhouse gases given out over the full life cycle of a product, service or event.

For example, the carbon footprint of a plastic bag made from poly(ethene) could include carbon dioxide released as fuels are burned to provide energy/heat to:

- drill for crude oil
- heat and vaporise crude oil in fractional distillation
- heat alkanes to crack them to make ethene
- provide heat and pressure to polymerise ethene to make poly(ethene)
- transport plastic bags to shops
- transport used to take plastic bags to waste disposal sites.

In addition, if the bag is burned in an incinerator after disposal then more carbon dioxide is produced.

KEY TERM

Carbon footprint The amount of carbon dioxide and other greenhouse gases given out over the full life cycle of a product, service or event.

Ways to reduce a carbon footprint

In order to prevent the amount of greenhouse gases in the atmosphere increasing further, it is important to reduce the carbon footprint of products, services and events. Some ways in which this can be done are shown in Table 9.1.

Table 9.1

Method of reducing carbon footprint	Comments
Increase the use of alternative energy sources	• We can reduce the use of fossil fuels and increase the use of alternative energy sources such as wind turbines, solar cells and nuclear power. These alternative sources have much lower carbon footprints than burning fossil fuels
Energy conservation	• There are many things we can do in a more energy efficient way that conserves energy. Some examples include: – using more energy efficient engines in cars and other vehicles – increasing insulation in homes – using more energy efficient boilers in heating systems – using low energy (LED) light bulbs instead of filament or halogen light bulbs – using better detergents so clothes can be washed at lower temperatures – switching off electrical devices instead of leaving them on standby
Carbon capture and storage (Figure 9.23)	• A significant amount of the carbon dioxide produced by burning fossil fuels in a power station can be captured to prevent it being released into the atmosphere. This carbon dioxide can then be moved by pipeline and stored deep underground in rocks. This is a new technology that is being developed but some plants are in operation and it is thought it could reduce carbon dioxide emissions from power stations by up to 90%
Carbon off-setting	• Carbon off-setting is where something is put in place to reduce the emission of greenhouse gases to compensate for the release of greenhouse gases elsewhere. Some examples of carbon off-sets include: – setting up renewable energy projects, e.g. wind farms – setting up plants to prevent the emission into the atmosphere of methane from landfill sites – planting more trees
Carbon taxes and licences	• Governments can put a carbon tax on the use of fossil fuels. For example, in many countries there is a large tax on petrol and diesel for cars. As this makes the fuel very expensive for drivers, car-makers are producing more fuel efficient engines so that less fuel is needed to meet the demands of car owners. Carbon licences can also be used where the amount of greenhouse gases a company can release is limited
Carbon neutral fuels (Figure 9.24)	• There are fuels and processes that are carbon neutral. This means that their use results in zero net release of greenhouse gases to the atmosphere. For example, ethanol made from fermentation of crops can be used as a fuel – it is carbon neutral because it releases the same amount of carbon dioxide when it burns as the crops it was made from took in for photosynthesis as they grew

Problems of reducing carbon footprints

There is a need to reduce carbon footprints, but there are some problems that make the reduction in greenhouse gas emissions difficult (Table 9.2).

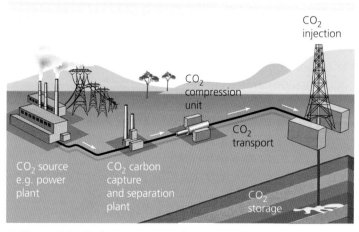

▲ **Figure 9.23** Carbon capture and storage.

▲ **Figure 9.24** 85% of the fuel in E85 fuel is ethanol which is a carbon neutral fuel.

Table 9.2

Problem	Explanation
Scientific disagreement	• Scientists do not all agree about the causes and consequences of global climate change. For example, there are some scientists who do not believe that the extra greenhouse gases in the air are responsible for global warming. This makes efforts to reduce greenhouse gas emissions more difficult as some scientists say there is no need to do so
Economic considerations	• Many of the methods of reducing carbon footprints are expensive. For example, carbon capture and storage significantly increases the cost of generating electricity by burning fossil fuels. There is some reluctance to put these costly methods into place while there is some disagreement about the cause of climate change. The extra cost would be particularly difficult for poor countries in the developing world

Incomplete international co-operation	• The Kyoto Protocol is a major international treaty agreed by many nations to reduce emissions of greenhouse gases. This has been successful in some countries with, for example, greenhouse gas emissions in the UK falling by about 20% since 1990 (Figure 9.25).
	• However, not all countries are keen to do this. For example:
	  – some countries did not sign up;
	  – developing nations such as China and India (countries where over a third of the world's population live) are exempt;
	  – the USA, which emits more greenhouse gases than any other nation, has not agreed to this which means it is not required to cut its emissions
	  – Canada has withdrawn from the agreement
	  – greenhouse gas emissions have actually increased in some countries that agreed to the treaty
	• The original Kyoto Protocol expired in 2012, but the Paris Agreement was negotiated by many countries in 2015 to cut greenhouse gas emissions. It is hoped that enough countries will sign up to this so that it has an impact.

Lack of public information and education	• Many people are confused and do not understand the issues of greenhouse gases and climate change. This problem is even greater in developing countries. The better educated and informed people are about the issues, the more likely that effective action will be taken to reduce the problem

Lifestyle changes	• The world's energy consumption is rapidly increasing. This is partly because there are more people on the planet but also due to our greater demand for energy as our standard of living increases and lifestyle changes. For example:
	  – we are using more electrical devices (e.g. smart phones, tablets, dishwashers, tumble driers)
	  – we expect to be warmer in our homes than in the past
	  – we are travelling more (e.g. longer commutes to work, travelling further on holiday, making more overseas business visits)

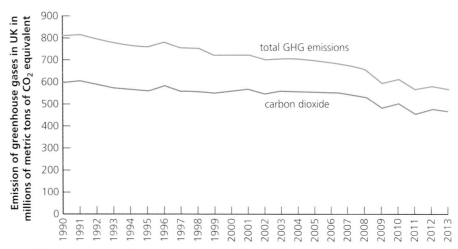

▲ **Figure 9.25** Greenhouse gas (GHG) emission in the UK 1990–2013.

Test yourself

16 What is meant by the term carbon footprint?

17 There are many ways to reduce our carbon footprint.

  a) One way is to use more alternative energy sources. Give some examples of these.

  b) Another way is to be more energy efficient. Give some examples of ways in which this can be done.

  c) Carbon capture and storage could reduce our carbon footprint a great deal. What happens in this process?

  d) Carbon off-setting can be used. Give some ways in which this can be done.

  e) Explain how carbon taxes on fuel for cars has led to cars with more fuel efficient engines being produced.

  f) Biodiesel is a carbon neutral fuel made from crops such as rapeseed. Explain what it means to be carbon neutral.

18 Explain why disagreement among scientists about the causes of global warming may limit steps to lower greenhouse gas emissions.

19 Worldwide energy consumption is increasing rapidly. Why might our improving lifestyles contribute to this increase?

Common atmospheric pollutants and their sources

○ What is formed when fuels burn?

Most fuels, including coal, oil and gas, contain carbon and/or hydrogen and may also contain some sulfur.

When fuels burn, they are oxidised which means that they react with oxygen and the atoms in their molecules combine with the oxygen. For example, when fuels containing hydrogen burn, the hydrogen combines with oxygen to form water vapour. When fuels containing sulfur burn, the sulfur combines with oxygen to form sulfur dioxide.

For fuels containing carbon, complete combustion takes place in a good supply of oxygen to form carbon dioxide. In a poor supply of oxygen, incomplete combustion takes place which gives carbon monoxide and/or soot (a form of carbon).

Table 9.3 gives some equations which are examples of reactions taking place when fuels burn.

Table 9.3

Substance in fuel	Products from combustion	Combustion reactions
Hydrogen, H_2 (an important fuel)	H_2 → oxygen → H_2O	$2H_2 + O_2 \longrightarrow 2H_2O$ hydrogen + oxygen $\longrightarrow$ water
Methane, CH_4 (the main compound in natural gas)	CH_4 → oxygen / oxygen → CO_2 or CO or C ; H_2O	$CH_4 + 2O_2 \longrightarrow CO_2 + 2H_2O$ (complete combustion) methane + oxygen $\longrightarrow$ carbon dioxide + water $2CH_4 + 3O_2 \longrightarrow 2CO + 4H_2O$ (incomplete combustion) methane + oxygen $\longrightarrow$ carbon monoxide + water $CH_4 + O_2 \longrightarrow C + 2H_2O$ (incomplete combustion) methane + oxygen $\longrightarrow$ carbon + water
Ethanol, C_2H_6O (a common alcohol used as a fuel)	C_2H_6O → oxygen / oxygen → CO_2 or CO or C ; H_2O	$C_2H_6O + 3O_2 \longrightarrow 2CO_2 + 3H_2O$ (complete combustion) ethanol + oxygen $\longrightarrow$ carbon dioxide + water $C_2H_6O + 2O_2 \longrightarrow 2CO + 3H_2O$ (incomplete combustion) ethanol + oxygen $\longrightarrow$ carbon monoxide + water $C_2H_6O + O_2 \longrightarrow 2C + 3H_2O$ (incomplete combustion) ethanol + oxygen $\longrightarrow$ carbon + water
Carbon disulfide, CS_2 (small amounts are found in oil)	CS_2 → oxygen / oxygen → CO_2 or CO or C ; SO_2	$CS_2 + 3O_2 \longrightarrow CO_2 + 2SO_2$ (complete combustion) carbon disulfide + oxygen $\longrightarrow$ carbon dioxide + sulfur dioxide $2CS_2 + 5O_2 \longrightarrow 2CO + 4SO_2$ (incomplete combustion) carbon disulfide + oxygen $\longrightarrow$ carbon monoxide + sulfur dioxide $CS_2 + 2O_2 \longrightarrow C + 2SO_2$ (incomplete combustion) carbon disulfide + oxygen $\longrightarrow$ carbon + sulfur dioxide
Ethanethiol, C_2H_6S (small amounts are found in oil)	C_2H_6S → oxygen / oxygen / oxygen → CO_2 or CO or C ; H_2O ; SO_2	$2C_2H_6S + 9O_2 \longrightarrow 4CO_2 + 6H_2O + 2SO_2$ (complete combustion) ethanethiol + oxygen $\longrightarrow$ carbon dioxide + water + sulfur dioxide $2C_2H_6S + 7O_2 \longrightarrow 4CO + 6H_2O + 2SO_2$ (incomplete combustion) ethanethiol + oxygen $\longrightarrow$ carbon monoxide + water + sulfur dioxide $2C_2H_6S + 5O_2 \longrightarrow 4C + 6H_2O + 2SO_2$ (incomplete combustion) ethanethiol + oxygen $\longrightarrow$ carbon + water + sulfur dioxide

○ Pollution from burning fuels

The burning of fuels is a major source of pollutants in the atmosphere. Table 9.4 shows some of these, how they are formed and how the problems they cause could be reduced.

Table 9.4

Product of combustion	How it is formed	Problems it causes	How the problem could be reduced
Carbon dioxide (CO_2)	Most fuels contain carbon and this produces carbon dioxide on complete combustion	It is a greenhouse gas and is thought to be causing global warming and climate change	Reduce the use of fossil fuels by using alternative energy sources and improving energy efficiency
Carbon monoxide (CO)	Most fuels contain carbon and this can produce carbon monoxide on incomplete combustion	It is toxic because it combines with haemoglobin in blood reducing the ability of the blood to carry oxygen It is also colourless and has no smell so can be hard to detect	Ensure there is a good supply of oxygen when fuel is burned
Soot (C) (Figure 9.26)	Most fuels contain carbon and this can produce soot (carbon particulates) on incomplete combustion	These carbon particulates pollute the air and blacken buildings They also cause global dimming which reduces the amount of sunlight reaching the Earth's surface They can damage the lungs and cause health problems to humans	Ensure there is a good supply of oxygen when fuel is burned
Water vapour (H_2O)	Many fuels contain hydrogen and this produces water vapour on combustion	It is not a problem as although it is a greenhouse gas, human activity does not appear to affect the amount of water vapour in the air as weather patterns prevent the amount of water vapour changing over time	It is not a problem
Sulfur dioxide (SO_2)	Many fuels contain some sulfur and this produces sulfur dioxide on combustion	This is a cause of acid rain which damages plants and stonework It also causes respiratory problems for humans	Remove the sulfur dioxide from the waste gases before they reach the atmosphere
Nitrogen oxides (NO and NO_2)	Nitrogen in the air reacts with oxygen in air at very high temperatures when fuels are burned	This is a cause of acid rain It also causes respiratory problems in humans	Remove the nitrogen oxides from the waste gases before they reach the atmosphere (this is done using catalytic converters in cars) Adjust the way the fuel is burned to reduce the amount produced
Unburned hydrocarbons	When hydrocarbon fuels are burned (e.g. petrol, diesel), some molecules may not burn and so unburned hydrocarbons are released into the atmosphere	This wastes fuel and they are greenhouse gases	Ensure efficient burning so that all the fuel is burned

▲ **Figure 9.26** Tower Bridge has been cleaned to remove layers of soot caused by air pollution.

Show you can...

Burning petrol in car engines can lead to the formation of acid rain

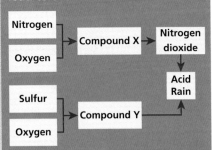

a) What is the formula of compound X?
b) What is the formula of nitrogen dioxide?
c) What is the source of the nitrogen and oxygen that react together to form compound X?
d) What does nitrogen dioxide react with to form acid rain?
e) What is the formula of compound Y?
f) State two ways in which acid rain can be prevented.

Test yourself

20 Write both a word equation and balanced equation to show what happens when each of the following substances undergoes complete combustion.
  a) hydrogen, H_2
  b) propane, C_3H_8
  c) butanol, $C_4H_{10}O$
  d) thiophene, C_4H_4S

21 Petrol is an important fuel used in cars.
  a) One of the chemicals in petrol is isooctane (C_8H_{18}). Under what conditions can the following gases be formed when isooctane burns:
    i) carbon dioxide
    ii) carbon monoxide
    iii) soot
    iv) water vapour
  b) Explain why some sulfur dioxide may be formed when petrol burns.
  c) Explain why some nitrogen monoxide and nitrogen dioxide may be formed when petrol burns.
  d) What two problems do the pollutants sulfur dioxide and nitrogen oxides cause in the air?
  e) Some unburned isooctane can be given off in exhaust fumes from cars. Why might this be a problem?

22 a) Carbon monoxide can be formed on the incomplete combustion of methane (CH_4).
    i) What is incomplete combustion?
    ii) Why is carbon monoxide toxic?
    iii) Why is it difficult to detect carbon monoxide?
  b) Soot can be formed on the incomplete combustion of methane (CH_4).
    i) What is soot?
    ii) What problems can soot cause?

Practical

Investigating the products of combustion of a fuel

The experiment shown in the diagram was carried out to investigate the products of the combustion of a hydrocarbon candle wax. In the experiment the products of the combustion were drawn through the apparatus by the vacuum pump.

Questions

1 Explain why a colourless liquid forms at the bottom of the U tube.

2 The colourless liquid in the U tube changes white anhydrous copper sulfate to blue hydrated copper sulfate. Suggest the name of the colourless liquid.

3 Liquid A was clear and colourless at the start of the experiment and it slowly became cloudy. Identify liquid A and explain why it became cloudy.

4 From the results of this experiment identify two of the combustion products formed when the hydrocarbon burns completely.

5 Some solid black particles were found on the apparatus at the end of the experiment. Suggest the identity of these particles and suggest how they are formed.

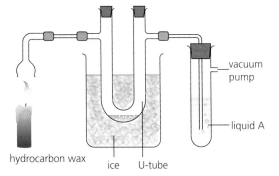

6 The experiment was repeated using a different fuel, which contained sulfur as an impurity.
  a) Name the gas formed when sulfur is burned in air.
  b) This gas is acidic. Suggest what happens to it as it passes through the apparatus.
  c) Suggest why burning sulfur containing fuels is a problem for the environment.

Chapter review questions

1 Copy and complete the gaps in the paragraph below.

The Earth is thought to be 4.6 _____ years old. There is _____ evidence about the atmosphere of the Earth when it was young. One theory is that there was a lot of _____ activity when the Earth was young and this gave out the gases that formed the early atmosphere.

2 a) The air today contains 78% nitrogen, 21% oxygen and 1% other gases. Draw a bar chart to show the composition of air today.

b) What happened on Earth that led to the formation of oxygen in the atmosphere?

3 a) What is a greenhouse gas?

b) Name three greenhouse gases and state where each one comes from.

c) The amount of greenhouse gases in the air is increasing. What impact is this having on the Earth's temperature?

d) State three ways in which this change in temperature could affect the global climate.

4 Some scientists believe that the gases that formed the atmosphere of the Earth when it was young came from volcanoes. The table below shows the proportions of gases given off by a present-day volcano.

Gas	Percentage
Water vapour	38
Carbon dioxide	50
Sulfur dioxide	10
Other gases	2

a) Draw a pie chart to show the gases given off by the present-day volcano.

b) Some scientists believe that the oceans were formed from water vapour given off by ancient volcanoes. How could the oceans have been formed from this water vapour?

5 Scientists believe that the main gas in the atmosphere when it was young was carbon dioxide. The evolution of life is thought to have changed the Earth's atmosphere when algae and plants evolved that photosynthesise.

a) Which two planets in the solar system have atmospheres that are mainly carbon dioxide?

b) Write a word equation and balanced equation for photosynthesis.

6 Copy and complete the gaps in the paragraph below.

Limestone and coal are both _____ rocks that contain carbon. Limestone is mainly the compound _____. Limestone was formed from the skeletons and _____ of marine organisms that died and fell into sediment.

7 When petrol burns in an engine, several pollutants are formed. For each one, explain how it is formed and one problem it can cause.

a) sulfur dioxide (SO_2)

b) nitrogen oxides (NO and NO_2)

c) carbon dioxide (CO_2)

d) carbon monoxide (CO)

e) soot (C)

8 a) Methane (CH$_4$) is a greenhouse gas. Describe two ways in which human activities are increasing the amount of methane in the atmosphere.

   b) Carbon dioxide (CO$_2$) is a greenhouse gas. Describe two ways in which human activities are increasing the amount of carbon dioxide in the atmosphere.

9 There is a lot of work taking place to reduce the carbon footprint of many processes.

   a) What does the term carbon footprint mean?

   b) Describe three ways in which carbon footprints can be reduced.

   c) Describe three problems that can prevent attempts to reduce carbon footprints.

10 Write a balanced equation for the complete combustion of each of the following fuels.

   a) hydrogen, H$_2$

   b) heptane, C$_7$H$_{16}$

   c) methanol, CH$_3$OH

11 There is a small amount of diethyl sulfide (C$_4$H$_{10}$S) found in the petrol fraction of crude oil used as a fuel in cars. Write a balanced equation for the complete combustion of diethyl sulfide.

12 The Sun's radiation is in the ultraviolet, visible and near infrared region of the electromagnetic spectrum. The radiation given out by the Earth is in the infrared region.

   a) How does the wavelength of the Earth's radiation compare to the Sun's?

   b) Why are carbon dioxide, methane and water vapour greenhouse gases?

Practice questions

1 Which one of the following makes up about 20% of the Earth's atmosphere? [1 mark]

A carbon dioxide   B helium

C nitrogen   D oxygen

2 A company has decided to reduce the amount of sulfur dioxide from the waste gases which it emits from its factories. Which one of the following substances in solution would most effectively remove the sulfur dioxide from the waste gases? [1 mark]

A calcium chloride   B calcium hydroxide

C sodium chloride   D sulfuric acid

3 Mars is often called the red planet due to the presence of haematite, which contains iron(III) oxide, on its surface. A recent study of the Huygens Crater on Mars has also shown the presence of iron(III) hydroxide and calcium carbonate.

a) i) Calcium carbonate and iron(III) hydroxide undergo thermal decomposition. What is meant by the term thermal decomposition? [2 marks]

   ii) Write a balanced equation for the thermal decomposition of iron(III) hydroxide into iron(III) oxide and water. [2 marks]

b) Atmosphere is the term used to describe the collection of gases that surround a planet. The suggested composition of the atmosphere of Mars is shown in the table below. Compare the composition of the Earth's atmosphere today, with that of the planet Mars. [4 marks]

Gas	Composition in %
Carbon dioxide	95.0
Nitrogen	3.0
Noble gases	1.6
Oxygen	Trace
Methane	Trace

c) Changes in the atmosphere of the Earth occurred slowly over millions of years due to photosynthesis and other processes.

   i) The equation for the production of glucose ($C_6H_{12}O_6$) in photosynthesis is shown below. Balance this equation. [1 mark]

   $$\_\_CO_2 + \_\_H_2O \rightarrow C_6H_{12}O_6 + \_\_O_2$$

   ii) Photosynthesis is an endothermic reaction. Explain why this reaction is endothermic in terms of breaking and making bonds. [3 marks]

   iii) State one other process which caused the composition of the Earth's atmosphere to change. [1 mark]

4 The table shows some information about different fossil fuels.

Fossil fuel	Appearance	% Carbon	% Moisture	Thermal energy released when burned
Natural gas	Colourless gas	75	0	High
Crude oil	Black liquid	80	0	High
Peat	Soft brown fibrous solid	20	70	Low
Lignite	Soft dark solid	55	35	Medium
Coal	Hard shiny black solid	90	0	High

a) Using the table state the relationship between thermal energy released when burned and

   i) % carbon [1 mark]

   ii) % moisture [1 mark]

b) State three differences between lignite and coal. [3 marks]

c) Name two products formed when coal burns incompletely. [2 marks]

d) If the coal contained sulfur, name one other product that may be formed during combustion, and state two environmental problems which may occur. [3 marks]

e) Some lignite is burned in a test tube and a gas given off. The gas can be condensed to give a liquid by passing the gas into a cooled test tube. The test tube is cooled in a beaker of iced water.

   i) Draw a labelled diagram to illustrate the assembled apparatus described above. [3 marks]

   ii) Name two products for the complete combustion of lignite. [1 mark]

5 a) There is much international concern that an increase in atmospheric concentrations of carbon dioxide and methane may lead to global warming and climate change.

   i) Give the formulae of carbon dioxide and methane. [2 marks]

   ii) What type of radiation is absorbed by carbon dioxide and methane molecules? [1 mark]

   iii) What is meant by global warming? [1 mark]

   iv) Suggest why scientists are more concerned about carbon dioxide as a greenhouse gas than methane. [1 mark]

b) Scientists in the Antarctic have measured the concentration of carbon dioxide in air bubbles in the ice there, and this has allowed them to estimate the atmospheric concentration of carbon dioxide over many thousands of years, as shown in graph 1. Graph 2 shows the change in average temperature of the Earth's surface over the same time period as graph 1.

Do the graphs show that an increase in atmospheric carbon dioxide concentration increases global warming? Explain your answer. [2 marks]

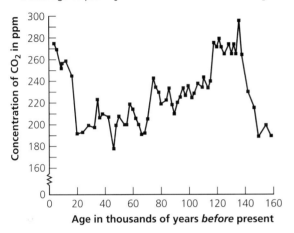

Graph 1

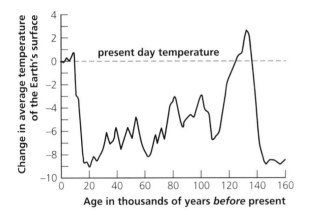

Graph 2

c) Chemists are developing methods to minimise climate change. State two methods which have been used to minimise climate change by removing carbon dioxide. [2 marks]

d) Governments have set targets to reduce emissions of carbon dioxide. One way of doing this is by issuing carbon licences to businesses. Describe how carbon licences could be used to reduce carbon dioxide emissions. [1 mark]

Working scientifically: Communicating scientific conclusions

When drawing conclusions from data about burning fossil fuels and air pollution, it is important to realise that a number of different influences can affect the data and these need to be taken into consideration. There are many things which a scientist must do to produce valid results and conclusions.

▶ A scientist may take a set of measurements or make some observations and draw conclusions from them. However to make valid conclusions the experiment must be repeated and similar data obtained.

▶ Getting other people to successfully repeat the experiment or achieving the same results with a slightly different method means that the results are reproducible. This increases the validity of the conclusion.

▶ Scientists publish their results in journals. Before their work can be published, it is checked and evaluated by other experts in the same field. This process is known as peer review. Scientists also present their ideas at conferences where other scientists can review and evaluate their ideas.

Cold fusion

> Nuclear fusion in a test tube developed by Utah professors – *Financial Times*

> Scientists pursue endless power source – *The Times*

> Scientists claim techniques to control nuclear fusion – *Boston Globe*

If you had read the newspaper headlines shown above in March 1989, you might have believed that the world's energy generation problems could be over for ever. Two scientists, Stanley Pons and Martin Fleischmann, called a press conference to announce that they had successfully caused nuclear fusion to happen using simple laboratory electrolysis apparatus at room temperature. This became known as "cold fusion".

Nuclear fusion involves two atomic nuclei overcoming the repulsion between them and joining together to make a larger nucleus. It is only known to happen at very high temperatures. This fusion releases huge amounts of energy and is the energy source in stars (Figure 9.27) where it takes place at about 15 million °C. Reports that fusion was occurring at room temperature received worldwide media attention and raised hopes of a cheap and abundant source of energy.

Pons and Fleischmann did not follow the normal route of publishing scientific work. They announced their findings at a press conference before

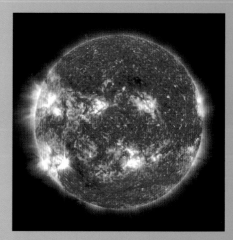

▲ Figure 9.27 Fusion reactions take place at high temperature inside the Sun.

other scientists had reviewed them or tried to repeat their work. It was only after their announcements that details of the experimental method were published. Other scientists tried to replicate the work of Fleischmann and Pons but they were unable to produce the same results. Soon, the work of Fleischmann and Pons was discredited and the hope for cold fusion gone.

Questions

1 What was unusual about the way in which Pons and Fleischmann announced their results?

2 What procedures should a scientist follow before presenting their findings?

3 The concentration of particulates in the air of a town centre was measured over several days. The number of patients seeking medical treatment for asthma was recorded over the same days. The results were plotted in the graph shown.

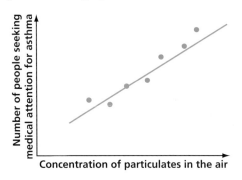

a) What correlation does the graph show?

b) Evaluate the validity of the data.

c) The results of this experiment were presented to a newspaper. The headline claims:

Asthma is caused by particulates in the air.

i) How much confidence can be placed in the newspaper claim?

ii) The particulates that pollute the air are made of carbon. Explain how carbon particulates get into the air.

10 Using the Earth's resources

The planet we live on provides all the raw materials for making the products we use in everyday life. Many of these products are made from resources such as rocks, crude oil, the oceans and the air. Some of these resources will run out if we keep using them as we do. This chapter looks at some ways in which we make use of the Earth's resources and how we can use them in a more sustainable way.

Specification coverage

This chapter covers specification points 4.10.1 to 4.10.4 and is called Using the Earth's resources.

It covers using the Earth's resources and obtaining potable water, life cycle assessment and recycling, using materials, the Haber process and the use of NPK fertilisers.

Previously you could have learned:

› Useful products are made from the raw materials found on the Earth.

› Metals are extracted from ores found in rocks; the method used to extract the metal depends on the reactivity of the metal.

› Alloys are mixtures of metals with small amounts of other metals or carbon; alloys are harder than pure metals.

› Polymers (plastics) are long chain molecules made from joining lots of short molecules together.

› Some chemical reactions are reversible and can reach a state of equilibrium in a closed system.

Test yourself on prior knowledge

1 What raw materials do we use to make each of the following?
   a) metals
   b) drinking water
   c) oxygen
   d) cotton

2 Some metals are extracted from their ores by heating with carbon and some are extracted by electrolysis.
   a) Identify two metals that could be extracted from their ores by heating with carbon.
   b) Identify two metals that could be extracted from their ores by electrolysis but not by heating with carbon.

3 a) What is an alloy?
   b) Give one property of an alloy that is different from the metal it is made from.

4 a) What is a polymer?
   b) Name two polymers.

5 Some reactions are reversible. What does this mean?

Using the Earth's resources

○ The Earth's resources

Everything that we need for life is provided by the Earth's resources. These resources include:

● rocks in the ground
● fuels such as coal, oil and natural gas found underground
● plants and animals (produced by agriculture)
● fresh water and sea water
● air
● sunlight
● wind.

Table 10.1 shows some examples of how we use these resources to provide energy and essential substances for everyday life.

Table 10.1

What we need	Where it comes from
Oxygen	Air
Water	Rain
Food	Plants and animals
Clothes	Fibres made from chemicals in oil (e.g. nylon, polyesters) or from natural fibres from plants (e.g. cotton) or animals (e.g. wool)
Shelter / buildings	Building materials include: stone, sand, bricks made from clay, cement made from limestone and clay, timber from trees
Warmth	Burning fuels (e.g. coal, oil, gas, biofuels from plants)
Electricity	Generated using fuels (e.g. coal, oil, gas, biofuels from plants), wind, sunlight, waves, decay of radioactive substances, etc.
Fuel for transport	Burning fuels (e.g. oil, natural gas, biofuels from plants)
Medicines	Mainly made from chemicals found in oil and/or plants
Fertilisers	From plant and animal waste or made from nitrogen in the air, water, natural gas and minerals from the ground
Metals	Made from ores in the ground
Polymers (plastics)	Mainly made from chemicals found in oil

Many substances, such as metals, medicines, plastics and some clothing fibres, are made from natural resources by chemical reactions. In fact, chemistry is all about the production of useful substances from the Earth's natural resources.

As time goes on, chemistry is playing an important role in improving agricultural and industrial processes. For example:

- new products with new uses are being made (e.g. new medicines)
- new products to supplement natural ones are being made (e.g. polyesters, nylon, Gore-Tex® and microfibre for clothing fibres to add to natural fibres such as wool, cotton and silk) and
- new products or improving processes (e.g. better fertilisers or more efficient processes to make them).

○ Sustainable development

Many of the Earth's resources are finite. This means that we cannot replace them once we have used them. For example, supplies of crude oil and many metal ores are limited and if we continue to use them as we do now we will run out. In contrast to this, some resources are renewable, which means we can replace them once we have used them. Biofuels, which are fuels made from plants, are good examples of renewable resources. Examples of biofuels include ethanol and biodiesel that can be made from crops and so can be replaced once used.

Sustainable development is where we use resources to meet the needs of people today without preventing people in the future from meeting theirs. Table 10.2 shows some ways in which we can meet our needs today in more sustainable ways.

Table 10.2

	Unsustainable way	Sustainable alternative
Metals	Throw away metals and then use metal ores to provide more metal	Recycle metals
Fuels for transport	Use fossil fuels (e.g. petrol and diesel from crude oil)	Use biofuels (e.g. biodiesel and ethanol made from crops)
Electricity generation	Use fossil fuels (e.g. coal and natural gas)	Use renewable energy sources (e.g. solar, wind, tidal) (Figure 10.1)

KEY TERMS

Finite resource A resource that cannot be replaced once it has been used.

Renewable resource A resource that we can replace once we have used it.

Sustainable development Using resources to meet the needs of people today without preventing people in the future from meeting theirs.

▲ **Figure 10.1** Using the wind to generate electricity is sustainable.

○ Ways to reduce the use of resources

We use large amounts of metals, glass, building materials, clay ceramics and many plastics, but they are all produced from limited raw materials. Much of the energy used in their production comes from resources that are also limited, such as fossil fuels. Many of the raw materials used to make these products, such as rocks and ores, are obtained by mining or quarrying which also have environmental impact.

If we, as the end-users of a product, reuse or recycle products then we make processes more sustainable as it will reduce:

- the use of the Earth's limited resources
- energy consumption and the use of fuels
- the impact of manufacturing new materials on the environment
- the impact of waste disposal on the environment.

Reuse

Some products, such as glass bottles, can be reused. For example, most glass milk bottles are collected, washed and reused (Figure 10.2). Some glass beer bottles are also reused in a similar way.

Recycling

Many products cannot be reused, but the materials they are made of can be recycled. This usually involves melting the products and then remoulding or recasting the materials into new products. Some examples are shown in Table 10.3.

▲ **Figure 10.2** Milk bottlee are reused.

Table 10.3

Materials	How they are recycled
Metals	• Separated (e.g. iron/steel can be removed with a large magnet) • Melted • Recast/reformed into new metal products
Glass	• Separated into different colours • Crushed • Melted • Remoulded into new glass products
Plastics	• Separated into different types • Melted • Remoulded into new plastic products

Materials for some uses have to be of very high quality. For example, glass used for making pans for cooking and metals used for making tools must be of very high quality to have the required properties. For other uses, such as plastic for making rulers, the quality of the material is less important. In general, when using recycled products to produce materials, the better the separation of the materials when they are recycled, the higher the quality of the final material produced.

When metals, glass or plastics are recycled, the different types are separated. For example:

- iron/steel, aluminium and copper are common metals that are separated from other metals
- colourless, brown and green glass are separated
- plastics such as high-density poly(ethene), low density poly(ethene), PET and PVC are separated (Figure 10.3).

| PET | HDPE | PVC | LDPE | PP | PS |
| PET | high density poly(ethene) | PVC | low density poly(ethene) | poly(propene) | polystyrene |

▶ **Figure 10.3** Some symbols on plastic products to help separation when recycling.

A good example of recycling is the use of scrap steel to make new steel (Figure 10.4). Steel is made using mainly iron which is extracted from iron ore in a blast furnace. By adding scrap steel to some iron from a blast furnace, the amount of iron that needs to be extracted to make steel is reduced.

◯ Life cycle assessment

A life cycle assessment is carried out to assess the impact of a product on the environment throughout its life. This includes the extraction of raw materials, its manufacture, its use and its disposal at the end of its useful life. Factors that are considered include the

- use of and sustainability of raw materials (including those used for the packaging of the product)
- use of energy at all stages
- use of water at all stages
- production and disposal of waste products (including pollutants) at all stages
- transportation and distribution at all stages.

Simple life cycle assessments of the use of plastic and paper to make shopping bags are shown in Table 10.4 for comparison.

▲ **Figure 10.4** Some scrap steel can be used to make new steel.

KEY TERM

Life cycle assessment An examination of the impact of a product on the environment throughout its life.

Table 10.4

		Plastic shopping bags	Paper shopping bags
Raw materials	What they are	Crude oil	Trees
	Sustainability	Not sustainable – crude oil cannot be replaced	Sustainable – more trees can be planted (but take a long time to grow)
	Obtaining raw materials	Extracting crude oil uses lots of energy	Habitats are destroyed as trees are cut down
	Transporting raw materials	Transport of crude oil uses up fuels and causes some pollution; potential damage to environment from spillages	Transport of logs uses up fuels and causes some pollution
Manufacture of bag		Much energy used to separate crude oil into fractions, for cracking and for polymerisation	Uses a lot of water; uses some harmful chemicals (leaks would damage the environment)
Use of bags	Transport to where used	Transport of bags uses up fuels and causes some pollution	Transport of bags uses up fuels and causes some pollution
Disposal options	Landfill	Does not rot and so remains in ground for many years	Rots releasing greenhouse gases including methane
	Incinerator	Gives off greenhouse gas carbon dioxide when burned	Gives off greenhouse gas carbon dioxide when burned
	Recycled	Transport of bags uses up fuels and causes some pollution; melting of plastic uses energy	Transport of bags uses up fuels and causes some pollution; relatively easy to recycle

In a life cycle assessment, the use of raw materials, energy and water plus the production of some waste can be quite easy to quantify. However, it can be difficult to quantify the effects of some pollutants. For example, it is hard to judge how much damage to the environment is caused by the release of the greenhouse gas methane from the rotting of paper bags in landfill sites. This means that a life cycle assessment involves some personal judgements to make a value judgement to decide whether use of a product is good or not. This means that the process is not completely objective and so judgements about the benefits or harm of the use of a product may be a matter of opinion.

Life cycle assessments can be produced that do not show all aspects of a product's manufacture, use and disposal. These could be misused, for example by the manufacturer of a product to support the use of their product.

Show you can...

The table shows data about plastic and glass milk bottles.

	Energy needed: non-reusable plastic milk bottle in MJ	Energy needed: reusable glass milk bottle in MJ
Manufacture	4.7	7.2
Washing, filling and delivering	2.2	2.5

a) How many times must the glass bottle be reused before there is an energy saving, compared to the plastic milk bottle?
b) Create a life cycle assessment for the glass bottle and for the plastic bottle.

Test yourself

4 a) Glass bottles can be reused or recycled. What is the difference?
  b) What are the stages in the recycling of glass?
5 Describe how some scrap steel can be used in the manufacture of steel using iron from a blast furnace.
6 How does the way in which waste plastics are separated before melting down to make recycled plastic affect the usefulness of the recycled plastic formed?
7 What is a life cycle assessment?
8 Create a life cycle assessment for copper water pipes that will be recycled at the end of their useful life.
9 The manufacturer of a product included a life cycle assessment on their website. Why should we be cautious about this life cycle assessment?

The use of water

○ Producing potable water

Types of water

Water is essential for life and humans need water that is safe to drink (Figure 10.5). Water that is safe to drink is called potable water. Potable water is not pure water as it contains some dissolved substances, but these are at low levels that are safe. There should also be safe levels of microbes in potable water and ideally none.

Potable water and other types of water are described in Table 10.5.

▲ **Figure 10.5** Humans need water that is safe to drink.

Table 10.5

Type	Description	Contents	
		Dissolved substances	Microbes
Pure water	Water that contains only water molecules and nothing else	✗	✗
Potable water	Water that is safe to drink	✓ low levels	✗ (or very low levels)
Fresh water	Water found in places such as lakes, rivers, the ice caps, glaciers and underground rocks and streams	✓ low levels	✓ (very low from some sources)
Ground water	Fresh water found underground streams and in porous rocks (aquifers)	✓ low levels	✓
Sea water	Water in the seas and oceans	✓ high levels	✓
Waste water	Used water from homes, industry and agriculture	✓ high levels	✓

Water treatment

In the UK we produce our potable water from fresh water that comes from rain. This rain collects in lakes, rivers and reservoirs. It also collects in the ground in underground streams and in the pores of some rocks which are called aquifers.

This fresh water can be treated to produce potable water. In some parts of the UK, fresh water comes mainly from rivers and reservoirs, but in other areas much water comes from ground water sources. There are many stages in making this water safe to drink, but two of the main stages are shown in Table 10.6.

Table 10.6

Stage	What it does	Details
Filtration	Removes solids	The water is passed through filter beds. These filter beds are usually made of sand. As the water flows through the sand, any solids in the water are removed
Sterilisation	Kills microbes	The most common way is to use small amounts of chlorine. However, the microbes can also be killed by treating the water with ozone or by passing ultraviolet light through the water

In some parts of the world, such as Saudi Arabia, the United Arab
Emirates (UAE) and parts of Spain, there is little fresh water but there
is lots of sea water. Potable water can be made from sea water by
desalination. Two methods of doing this are distillation and reverse
osmosis (Table 10.7).

Table 10.7

Method	How it works	Diagram
Distillation	Sea water is heated so that it boils. The water molecules are turned to steam leaving behind the dissolved substances. The water vapour is then cooled and condensed	
Reverse osmosis	Sea water is passed through a semipermeable membrane using pressure. The water molecules pass through the membrane but many of the dissolved substances cannot. This is the opposite of normal osmosis where water would move in the opposite direction	

Both methods require a lot of energy. In distillation the energy is
needed to boil the water. In reverse osmosis the energy is needed
to create the pressure. Reverse osmosis is becoming more popular
in places with sea water but little fresh water as it is cheaper
than distillation. However, due to the high energy costs both
methods are more expensive than producing potable water from
fresh water.

Test yourself

10 What is potable water?

11 a) In the UK, potable water is produced from fresh water. What is fresh water?

b) A key step in the production of potable water is filtration. Why and how is this done?

c) Another key step in the production of potable water is sterilisation. Why is this done? Give three ways in which this is done.

12 a) What is desalination?

b) More potable water is produced by desalination in Saudi Arabia than in any other country. Why is water produced this way in Saudi Arabia?

c) Describe how potable water is produced from sea water by distillation.

d) Describe how potable water is produced from sea water by reverse osmosis.

13 a) Give three sources of waste water.

b) For each of the following stages of sewage treatment, describe what happens.

i) screening

ii) sedimentation

iii) aerobic treatment of effluent

iv) anaerobic treatment of sludge

Show you can...

	Waste water	Ground water	Salt water
Method of producing potable water			Desalination by distillation or reverse osmosis

a) Copy and complete the table to give details of the methods of converting the different types of water into potable water. One column has been completed for you.

In 2010 the Thames Water desalination plant, the first and only desalination plant in the UK, opened and is able to produce 1.5% of the UK's water requirement. Israel has many desalination plants and 40% of its water requirement is supplied in this way. Singapore, once heavily dependent on Malaysia for its water supply, now recycles sewage and waste water and uses it as a primary supply of potable water.

b) State and explain which method of obtaining potable water is primarily used in the UK.

c) Suggest reasons why there is a difference in the percentage use of desalination in the UK and in Israel.

d) Suggest why using waste water has been important for Singapore.

Metals and other materials

○ Metals

Metals are very important materials that have many uses. For example:

- cars, trains and planes are made from metal
- large buildings have metal frameworks
- all electrical devices have metal circuits and wires
- tools are made from metal.

Metals are extracted from compounds found in rocks (see Chapter 4). The metal compounds are found in rocks are called ores. An ore is a rock from which a metal can be extracted for profit. There are several ways to extract the metals from rocks but the method used depends on the reactivity of the metal.

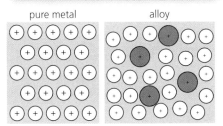

▲ **Figure 10.7** Alloys contain atoms of a different size.

Alloys

Pure metals such as aluminium, copper, gold and iron are too soft for many uses. They can be made harder and so more useful by turning them into alloys. An alloy is a mixture of a metal with small amounts of other metals or carbon.

Pure metals are soft because all the atoms are the same size and so layers of atoms can slide over each other. Alloys are harder because there are some different sized atoms present which makes it much more difficult for atoms to slide (Figure 10.7).

Some examples of alloys are described in Table 10.8.

Table 10.8

Alloy	Main metal	Other elements	Uses	Comments
Bronze	Copper	Tin	To make statues and decorative objects	2p and 1p coins were made from bronze until 1992
Brass	Copper	Zinc	To make water taps, door fittings and musical instruments (e.g. trumpet)	The copper in brass kills bacteria and so brass is used for many door handles and hand rails in public places
Gold alloys	Gold	Silver / copper / zinc	To make jewellery	The colour of gold alloys depends on which metals are mixed with the gold
Steels	Iron	Carbon (usually)	To make cars, building frameworks, cutlery, tools	There are many different types of steel with different properties and uses
Aluminium-magnesium alloys	Aluminium	Magnesium	To make aeroplanes (Figure 10.8)	These alloys have a low density

▲ **Figure 10.8** Many aeroplanes are made from aluminium-magnesium alloys.

▲ **Figure 10.9** Gold jewellery is made from gold alloys.

Gold alloys

The purity of gold alloys is measured in carats. Pure gold is 24 carat. An alloy containing 75% gold would be 18 carat. Jewellery is not made from pure gold because it is too soft and would lose its shape and any engravings (Figure 10.9)

$$\% \text{ of gold in alloy} = \frac{\text{number of carats}}{24} \times 100$$

Example

An alloy of gold is found to be 21 carats. Calculate the percentage of gold in this alloy.

Answer

$\% \text{ of gold in alloy} = \frac{\text{number of carats}}{24} \times 100 = \frac{21}{24} \times 100 = 87.5\%$

Example

An alloy of gold is found to contain 90% gold. Calculate how many carats this alloy has.

Answer

$\text{number of carats} = \frac{24 \times \% \text{ of gold in alloy}}{100} = \frac{24 \times 90}{100} = 21.6 \text{ carats}$

Steels

Steels are alloys of iron mixed with small amounts of carbon and/or other metals. Steels are used more than other alloys due to their low cost and high strength.

▲ **Figure 10.10** Car body panels are made from low carbon steel.

There are many different steel alloys, each with its own properties and containing specific amounts of carbon and/or other metals. Three common types of steels are shown in the Table 10.9.

Table 10.9

Type of steel	Contents	Properties	Uses
High carbon steels	Iron + 2% carbon	Hard but brittle	Saw blades
Low carbon steels	Iron + 0.1% carbon	Soft and easily shaped	Car bodies (Figure 10.10)
Stainless steels	Iron + 11% chromium + some nickel	Hard and does not corrode	Sinks, cutlery

Test yourself

14 From where are metals extracted?

15 **a)** What is an alloy?

   b) Why are pure metals made into alloys?

16 Identify each of the following alloys.

   a) a mixture of iron with carbon

   b) mixture of copper with zinc

   c) mixture of copper with tin

   d) mixture of iron with chromium and some nickel

17 **a)** What percentage of gold is there in a 20 carat gold alloy?

   b) What is the carat rating of a gold alloy containing 80% gold?

   c) Why is jewellery not made from pure gold?

18 **a)** Why are sinks made from stainless steel?

   b) Why are saw blades made from high carbon steel?

   c) Why are car body panels made from low carbon steel?

KEY TERMS

Corrosion The destruction of materials by chemical reactions with substances in the environment.

Rusting The corrosion of iron or steel.

Show you can...

The table shows five different alloys A, B, C, D and E.

Alloy	Elements present in the alloy	Properties
A	Lead and tin	Melts at 203°C
B	Bismuth, cadmium, lead, tin	Melts at 70°C
C	Carbon, iron, tungsten	Hard, unaffected at high temperatures
D	Copper and zinc	Golden colour, does not tarnish
E	Aluminium and lithium	Low density, high strength

Which alloy A, B, C, D or E is best used in each case. Explain your answer.

a) joining electric wire
b) making jewellery
c) making aircraft bodywork
d) making a drill for bricks and stone

TIP

Most metals corrode but only iron and steel rust. This is because rusting is defined as the corrosion of iron and steel. It is wrong therefore to say that any other metals rust when they corrode.

▲ **Figure 10.11** A patch of rust on a car.

Corrosion of metals

Corrosion is the destruction of materials by chemical reactions with substances in the environment. Many metals corrode when they come into contact with oxygen and/or water.

What is rusting?

Many metals corrode, but **rusting** is specifically the corrosion of iron or steel (Figure 10.11).

When iron (or the iron in steel) corrodes, it reacts with oxygen and water to produce rust.

$$\text{iron} + \text{water} + \text{oxygen} \rightarrow \text{rust}$$

This can be shown by experiment where nails are placed in a series of boiling tubes under different conditions (Table 10.10).

Table 10.10

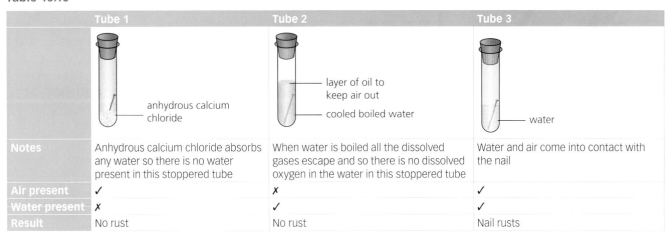

	Tube 1	Tube 2	Tube 3
	anhydrous calcium chloride	layer of oil to keep air out / cooled boiled water	water
Notes	Anhydrous calcium chloride absorbs any water so there is no water present in this stoppered tube	When water is boiled all the dissolved gases escape and so there is no dissolved oxygen in the water in this stoppered tube	Water and air come into contact with the nail
Air present	✓	✗	✓
Water present	✗	✓	✓
Result	No rust	No rust	Nail rusts

The nail does not rust in tube 1 where there is no water present. It also does not rust in tube 2 when there is no air present. However, when there is both air and water present the nail rusts. It can also be shown by experiment that it is the oxygen in the air that reacts when a nail rusts in the presence of air and water.

Preventing corrosion

Two basic ways to prevent the corrosion of metals are to put a coating on the surface of the metal to act as a protective layer or to use sacrificial protection where a more reactive metal is attached to the metal (Table 10.11).

(a)

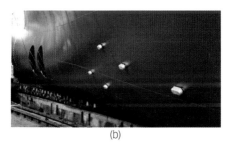
(b)

▲ **Figure 10.12** (a) Layers of paint help to protect a car from corrosion. (b) Blocks of zinc or magnesium are attached to ships to prevent corrosion.

Table 10.11

Method	How it works	Examples
Surface coating (Figure 10.12a)	It acts as a barrier preventing chemicals that cause corrosion coming into contact with the metal	Paint (e.g. car)
		Grease/oil (e.g. bike chain)
		Plastic coating (e.g. fridge shelf)
		Metal coating – applied by electroplating which is a form of electrolysis (e.g. 1p and 2p coins are made of steel which is coated with a layer of electroplated copper)
		Oxide layer (e.g. aluminium has a layer of aluminium oxide on its surface that protects the aluminium from further corrosion)
Sacrificial protection (Figure 10.12b)	The metal is joined to a more reactive metal – the more reactive metal corrodes instead	Zinc or magnesium blocks are attached to the hulls of steel ships to prevent the steel corroding
		Magnesium is attached to steel pipelines and railway lines to prevent corrosion

▲ **Figure 10.13** Steel fire escapes are galvanised to prevent corrosion.

Steel can be galvanised, which combines both of these methods (Figure 10.13). The steel is coated in a layer of zinc. Surface coatings often stop working if the surface is scratched and the metal exposed. However, if the layer of zinc is scratched the steel is still protected by sacrificial protection as the zinc is more reactive.

Test yourself

19 **a)** What is corrosion?
   b) What is rusting?
20 **a)** Write a word equation to show what happens when iron rusts.
   b) Describe an experiment you could do to show that it is the oxygen rather than the nitrogen or argon in air that is required for iron to rust. Describe the results you would expect.
21 Explain how each of the following prevents steel from rusting.
   a) connecting the steel to some magnesium
   b) painting the steel
   c) coating the steel with a layer of chromium by electroplating
   d) galvanising the steel
22 Aluminium does not appear to corrode. Explain why.

Show you can...

Iron can be protected from rusting by sacrificial protection. In an experiment to investigate sacrificial protection, different metals are wrapped around iron nails and left in water for one week.

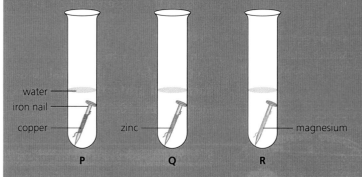

a) State two ways in which you could ensure that this experiment was a fair test.
b) At the end of the week in which tube(s) would rusting have occurred? Explain your answer.

Extraction of copper

The Earth's resources of metal ores are limited. We have already reached a point with copper metal where there are only low-grade ores remaining. Low-grade ores are ones which only contain a small percentage of metal compounds. Due to this, scientists are having to develop new methods to extract copper.

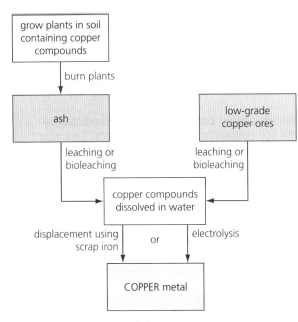

▲ **Figure 10.14** Phytomining.

KEY TERMS

Bioleaching The use of bacteria to produce soluble metal compounds from insoluble metal compounds.

Leachate A solution produced by leaching or bioleaching.

Leaching The use of dilute acid to produce soluble metal compounds from insoluble metal compounds.

Phytomining The use of plants to absorb metal compounds from soil as part of metal extraction.

One way of doing this is to use phytomining to extract copper from soil that contains copper compounds (Figure 10.14). Some plants are very good at absorbing copper compounds from the ground through their roots. These plants can be grown in land containing copper compounds. These plants can then be burned leaving ash that is rich in copper compounds.

Copper can be extracted from this plant ash or from remaining low-grade ores using bioleaching or leaching. Bioleaching uses bacteria whereas leaching uses an acid such as sulfuric acid. In each case, insoluble copper compounds react to produce a solution containing soluble copper compounds. This solution is known as a leachate. Some bacteria are very efficient at this and so bioleaching can be very effective.

The copper can be extracted from the solution containing copper compounds. This can be done by electrolysis or using scrap iron in a displacement reaction (Table 10.12).

Traditional methods of metal extraction involve mining with digging, moving and disposing of large amounts of rock. These new methods for copper extraction are very different and do not involve this and so are better for the environment.

Table 10.12

	Electrolysis	Displacement reaction
What happens	Two inert electrodes (e.g. graphite) are placed in the solution of copper compounds in an electrolysis circuit – copper forms at the negative electrode	Scrap iron/steel is placed in the solution of copper compounds – iron is more reactive than copper and so the copper is displaced by the iron
Equations	Negative electrode: $Cu^{2+}(aq) + 2e^- \rightarrow Cu(s)$	$Cu^{2+}(aq) + Fe(s) \rightarrow Cu(s) + Fe^{2+}(aq)$

Thermosoftening and thermosetting polymers

The atoms within polymer chains are joined together by covalent bonds. In thermosoftening polymers, these polymer chains are not joined together. In thermosetting polymers, the polymer chains are joined to each other by covalent bonds (often called cross-links) (Table 10.15).

Table 10.15

	Thermosoftening polymer	**Thermosetting polymer**
Structure	Contains long polymer chains	Contains long polymer chains
	Chains are not joined to each other (but they are tangled up with each other)	Chains are joined to each other by covalent bonds (often called cross-links)
	polymer chains	polymer chains cross-links between chains
What happens when they are heated	Soften and then melt when heated	Do not soften or melt when heated
Examples	Poly(ethene)	Superglue
	PVC	Epoxy resins
	Poly(propene)	Melamine
	Perspex	Bakelite
	Teflon	
	Polyesters	
	Nylon	

Glass

Glass is used for windows and containers such as bottles. It is a very useful material that is hard, see-through and unreactive. There are many different types of glass with slightly different properties and uses. Two of these are shown in the Table 10.16.

Table 10.16

Type of glass	Soda-lime glass	Borosilicate glass
What it is made from	Sand	Sand
	Sodium carbonate	Boron trioxide
	Calcium carbonate	
Uses	Windows	Laboratory glassware
	Containers (e.g. bottles)	Cookware (e.g. glass pans) (Figure 10.15)
Comment	Most of the glass we use is soda-lime glass	Melts at a higher temperature than soda-lime glass

▲ **Figure 10.15** Cooking pans are made of borosilicate glass.

▲ **Figure 10.16** Bricks are ceramic materials made from baking clay in a kiln.

Clay ceramics

Bricks and pottery such as plates are examples of clay ceramics. These are made by shaping wet clay that is then baked in a furnace or kiln (Figure 10.16). Clay ceramics are very hard, unreactive and resistant to heat.

Composites

Most composite materials are made from two or more different materials. The composite material has different properties from the materials it is made from. In composite materials, fibres or fragments of one material (called the reinforcement) are surrounded by a binder/matrix material that binds them together. For example, in fibreglass, glass fibres are bound together in a polymer. This and some other composites are described in Table 10.17.

Table 10.17

Composite material	Matrix / binder	Reinforcement fibres / fragments	Uses
Fibreglass	Polymer	Glass fibres	Building material, storage tanks
Concrete	Cement and water	Sand and crushed rock	Building material
Natural wood	Lignin	Cellulose fibres	Building material, furniture
Composite wood (e.g. MDF, plywood)	Adhesives	Wood fibres	Building material, furniture
Carbon fibre composites	Polymer	Carbon fibres or carbon nanotubes	Sports equipment (e.g. bikes, tennis racquets, golf clubs) (Figure 10.17)

▲ **Figure 10.17** Many golf clubs are made using carbon fibre composites.

Test yourself

27 a) What is a polymer?
   b) Give two factors that can be changed to produce polymers with different properties.
28 a) What happens when a thermosoftening polymer is heated?
   b) What happens when a thermosetting polymer is heated?
   c) Explain why thermosoftening and thermosetting polymers act differently when heated.
29 a) What is borosilicate glass made from?
   b) What is soda-lime glass made from?
   c) Which is the main type of glass that is used?
   d) Which glass can be used at higher temperatures?
30 a) Give two examples of ceramic materials made from clay.
   b) How are ceramics made from clay?
31 Composite materials are made from a matrix or binder and a reinforcement material.
   a) Fibreglass and carbon fibre composites both use polymers as the binder. What does each composite use as the reinforcement material?
   b) Give two more examples of composite materials.

Show you can...

Select a suitable material from the word box for each of the following uses and give two reasons in each case why it is a suitable material.

thermosetting polymer   thermosoftening polymer   soda-lime glass
borosilicate glass   clay ceramic   carbon fibre composite

a) boiling tube
b) dining plate
c) casing of hair straighteners

Making fertilisers

Fertilisers are used to help crops grow. They supply nutrients that are essential for the growth of plants. The use of fertilisers increases the yield of crops significantly.

The world's population reached 3 billion people in 1959. In 2011, it reached 7 billion people and it is forecast to keep rising in the future. We need to be able to provide enough food to feed everyone on the planet and many scientists argue that we could not do this without the use of fertilisers.

◯ The Haber process

Most fertilisers contain nitrogen compounds which are made from ammonia. Ammonia has the formula NH_3 and is made by the Haber process (Figure 10.18).

In the Haber process, nitrogen reacts with hydrogen. The nitrogen is obtained from the air. The hydrogen is made by reaction of steam with methane from natural gas or other sources (Figure 10.19).

▲ **Figure 10.18** Fritz Haber (1868–1934) invented the Haber process.

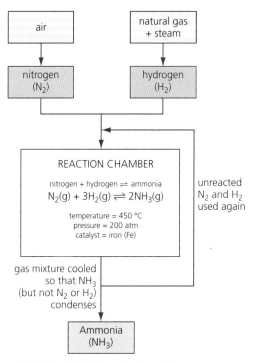

▲ **Figure 10.19** The Haber process.

The purified nitrogen and hydrogen pass into a reaction chamber where they react to form ammonia. However, the reaction is reversible which means that some of the ammonia breaks back down into nitrogen and hydrogen. The reaction is also slow. Due to the reaction being slow and reversible, it is carried out at a temperature of 450°C, a pressure of 200 atmospheres and with an iron catalyst. These conditions are used so that a good yield of ammonia is produced at a good rate.

The mixture of gases that leaves the reaction chamber is cooled so that the ammonia condenses and is removed as a liquid. This separates the ammonia from the leftover nitrogen and hydrogen which can then be used again.

The reaction between nitrogen and hydrogen reaches a state of dynamic equilibrium. The forward reaction is exothermic. In the equation, there are four reactant gas molecules but only two product gas molecules (Figure 10.20).

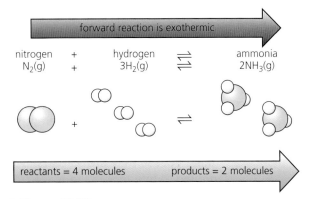

▲ Figure 10.20

The conditions used are a compromise between the yield of ammonia (which depends on the position of the equilibrium), the reaction rate and the cost of raw materials and energy (Table 10.18).

Table 10.18

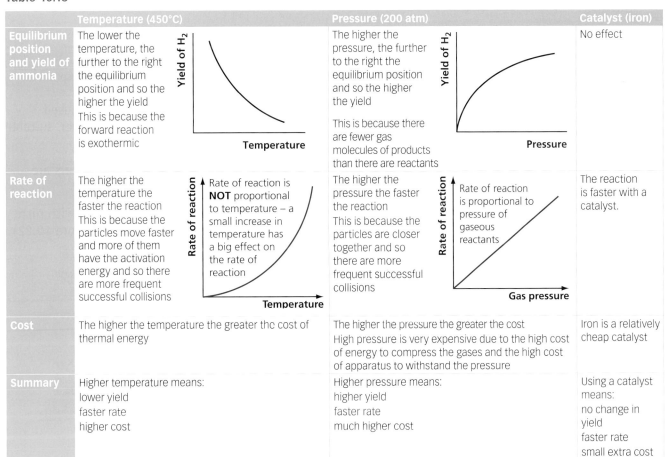

	Temperature (450°C)	Pressure (200 atm)	Catalyst (iron)
Equilibrium position and yield of ammonia	The lower the temperature, the further to the right the equilibrium position and so the higher the yield This is because the forward reaction is exothermic	The higher the pressure, the further to the right the equilibrium position and so the higher the yield This is because there are fewer gas molecules of products than there are reactants	No effect
Rate of reaction	The higher the temperature the faster the reaction This is because the particles move faster and more of them have the activation energy and so there are more frequent successful collisions Rate of reaction is **NOT** proportional to temperature – a small increase in temperature has a big effect on the rate of reaction	The higher the pressure the faster the reaction This is because the particles are closer together and so there are more frequent successful collisions Rate of reaction is proportional to pressure of gaseous reactants	The reaction is faster with a catalyst.
Cost	The higher the temperature the greater the cost of thermal energy	The higher the pressure the greater the cost High pressure is very expensive due to the high cost of energy to compress the gases and the high cost of apparatus to withstand the pressure	Iron is a relatively cheap catalyst
Summary	Higher temperature means: lower yield faster rate higher cost	Higher pressure means: higher yield faster rate much higher cost	Using a catalyst means: no change in yield faster rate small extra cost

In addition, by recycling the leftover nitrogen and hydrogen, less raw materials are needed which saves costs.

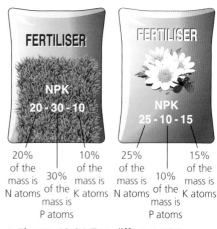

20% of the mass is N atoms | 30% of the mass is P atoms | 10% of the mass is K atoms | 25% of the mass is N atoms | 10% of the mass is P atoms | 15% of the mass is K atoms

▲ **Figure 10.21** Two different NPK fertilisers.

Production and uses of NPK fertilisers

Some water-soluble compounds of nitrogen (N), phosphorus (P) and potassium (K) are used as fertilisers to increase agricultural productivity. This means that they help crops grow and so increase the yield of the crops. Fertilisers that contain compounds of each of these three elements are called NPK fertilisers.

NPK fertilisers are a mixture of compounds in the correct proportions to give the desired N : P : K ratio. There are many different formulations of NPK fertilisers which contain different percentages of nitrogen, phosphorus and potassium. These different formulations are used for different crops and for different soil conditions (Figure 10.21).

Some of the compounds used in NPK fertilisers are shown in Table 10.19. They are all salts. Some of these salts are mined from the ground, while others are made from reactions between acids and bases.

Table 10.19

Name of salt	Contains N	Contains P	Contains K	Source
Ammonium nitrate	✓			Reaction of ammonia with nitric acid
Ammonium sulfate	✓			Reaction of ammonia with sulfuric acid
Ammonium phosphate	✓	✓		Reaction of ammonia with phosphoric acid
Potassium chloride			✓	Mined
Potassium sulfate			✓	Mined
Calcium nitrate	✓			Reaction of phosphate rocks with nitric acid
Single superphosphate (calcium phosphate + calcium sulfate)		✓		Reaction of phosphate rocks with sulfuric acid
Triple superphosphate (calcium phosphate)		✓		Reaction of phosphate rocks with phosphoric acid

Phosphate rock is often used as a raw material. It cannot be used directly as a fertiliser because it is insoluble in water. However, suitable salts can be made when the rock reacts with acids.

Many of the processes to make the compounds in NPK fertilisers are integrated. For example,

- Ammonium nitrate is made from the reaction of ammonia with nitric acid, but nitric acid itself is also made from ammonia (Figure 10.22).

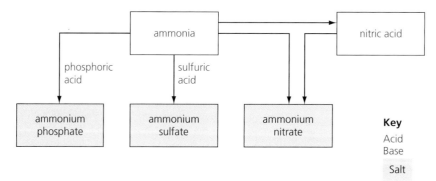

▲ **Figure 10.22** Salts for fertilisers made from ammonia.

● Triple superphosphate fertiliser (calcium phosphate) is made from reaction of phosphate rock with phosphoric acid, which itself is made from reaction of phosphate rock with nitric acid, which is itself made from ammonia (Figure 10.23).

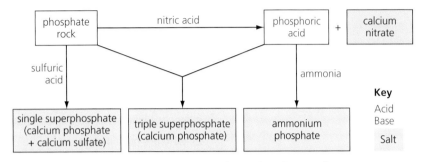

▲ **Figure 10.23** Salts for fertilisers made from phosphate rock.

Test yourself

32 Ammonia is made from reaction of nitrogen with hydrogen in the Haber process.
  a) Give the formula of ammonia.
  b) Write a word equation for this reaction.
  c) Write a balanced equation for this reaction.
  d) This reaction is reversible. What does this mean?
  e) Why is a catalyst used?
  f) From what raw material is the nitrogen obtained?
  g) From what raw materials is the hydrogen made?
  h) How is the ammonia removed from the mixture of gases produced?
  i) What is done with the unreacted nitrogen and hydrogen?

33 Ammonia is made from reaction of nitrogen with hydrogen at a temperature of 450°C, a pressure of 200 atmospheres and in the presence of an iron catalyst. The reaction reaches a state of dynamic equilibrium.
  a) Explain clearly why the temperature used of 450°C is a compromise.
  b) Explain clearly why the pressure used of 200 atmospheres is a compromise.

34 a) What are fertilisers used for?
  b) What are NPK fertilisers?
  c) Ammonium nitrate is in many NPK fertilisers. It is made by reaction of ammonia with nitric acid. Write a word equation and a balanced equation for this reaction.
  d) Many of the compounds in NPK fertilisers are made from phosphate rock. Why can the rock not be crushed up and used as a fertiliser as it is?

Show you can...

Four fertilisers are listed below:
A Easy grow NH_4NO_3
B Epsom salts $MgSO_4$
C Saltpetre KNO_3
D Superphosphate $Ca(H_2PO_4)_2$

a) Give the letter of one fertiliser which contains the element
   i)   nitrogen   ii) potassium
   iii) phosphorus   iv) magnesium
b) How many different elements are present in fertiliser A?
c) Compare the percentage of nitrogen present in fertiliser A and fertiliser C using the formula

$$\text{percentage of nitrogen} = 100 \times \frac{\text{number of nitrogen atoms} \times A_r \text{ of nitrogen}}{M_r \text{ of compounds}}$$

Chapter review questions

1 Here is a list of some of the Earth's raw materials.

ores, crude oil, sea water, air, plants

From which of these raw materials is each of the following products made?

a) oxygen gas　　**b)** plastics　　　**c)** iron　　　　**d)** biofuels

2 State whether each of the following raw materials is finite or renewable.

a) crude oil　　　　　**b)** ores　　　　　　**c)** rain water

d) cotton plants　　　**e)** limestone

3 24 carat gold is 100% pure gold and is too soft to be useful. 18 carat gold is an alloy containing 75% gold and is harder and more useful.

a) Why are pure metals soft?

b) What is an alloy?

c) Why are alloys harder than pure metals?

d) Calculate the percentage of gold in 22 carat gold.

4 Ammonia (NH_3) is made by reaction of nitrogen with hydrogen. Much of the ammonia is used to make fertilisers.

a) From what raw material is nitrogen obtained?

b) From what raw materials is hydrogen obtained?

c) The reaction between nitrogen and hydrogen is reversible. What does this mean?

d) How is ammonia removed from the reaction mixture?

e) What are fertilisers used for?

f) What elements do NPK fertilisers contain?

5 **a)** What is meant by a process that is sustainable?

b) Suggest a way in which each of the following products could be made in a more sustainable way than the way shown here.

i) extracting aluminium from the ore bauxite

ii) using petrol to fuel a car

iii) making a glass wine bottle

iv) generating electricity by burning natural gas (methane)

c) Some products can be reused while others can be recycled. Explain the difference.

6 **a)** Plates are clay ceramics. Describe how plates are made.

b) Produce a life cycle assessment for a plate that is eventually disposed of in a landfill site.

7 **a)** Water that is safe to drink is called potable water. Name two important stages in the production of potable water from fresh water in the UK. For each stage, explain why it is done.

b) In some countries potable water is produced from sea water by desalination. Outline how this can be done by

i) distillation　　　　　　　　　**ii)** reverse osmosis

c) Explain why distillation and reverse osmosis are expensive ways to produce potable water.

8 Copper is a very important metal used for making water pipes and electrical cables. There is a shortage of copper-rich ores and new methods are being developed to extract copper from low-grade ores.

a) Solutions containing copper compounds can be produced from low-grade ores by bioleaching. What is used to bioleach copper compounds in this way?

b) Copper metal can be extracted from solutions containing copper compounds by electrolysis or a displacement reaction with scrap iron.

i) Write a half equation to show the formation of copper at the negative electrode in the electrolysis of copper sulfate.

ii) Write a balanced equation for extraction of iron from copper sulfate solution using iron.

c) Copper can be produced by phytomining. Outline what happens in phytomining.

9 The diagrams below represent the structures of a thermosetting polymer and a thermosoftening polymer.

polymer A   polymer B

a) What is a polymer?

b) Which diagram represents which type of polymer?

c) By considering their structure and bonding, explain why thermosetting polymers do not melt on heating but thermosoftening polymers do.

d) Which type of polymer could be recycled? Explain your answer.

10 In each case below, what has been done differently to produce the two polymers which have different properties?

a) high-density poly(ethene) and low density poly(ethene)

b) PVC and polystyrene

11 Waste water from domestic, industrial and agricultural uses of water has to be treated before the water can safely be returned to the environment. Describe what happens in each of the following stages.

a) screening

b) sedimentation

c) aerobic treatment of effluent

d) anaerobic treatment of sludge

12 Ammonia is made from reaction of nitrogen and hydrogen in the Haber process. This reaction is done at a temperature of 450°C, 200 atmospheres pressure and with an iron catalyst. Explain clearly why each condition is used and why these are compromise conditions.

271

Practice questions

1 Which one of the following is NOT true of the conditions under which the Haber process is conducted? [1 mark]

  A the temperature is approximately 450°C

  B the reactants are at a high pressure

  C the reactants are mixed in the ratio by volume of 3 parts nitrogen to 1 part hydrogen

  D unused reactants are reused to bring about further reaction

2 Which of these methods could be used to obtain a sample of pure water from sea water: [1 mark]

  A chromatography   B crystallisation

  C distillation   D filtration

3 Nitrogenous fertilisers contain ammonium compounds such as ammonium nitrate which is produced when ammonia reacts with nitric acid.

  a) i) Write a balanced symbol equation for the reaction of ammonia with nitric acid to produce ammonium nitrate. [2 marks]

   ii) Suggest why ammonium nitrate is not an NPK fertiliser. [1 mark]

  b) In industry ammonia gas is produced by the Haber process which involves a reversible reaction between the gases nitrogen and hydrogen.

   $$N_2 + 3H_2 \rightleftharpoons 2NH_3$$

   i) State what you understand by the term reversible reaction. [1 mark]

   ii) Where is the nitrogen obtained for this process? [1 mark]

   iii) Name the catalyst used in the Haber process. [1 mark]

   iv) Describe how ammonia is separated from the unreacted nitrogen and hydrogen. [2 marks]

  c) The graph below shows how the percentage yield of ammonia changes with pressure and temperature. Use the graph to answer the following questions.

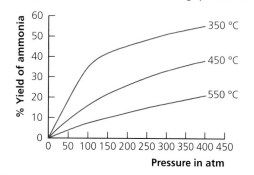

i) State the effect of increasing temperature on the yield of ammonia at constant pressure. [1 mark]

ii) 450°C and 200 atm are commonly used conditions for the Haber Process. What is the percentage yield of ammonia using these conditions? [1 mark]

iii) Which temperature shown on the graph produces the highest yield of ammonia? [1 mark]

iv) Suggest why industry uses 450°C and 200 atm when it is possible to obtain a higher yield of ammonia using a lower temperature and a higher pressure. [4 marks]

4 In the United Kingdom 45 million mobile phone users discard about 15 million handsets every year. Only about 2% of these handsets are recycled, with the remainder going to landfill dumps. Recycling mobile phones means an overall reduction in carbon dioxide emissions as well as saving precious metals such as gold, silver and copper

  a) i) Suggest why it is important to recycle phones to recover metals such as gold, silver and copper. [1 mark]

   ii) Suggest why metals such as gold are used in the circuit boards and wiring of mobile phones. [1 mark]

  b) In the recycling process the phones are first placed in a shredder, and then the shredded material is heated in ovens to incinerate the plastic. The remains from this process enter a melting furnace to produce a metal alloy covered with slag, which consists mainly of silicates. The metal alloy bars from one tonne of mobile phones can contain 2.3 kg of silver and 227 g of gold.

   i) State one advantage and one disadvantage of incineration as a method of disposal of plastics. [2 marks]

   ii) What is meant by the term alloy? [1 mark]

   iii) The silicate ion found in slag has the formula SiO_3^{2-}. Write the formula for copper(II) silicate. [1 mark]

   iv) Calculate the number of moles of silver in one tonne of mobile phones. [1 mark]

   v) The metal alloy bars can be used to produce jewellery. Suggest and explain why alloys are used to make jewellery rather than pure metals. [2 marks]

5 Copper can be found in the Earth's crust as ores containing copper sulfide. Copper ores are becoming scarce and many quarries are exhausted. Large areas of land around exhausted quarries contain low percentages of copper sulfide.

a) i) State one reason why extracting copper from this land using traditional methods is too expensive. [1 mark]

ii) Explain why extracting copper from this land would have a major environmental impact. [1 mark]

b) New methods such as phytomining and bioleaching can be used to extract copper from land containing low percentages of copper sulfide.

i) Describe the process of phytomining. [4 marks]

ii) State two advantages of phytomining over traditional methods of extracting copper. [2 marks]

c) Bioleaching uses bacteria to produce a solution of copper sulfate. It is possible to extract copper from copper sulfate solution using scrap iron.

i) Suggest why it is economical to extract copper using scrap iron. [1 mark]

ii) Write a symbol equation for the reaction of iron and copper sulfate to produce copper. Explain why this reaction occurs. [3 marks]

6 Polymers are produced from crude oil.

a) Name two different types of polymer made from the monomer ethene. [2 marks]

b) What is meant by the term monomer? [1 mark]

c) Compare and contrast thermosoftening polymers and thermosetting polymers by describing the structure, bonding and the effect of heat. [6 marks]

7 There are different types of water. Some of these are listed below.

fresh water, ground water, potable water, waste water

Match each of these types of water to the descriptions below. You should use each answer once.

a) used water from homes, industry and agriculture [1 mark]

b) water found in underground streams and in porous rocks [1 mark]

c) water found in places such as lakes, rivers, the ice caps, glaciers and underground rocks and streams [1 mark]

d) water that is safe to drink [1 mark]

8 Composite materials are made from two or more different materials. They contain a reinforcement material in a matrix material.

a) Identify the matrix and reinforcement material in each of the following composites.

i) carbon fibre composites

ii) fibreglass [2 marks]

b) Identify the following composites.

i) matrix = lignin, reinforcement = cellulose fibres

ii) matrix = adhesives, reinforcement = wood fibres [2 marks]

9 Most metals corrode as they react with oxygen and/or water. The corrosion of iron and steel is called rusting.

a) Explain how a surface layer such as a layer of paint prevents steel from rusting. [2 marks]

b) Explain how attaching a block of magnesium to the bottom of a steel ship prevents it from rusting. [2 marks]

c) Give an example where each of the following methods of rust prevention are used.

i) oil

ii) electroplating

iii) galvanising [3 marks]

10 Phosphoric acid has the formula H_3PO_4 and contains the phosphate ion PO_4^{3-}.

a) Give the formula of the fertiliser ammonium phosphate. [1 mark]

b) Write a balanced equation for the reaction of ammonia with phosphoric acid to produce ammonium phosphate. [2 marks]

c) Single superphosphate fertiliser contains calcium phosphate and calcium sulfate. Give the formula of both compounds. [2 marks]

273

Working scientifically:
Evaluating results and procedures

In everyday life, we are constantly evaluating situations and drawing conclusions. For example when baking muffins, we might evaluate how the finished muffins look and decide how to improve on our method. If the muffins are:

▶ **too pale** – then they probably should have been left in the oven for a longer time.

▶ **burned around the edges** – then a cooler oven should have been used or perhaps the muffins should have been removed from the heat sooner.

▶ **not moist enough** – then perhaps some extra liquid ingredients should have been added or maybe less flour should have been used.

▶ **very heavy** – then it is likely that the ingredients should have been stirred for longer to introduce more air to the mixture.

Evaluating an experimental procedure allows us to assess its effectiveness, to plan for future modifications and to judge whether an alternative method might be more suitable. Part of the evaluation process includes asking questions such as those in the table below.

Questions to consider	Explanation
Are the results **accurate**?	Accurate results are those which are close to the true value
Are the results **repeatable**?	Results are repeatable if similar results are obtained when the same person carries out the same experiment several times
Are the results **reproducible**?	Results are reproducible if similar results are obtained using a different technique or by someone else doing the same experiment
Are the results **precise**?	Precise results are those in which there is little spread around the mean value, i.e. all the results are close to the mean value
Are there any significant **random errors**?	There will be some small variations every time an experiment is done that lead to slightly different results each time. In a well designed, well carried out experiment, these random errors will be small. The impact of random errors is reduced by doing several repeats and finding a mean value
Are there any **systematic errors**?	Systematic errors give results that differ from the true value by a similar amount each time. This is because the same problem is affecting the result each time. Finding a mean value does not reduce the impact of systematic errors
Are the results **valid**?	Valid results are those from a fair test in which only one variable is changed
Was the method suitable?	Was the method easy to carry out? Did it give results that were accurate, repeatable, reproducible and precise, with minimal random errors and no systematic errors?

KEY TERMS ⭐

Accurate Result close to the true value

Repeatable Similar results are achieved when an experiment is repeated by the same person

Reproducible Similar results are achieved when an experiment is repeated by another person or by a different method is used

Precise Measurements which have little spread around the mean value

Systematic error Errors that cause a measurement to differ from the true value by a similar amount each time

Random error Errors that cause a measurement to differ from the true value by different amounts each time

Valid results Results from a fair test in which only one variable is changed

Use the questions above to help you evaluate the following experiments.

Questions

Student A	21.4%	21.2%	21.1%	21.3%	21.4%
Student B	22.5%	22.4%	22.6%	22.5%	22.5%

1 Two students carried out experiments to find the percentage by mass of nitrogen in ammonium sulfate. The true value is 21.2%.

   a) Calculate the mean value for each student including the uncertainty as a ± value.

   b) Comment on the accuracy of the result for each student.

   c) Comment on the repeatability of the result for each student.

   d) Why did the results differ when each student repeated them?

   e) Did either of the students have a systematic error in their experiment? Explain your answer.

2 Susie measured out 25 cm^3 of sodium hydroxide solution into a conical flask using a 25 cm^3 measuring cylinder. She added three drops of indicator and added sulfuric acid rapidly from a burette until the colour of the indicator changed. She found it difficult to see the colour of the indicator in the conical flask placed on the bench. The result of the titration was recorded and the experiment was repeated several times. The results were 23.3, 27.8, 25.4, 25.9 and 22.6 cm^3.

Evaluate the procedure used, calculate the mean volume (including the uncertainty as a ± value) and comment on the precision of the measurements.

3 In an experiment 25 cm^3 of 0.1 mol/dm^3 hydrochloric acid was measured into a flask, placed on a pencil drawn cross on a piece of paper and 25 cm^3 of sodium thiosulfate solution added. A stopwatch was started and the time taken for the cross to disappear from sight was recorded. The experiment was repeated using five different concentrations of hydrochloric acid. A different pencil cross was used each time.

Evaluate the experimental procedure.

4 In an experiment 2.0 g of calcium carbonate chips were added to hydrochloric acid in a conical flask. A stopper with a delivery tube was placed in the flask, a stopwatch started and the volume of carbon dioxide collected under water in an upturned measuring cylinder was recorded every 2 minutes. The experiment was repeated using 2.0 g of calcium carbonate powder.

State any source of error in this experiment and state any improvements which could be made.

Chemists use formulae and equations as a quick way of identifying substances and showing what happens in chemical reactions. Being formula literate is vital for any chemist. This chapter looks at writing formulae and understanding what they mean, as well as how to write chemical, ionic and half equations.

Specification coverage

This chapter is called Formulae and Equations and brings together points from throughout the specification including 4.1.1, 4.2.2, 4.3.1, 4.3.2, 4.4.1, 4.4.2, 4.4.3, 4.5.2, 4.7.1 and 4.8.3.

This chapter can be used when required during the course to teach about formulae and equations, as well as to review all the work later on.

Writing formulae

Chemists use formulae a lot and it is important that you are formula literate meaning that you can write and recognise formulae.

○ Elements

The formula for most elements is just its symbol. For example, the formula of argon is Ar and that of magnesium is Mg (Table 11.1).

However, this is not the case for elements made of molecules. Many of these molecules contain two atoms (called diatomic molecules) such as hydrogen (H_2) and oxygen (O_2). Some elements that are made of molecules contain more than two atoms in their molecules, such as phosphorus molecules which contain four atoms (P_4).

KEY TERM

Diatomic molecule A molecule containing two atoms.

Table 11.1

Common elements whose formula is the symbol				Common elements whose formula is not the symbol			
Al	aluminium	Pb	lead	Br_2	bromine	I_2	iodine
Ar	argon	Li	lithium	C_{60}	carbon (buckminsterfullerene)	N_2	nitrogen
Be	beryllium	Mg	magnesium	Cl_2	chlorine	O_2	oxygen
B	boron	Ne	neon	F_2	fluorine	P_4	phosphorus
Ca	calcium	Ni	nickel	H_2	hydrogen	S_8	sulfur
C	carbon (diamond)	K	potassium				
C	carbon (graphite)	Si	silicon				
Cu	copper	Ag	silver				
Au	gold	Na	sodium				
He	helium	Sn	tin				
Fe	iron	Zn	zinc				

○ Compounds

Some common compounds

It is very useful to know the formula of some common compounds. Some are listed in Table 11.2.

Table 11.2

Common compounds	
NH_3	ammonia
CO_2	carbon dioxide
CO	carbon monoxide
CH_4	methane
NO	nitrogen monoxide
NO_2	nitrogen dioxide
SO_2	sulfur dioxide
SO_3	sulfur trioxide
H_2O	water

Ionic compounds

Compounds made from metals combined with non-metals have an ionic structure. The formula of each of these compounds can be worked out using ion charges. The charges of common ions are shown in Tables 11.3 and 11.4.

Table 11.3 Positive ions.

Group 1 ions (form 1+ ions)		Group 2 ions (form 2+ ions)		Group 3 ions (form 3+ ions)		Others	
Li^+	lithium	Mg^{2+}	magnesium	Al^{3+}	aluminium	NH_4^+	ammonium
Na^+	sodium	Ca^{2+}	calcium			Cu^{2+}	copper(ıı)
K^+	potassium	Ba^{2+}	barium			H^+	hydrogen
						Fe^{2+}	iron(ıı)
						Fe^{3+}	iron(ııı)
						Pb^{2+}	lead
						Ag^+	silver
						Zn^{2+}	zinc

Table 11.4 Negative ions.

Group 6 ions (form 2– ions)		Group 7 ions (form 1– ions)		Others	
O^{2-}	oxide	F^-	fluoride	CO_3^{2-}	carbonate
S^{2-}	sulfide	Cl^-	chloride	OH^-	hydroxide
		Br^-	bromide	NO_3^-	nitrate
		I^-	iodide	SO_4^{2-}	sulfate

In an ionic substance the total number of positive charges must equal the total number of negative charges. This allows us to work out the formula of ionic substances.

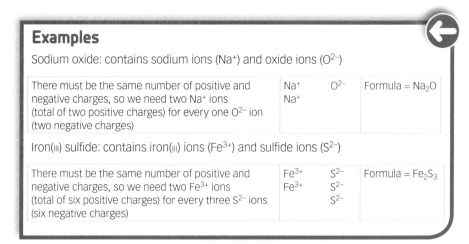

Examples

Sodium oxide: contains sodium ions (Na^+) and oxide ions (O^{2-})

There must be the same number of positive and negative charges, so we need two Na^+ ions (total of two positive charges) for every one O^{2-} ion (two negative charges)	Na^+ Na^+	O^{2-}	Formula = Na_2O

Iron(ııı) sulfide: contains iron(ııı) ions (Fe^{3+}) and sulfide ions (S^{2-})

There must be the same number of positive and negative charges, so we need two Fe^{3+} ions (total of six positive charges) for every three S^{2-} ions (six negative charges)	Fe^{3+} Fe^{3+}	S^{2-} S^{2-} S^{2-}	Formula = Fe_2S_3

Some ions contain atoms of different elements. Examples include sulfate (SO_4^{2-}), hydroxide (OH^-) and nitrate (NO_3^-). These are sometimes called compound ions or molecular ions. If you need to write more than one of these in a formula, then these ions should be placed in a bracket.

Example

Magnesium hydroxide: contains magnesium ions (Mg^{2+}) and hydroxide ions (OH^-)

There must be the same number of positive and negative charges, so we need one Mg^{2+} ion (total of two positive charges) for every two OH^- ions (two negative charges)	Mg^{2+}	OH^- OH^-	Formula = $Mg(OH)_2$

Table 11.5

Name	+ Ions		– Ions		Formula
Sodium chloride	Na^+	(1+ charge)	Cl^-	(1– charge)	NaCl
Magnesium chloride	Mg^{2+}	(2+ charges)	Cl^- Cl^-	(2– charges)	$MgCl_2$
Magnesium sulfide	Mg^{2+}	(2+ charges)	S^{2-}	(2– charges)	MgS
Copper(ɪɪ) sulfate	Cu^{2+}	(2+ charges)	SO_4^{2-}	(2– charges)	$CuSO_4$
Sodium carbonate	Na^+ Na^+	(2+ charges)	CO_3^{2-}	(2– charges)	Na_2CO_3
Ammonium sulfate	NH_4^+ NH_4^+	(2+ charges)	SO_4^{2-}	(2– charges)	$(NH_4)_2SO_4$
Calcium nitrate	Ca^{2+}	(2+ charges)	NO_3^- NO_3^-	(2– charges)	$Ca(NO_3)_2$
Aluminium oxide	Al^{3+} Al^{3+}	(6+ charges)	O^{2-} O^{2-} O^{2-}	(6– charges)	Al_2O_3
Iron(ɪɪɪ) hydroxide	Fe^{3+}	(3+ charges)	OH^- OH^- OH^-	(3– charges)	$Fe(OH)_3$

Test yourself

1 Write the formula of each of the following elements and compounds.
   a) copper
   b) hydrogen
   c) carbon dioxide
   d) argon
   e) silver
   f) oxygen
   g) ammonia
   h) chlorine
   i) carbon (diamond)
   j) carbon (buckminsterfullerene)
   k) sulfur dioxide
   l) methane

2 Write the formula of each of the following ionic compounds.
   a) potassium oxide
   b) sodium sulfate
   c) aluminium fluoride
   d) iron(ɪɪɪ) sulfide
   e) copper(ɪɪ) nitrate
   f) lithium carbonate
   g) ammonium bromide
   h) barium hydroxide
   i) silver nitrate
   j) aluminium sulfate
   k) strontium oxide
   l) potassium selenide

3 Name each of the following substances.
   a) Br_2
   b) Na
   c) Cu
   d) CO
   e) SO_3
   f) CaO
   g) AlF_3
   h) CuS
   i) KNO_3
   j) $(NH_4)_2CO_3$
   k) FeO
   l) Fe_2O_3

Classifying substances

○ Structure types

It is very useful to be able to identify what type of structure a substance has from its name or formula. Table 11.6 gives some general guidance on this.

Table 11.6

Structure type	Description of structure	Which substances have this structure
Monatomic	Made of individual atoms	Group 0 elements
Molecular	Made of individual molecules	Some non-metal elements (e.g. H_2, C_{60}, N_2, O_2, F_2, P_4, Cl_2, Br_2, I_2) Compounds made from non-metals (e.g. CH_4, CO_2, H_2O, NH_3, $C_6H_{12}O_6$)
Giant covalent	Lattice of atoms joined by covalent bonds	Diamond (C), graphite (C), graphene (C), silicon (Si), silicon dioxide (SiO_2)
Ionic	Lattice of positive and negative ions	Compounds made from metals combined with non-metals (e.g. NaCl, Fe_2O_3, $CuSO_4$)
Metallic	Lattice of metal atoms in a cloud of delocalised outer shell electrons	Metals (e.g. Cu, Fe, Al, Na, Ca, Mg, Au, Ag, Pt)

○ Acids, bases, alkalis and salts

Some compounds act as acids, bases, alkalis or salts. It is very useful if you can identify an acid, base, alkali or salt although not all substances are one of these (Table 11.7).

Table 11.7

Acids	Bases	Salts
Substances that react with water to release H^+ ions	Substances that react with acids to form a salt and water (and sometimes carbon dioxide as well)	Ionic substances formed when acids react with bases
Common acids: H_2SO_4 sulfuric acid HCl hydrochloric acid HNO_3 nitric acid H_3PO_4 phosphoric acid CH_3COOH ethanoic acid	*Common bases:* Metal oxides e.g. CaO, Na_2O Metal hydroxides e.g. $Ca(OH)_2$, NaOH Metal carbonates e.g. $CaCO_3$, Na_2CO_3 **Alkalis** Substances that react with water to release OH^- ions (they are a special type of water-soluble base) *Common alkalis:* NH_3 ammonia *plus water-soluble metal hydroxides:* NaOH sodium hydroxide KOH potassium hydroxide $Ca(OH)_2$ calcium hydroxide	*Common salts:* Sulfates from sulfuric acid Chlorides from hydrochloric acid Nitrates from nitric acid Phosphates from phosphoric acid Ethanoates from ethanoic acid Citrates from citric acid

Alkalis are a special type of base and so any substance that is an alkali is also a base.

○ Acid–base character of oxides

Most metal oxides are basic (Table 11.8). For example, calcium oxide (CaO) is used as a base to neutralise acidic soil on farms.

Most non-metal oxides are acidic. For example, carbon dioxide (CO_2) dissolves in rain water to make rain naturally slightly acidic.

Table 11.8

Type of oxide	Metal oxides	Non-metal oxides
Acidic or basic	Basic (react with acids)	Acidic (react with bases)
Examples	Calcium oxide (CaO) Sodium oxide (Na_2O) Copper oxide (CuO)	Carbon dioxide (CO_2) Sulfur dioxide (SO_2) Phosphorus oxide (P_4O_{10})

Test yourself

4 What type of structure does each of the following substances have?
  a) lead (Pb)
  b) argon (Ar)
  c) potassium iodide (KI)
  d) oxygen (O_2)
  e) diamond (C)
  f) methane (CH_4)
  g) ethanol (C_2H_5OH)
  h) aluminium oxide (Al_2O_3)
  i) chromium (Cr)
  j) silicon dioxide (SiO_2)
  k) sulfur dioxide (SO_2)
  l) potassium nitrate (KNO_3)

5 Classify each of these substances as an acid, base, alkali or salt.
  a) Fe_2O_3
  b) Na_2SO_4
  c) KOH
  d) $ZnCO_3$
  e) HNO_3
  f) $Ca(NO_3)_2$
  g) NH_3
  h) K_2O
  i) HCl
  j) $MgCl_2$
  k) NaBr
  l) H_2SO_4

6 Classify each of the following oxides as acidic or basic.
  a) NO_2
  b) K_2O
  c) MgO
  d) SiO_2

Common reactions

There are some general reactions that are useful to know. Many of these involve acids and/or metals. Remember that hydrochloric acid produces chloride salts, sulfuric acid produces sulfate salts and nitric acid produces nitrate salts.

Examples

element + oxygen → oxide of element
  e.g. calcium + oxygen → calcium oxide

compound + oxygen → oxides of each element in compound
  e.g. methane (CH_4) + oxygen → carbon dioxide + water

water + metal → metal hydroxide + hydrogen (for metals that react with water)
  e.g. water + sodium → sodium hydroxide + hydrogen

acid + metal → salt + hydrogen (for metals that react with dilute acids)
  e.g. hydrochloric acid + magnesium → magnesium chloride + hydrogen

acid + metal oxide → salt + water
  e.g. sulfuric acid + copper oxide → copper sulfate + water

acid + metal hydroxide → salt + water
  e.g. nitric acid + potassium hydroxide → potassium nitrate + water

acid + metal carbonate → salt + water + carbon dioxide
  e.g. hydrochloric acid + calcium carbonate → calcium chloride + water + carbon dioxide

acid + ammonia → ammonium salt
  e.g. nitric acid + ammonia → ammonium nitrate

Chemical reactions

Chemical reactions take place in one of the three ways shown in Table 11.9.

Table 11.9

Way in which the reaction takes place	Examples
1 Transfer of electrons	Metals reacting with non-metals e.g. sodium + chlorine → sodium chloride $2Na + Cl_2 → 2NaCl$ Sodium atoms lose electrons to form sodium ions. These electrons are transferred to chlorine atoms which form chloride ions. This forms the ionic compound sodium chloride Displacement reactions e.g. zinc + copper sulfate → zinc sulfate + copper $Zn + CuSO_4 → ZnSO_4 + Cu$ Zinc atoms lose electrons to form zinc ions. These electrons are transferred to copper ions in copper sulfate forming copper atoms.
2 Sharing of electrons	Non-metals reacting with non-metals e.g. hydrogen + oxygen → water $H_2 + O_2 → 2H_2O$ Hydrogen atoms share electrons with oxygen atoms to form covalent bonds in water
3 Transfer of protons	All acids contain H^+ ions. An H^+ ion is simply a proton. When acids react, they transfer H^+ ions to the substance they react with Acids reacting with alkalis e.g. hydrochloric acid + sodium hydroxide → sodium chloride + water $HCl + NaOH → NaCl + H_2O$ H^+ ions (protons) are transferred from the hydrochloric acid to the OH^- ions in sodium hydroxide to form water Acids reacting with metal oxides e.g. sulfuric acid + copper oxide → copper sulfate + water $H_2SO_4 + CuO → CuSO_4 + H_2O$ H^+ ions (protons) are transferred from the sulfuric acid to the O^{2-} ions in copper oxide to form water

Test yourself

7 Write a word equation for each of the following reactions with oxygen.
   a) magnesium (Mg) + oxygen
   b) hydrogen sulfide (H_2S) + oxygen
   c) phosphorus (P_4) + oxygen
   d) silane (SiH_4) + oxygen
   e) propane (C_3H_8) + oxygen
   f) methanol (CH_3OH) + oxygen

8 Write a word equation for each of the following reactions.
   a) potassium + water
   b) nitric acid + zinc
   c) sulfuric acid + nickel oxide
   d) hydrochloric acid + potassium hydroxide
   e) nitric acid + sodium carbonate
   f) hydrochloric acid + ammonia
   g) magnesium hydroxide + sulfuric acid
   h) calcium + water
   i) copper carbonate + nitric acid
   j) ammonia + sulfuric acid
   k) magnesium oxide + nitric acid
   l) cobalt + hydrochloric acid

9 Which of the following reactions involves the
   ● transfer of electrons?
   ● sharing of electrons?
   ● transfer of protons?
   a) nitric acid + sodium oxide → sodium nitrate + water
   b) aluminium + iron oxide → aluminium oxide + iron
   c) hydrogen + sulfur → hydrogen sulfide
   d) aluminium + bromine → aluminium bromide
   e) hydrochloric acid + sodium carbonate → sodium chloride + water + carbon dioxide

Balancing equations

Word equations show the names of the reactants and products in a reaction. A balanced equation shows the formula of each substance and how many particles of each are involved in the reaction. An example of this is shown in Table 11.10.

Table 11.10

Type of equation	Word equation	Balanced equation
Equation	nitrogen + hydrogen → ammonia	$N_2 + 3H_2 \rightarrow 2NH_3$
What it tells us	Nitrogen reacts with hydrogen to form ammonia	One molecule of nitrogen (N_2) reacts with three molecules of hydrogen (H_2) to form two molecules of ammonia (NH_3)

In a balanced equation, the total number of atoms of each element on both sides of the equation must be the same. This is because atoms cannot be created or destroyed. In the equation for the reaction between nitrogen and hydrogen above, there are two nitrogen atoms and six hydrogen atoms in both the reactants and products.

You are often required to write a balanced equation. Here are some steps to follow plus two examples.

Step 1   Write the word equation.

Step 2   Rewrite the equation with formulae (be very careful to ensure the formulae are correct).

Step 3   Count the number of atoms of each element on each side of the equation. If they are the same then the equation is already balanced and nothing more needs to be done.

TIP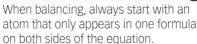

When balancing, always start with an atom that only appears in one formula on both sides of the equation.

Step 4   If the equation is not balanced, then add in extra molecules to try and balance it. You must never change the formulae themselves. For example, you could not change the formula of water from H_2O to H_4O in example 1 to balance the H atoms because it is water that is formed and that has the formula H_2O and not H_4O.

Step 5   Write out the final balanced equation.

Examples

	Example 1	Example 2
Step 1	methane + oxygen → carbon dioxide + water	aluminium hydroxide + nitric acid → aluminium nitrate + water
Step 2	$CH_4 + O_2 \rightarrow CO_2 + H_2O$	$Al(OH)_3 + HNO_3 \rightarrow Al(NO_3)_3 + H_2O$
Step 3	reactants products C = 1 C = 1 H = 4 H = 2 O = 2 O = 3 The equation is not balanced	reactants products Al = 1 Al = 1 O = 6 O = 10 H = 4 H = 2 N = 1 N = 3 The equation is not balanced
Step 4	Add another H_2O to the products (so there are now $2H_2O$) to balance the H atoms: $CH_4 + O_2 \rightarrow CO_2 + 2H_2O$ reactants products C = 1 C = 1 H = 4 H = 4 O = 2 O = 4 Then add another O_2 to the reactants (so there are now $2O_2$) to balance the O atoms: $CH_4 + 2O_2 \rightarrow CO_2 + 2H_2O$ reactants products C = 1 C = 1 H = 4 H = 4 O = 4 O = 4 The equation is now balanced	It is easiest to start with N as it is the only unbalanced atom that is in just one substance on both sides of the equation Add two more HNO_3 to the reactants (so there are now $3HNO_3$) to balance the N atoms $Al(OH)_3 + 3HNO_3 \rightarrow Al(NO_3)_3 + H_2O$ reactants products Al = 1 Al = 1 O = 12 O = 10 H = 6 H = 2 N = 3 N = 3 Then add two more H_2O to the products (so there are now $3H_2O$) to balance the O and H atoms $Al(OH)_3 + 3HNO_3 \rightarrow Al(NO_3)_3 + 3H_2O$ reactants products Al = 1 Al = 1 O = 12 O = 12 H = 6 H = 6 N = 3 N = 3 The equation is now balanced
Step 5	$CH_4 + 2O_2 \rightarrow CO_2 + 2H_2O$	$Al(OH)_3 + 3HNO_3 \rightarrow Al(NO_3)_3 + 3H_2O$

Balanced equations sometimes include state symbols to show the state of each substance:

- (s) = solid
- (l) = liquid
- (g) = gas
- (aq) = aqueous (dissolved in water).

For example, the equation:

$$CaCO_3(s) + 2HCl(aq) \rightarrow CaCl_2(aq) + H_2O(l) + CO_2(g)$$

means that calcium carbonate solid reacts with an aqueous solution of hydrochloric acid to form an aqueous solution of calcium chloride, water liquid and carbon dioxide gas.

Test yourself

10 Magnesium reacts with sulfuric acid as shown:

$Mg(s) + H_2SO_4(aq) \rightarrow MgSO_4(aq) + H_2(g)$

   a) What does the (s) mean?   **b)** What does the (aq) mean?

   c) What does the (g) mean?

11 Balance the following equations.

   a) $K + I_2 \rightarrow KI$ **b)** $Na + H_2O \rightarrow NaOH + H_2$

   c) $CuCO_3 \rightarrow CuO + CO_2$ **d)** $Al + O_2 \rightarrow Al_2O_3$

   e) $Ca + HCl \rightarrow CaCl_2 + H_2$ **f)** $KOH + H_2SO_4 \rightarrow K_2SO_4 + H_2O$

   g) $C_5H_{12} + O_2 \rightarrow CO_2 + H_2O$ **h)** $H_3PO_4 + NaOH \rightarrow Na_3PO_4 + H_2O$

   i) $NH_3 + H_2SO_4 \rightarrow (NH_4)_2SO_4$ **j)** $NO + H_2O + O_2 \rightarrow HNO_3$

12 Write a balanced equation for each of the following reactions.

   a) sodium + oxygen → sodium oxide

   b) propane (C_3H_8) + oxygen → carbon dioxide + water

   c) calcium + water → calcium hydroxide + hydrogen

   d) chlorine + sodium bromide → sodium chloride + bromine

   e) magnesium oxide + nitric acid → magnesium nitrate + water

Ionic equations

In a solid ionic compound, the positive and negative ions are bonded to each other strongly in a lattice. When it dissolves in water, the ions separate and become surrounded by water molecules (Figure 11.1).

When ionic compounds dissolved in water react, it is usual for some of the ions not to react and remain unchanged in the water. These are often called spectator ions as they are present but do not take part in the reaction.

We can write ionic equations for reactions involving ions. These ionic equations only show what happens to the ions that react. We do not include the spectator ions in these ionic equations.

The overall electric charge of the ions in the reactants must equal the overall electric charge of the ions in the products. This can sometimes be useful to help you check that the ionic equation is balanced or actually to help you balance the ionic equation.

Reaction of acids with alkalis

When sulfuric acid reacts with sodium hydroxide solution, it is only the hydrogen ions from the sulfuric acid and the hydroxide ions from the sodium hydroxide that react. These hydrogen ions and hydroxide ions

KEY TERMS

Spectator ions Ions that do not take part in a reaction and do not appear in the ionic equation for the reaction.

Ionic equation Balanced equation for reaction that omits any spectator ions.

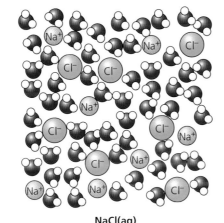

NaCl(s)   NaCl(aq)

▲ **Figure 11.1** Sodium chloride (NaCl) as a solid and when dissolved in water.

react to form water (Figure 11.2). The sulfate ions and sodium ions remain unchanged as they do not react and are left out of the ionic equation as they are spectator ions.

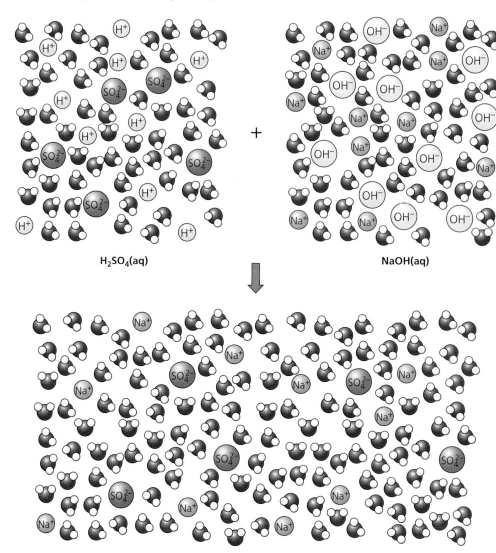

H₂SO₄(aq)   **NaOH(aq)**

▲ Figure 11.2

The ionic equation for this reaction can be written as:

$$H^+(aq) + OH^-(aq) \rightarrow H_2O(l)$$
from the acid   from the alkali

When any acid reacts with any alkali, the ionic equation is the same. Some examples are shown in Table 11.11.

Table 11.11

Examples	What reacts	Ions that do not react	Ionic equation
sulfuric acid (aq) + sodium hydroxide (aq)	H^+ ions from H_2SO_4 OH^- ions from NaOH	SO_4^{2-} ions from H_2SO_4 Na^+ ions from NaOH	$H^+(aq) + OH^-(aq) \rightarrow H_2O(l)$
hydrochloric acid (aq) + potassium hydroxide (aq)	H^+ ions from HCl OH^- ions from KOH	Cl^- ions from HCl K^+ ions from KOH	$H^+(aq) + OH^-(aq) \rightarrow H_2O(l)$
nitric acid (aq) + calcium hydroxide (aq)	H^+ ions from HNO_3 OH^- ions from $Ca(OH)_2$	NO_3^- ions from HNO_3 Ca^{2+} ions from $Ca(OH)_2$	$H^+(aq) + OH^-(aq) \rightarrow H_2O(l)$

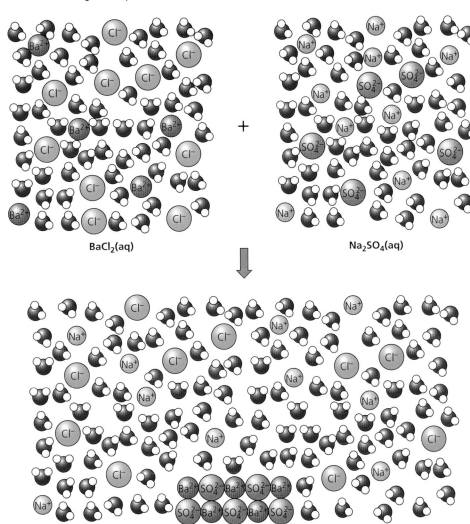

KEY TERM

Precipitate A solid formed when solutions are mixed together.

Precipitation reactions

A **precipitate** is sometimes formed when two solutions of ionic compounds are mixed. For example, when barium chloride solution is mixed with sodium sulfate solution, the barium ions from one solution combine with the sulfate ions from the other solution to form barium sulfate which is insoluble in water (Figure 11.3). The chloride ions and the sodium ions do not react and are not included in the ionic equation as they are spectator ions.

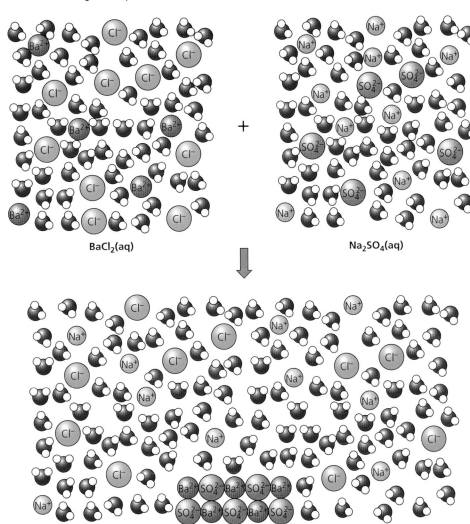

▲ Figure 11.3

The ionic equation for this reaction can be written as:

$$Ba^{2+}(aq) \quad + \quad SO_4^{2-}(aq) \quad \rightarrow \quad BaSO_4(s)$$

from the barium   from the sodium
chloride   sulfate

There are many examples of precipitation reactions, including reactions that are used to test for ions in solution. When writing the ionic equations it is very important to get the formula of the precipitate correct. This also helps you to balance the equation. In the final example in Table 11.12, the formula of copper hydroxide is $Cu(OH)_2$ and so one Cu^{2+} ion and two OH^- ions are needed to balance the equation.

Table 11.12

Examples	What reacts	Ions that do not react	Ionic equation
barium chloride (aq) + sodium sulfate (aq)	Ba^{2+} ions from $BaCl_2$ SO_4^{2-} ions from Na_2SO_4	Cl^- ions from $BaCl_2$ Na^+ ions from Na_2SO_4	$Ba^{2+}(aq) + SO_4^{2-}(aq) \rightarrow BaSO_4(s)$
silver nitrate (aq) + potassium iodide (aq)	Ag^+ ions from $AgNO_3$ I^- ions from KI	NO_3^- ions from $AgNO_3$ K^+ ions from KI	$Ag^+(aq) + I^-(aq) \rightarrow AgI(s)$
copper sulfate (aq) + sodium hydroxide (aq)	Cu^{2+} ions from $CuSO_4$ OH^- ions from NaOH	SO_4^{2-} ions from $CuSO_4$ Na^+ ions from NaOH	$Cu^{2+}(aq) + 2OH^-(aq) \rightarrow Cu(OH)_2(s)$

Displacement reactions

Ionic equations can also be written for displacement reactions that take place when a more reactive metal displaces a less reactive metal from a metal compound.

For example, when zinc reacts with copper sulfate solution, the zinc atoms in the zinc metal react with the copper ions in the copper sulfate (Figure 11.4). The sulfate ions are spectator ions and so do not appear in the ionic equation.

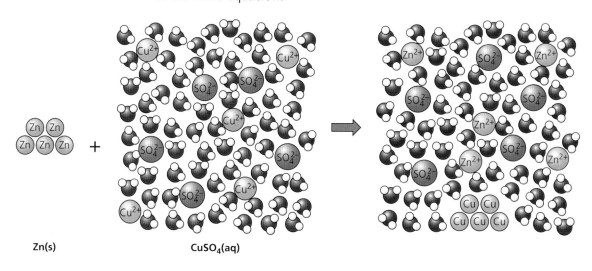

Zn(s) $CuSO_4$(aq)

▲ Figure 11.4

The ionic equation for this reaction can be written as:

$$Zn(s) \quad + \quad \underset{\substack{\text{from the} \\ \text{copper sulfate}}}{Cu^{2+}(aq)} \quad \rightarrow \quad Zn^{2+}(aq) + Cu(s)$$

This and some other examples of displacement reactions are shown in Table 11.13. In these reactions it helps to use the electric charges of the ions to balance the equation. For example, in the second example in the table, two Ag^+ ions are needed giving an overall 2+ charge on the left side of the equation to balance with the 2+ charge on the Fe^{2+} ion on the right side of the equation.

Table 11.13

Examples	What reacts	Ions that do not react	Ionic equation
zinc (s) + copper sulfate (aq)	Zn atoms in Zn metal Cu^{2+} ions from CuSO$_4$	SO$_4^{2-}$ ions from CuSO$_4$	Zn(s) + Cu^{2+}(aq) → Zn^{2+}(aq) + Cu(s)
iron (s) + silver nitrate (aq)	Fe atoms in Fe metal Ag$^+$ ions from AgNO$_3$	NO$_3^-$ ions from AgNO$_3$	Fe(s) + 2Ag$^+$(aq) → Fe^{2+}(aq) + 2Ag(s)
aluminium(s) + copper chloride (aq)	Al atoms in Al metal Cu^{2+} ions from CuCl$_2$	Cl$^-$ ions from CuCl$_2$	2Al(s) + 3Cu^{2+}(aq) → 2Al^{3+}(aq) + 3Cu(s)

Test yourself

13 Write an ionic equation (including state symbols) for each of the following reactions between acids and alkalis.
  a) hydrochloric acid (aq) + sodium hydroxide (aq)
  b) nitric acid (aq) + potassium hydroxide (aq)
  c) sulfuric acid (aq) + calcium hydroxide (aq)
  d) phosphoric acid (aq) + sodium hydroxide (aq)

14 Write an ionic equation (including state symbols) for each of the following reactions that produce a precipitate.
  a) Formation of calcium hydroxide (s) from reaction of calcium nitrate (aq) and sodium hydroxide (aq).
  b) Formation of silver bromide (s) from reaction of silver nitrate (aq) and sodium bromide (aq).
  c) Formation of iron(III) hydroxide (s) from reaction of iron(III) sulfate (aq) and sodium hydroxide (aq).
  d) Formation of barium sulfate (s) from reaction of barium nitrate (aq) and zinc sulfate (aq).

15 Write an ionic equation (including state symbols) for each of the following displacement reactions.
  a) Displacement of copper from copper(II) sulfate (aq) by magnesium.
  b) Displacement of silver from silver nitrate (aq) by magnesium.
  c) Displacement of zinc from zinc sulfate (aq) by aluminium.

16 Write an ionic equation (including state symbols) for each of the following reactions.
  a) Precipitation of lead(II) iodide (s) from reaction of lead(II) nitrate (aq) and potassium iodide (aq).
  b) Reaction of barium hydroxide (aq) with hydrochloric acid (aq).
  c) Precipitation of copper(II) hydroxide (s) from reaction of copper(II) sulfate (aq) and sodium hydroxide (aq).
  d) Displacement of silver from silver nitrate (aq) by zinc (s).
  e) Reaction of sulfuric acid (aq) with potassium hydroxide (aq).
  f) Displacement of nickel from nickel(II) sulfate (aq) by zinc (s).

Half equations

Many chemical reactions involve the transfer of electrons and half equations can be written for these reactions. These equations show the number of electrons that are gained or lost.

In these half equations:

- positive ions gain electrons
  (e.g. a 3+ ion gains 3 electrons, e.g. $Al^{3+} + 3e^- \rightarrow Al$)
- negative ions lose electrons
  (e.g. a 2– ion loses 2 electrons, e.g. $S^{2-} - 2e^- \rightarrow S$).

Half equations for the loss of electrons can be written in two ways. The electrons can be shown being taken away from the left-hand side, or shown on the right-hand side with the products.

For example, the half equation for the loss of electrons from S^{2-} ions to form S can be written as:

$$S^{2-} - 2e^- \rightarrow S \qquad \text{or} \qquad S^{2-} \rightarrow S + 2e^-$$

Some elements contain diatomic molecules (e.g. H_2, O_2, Cl_2, Br_2, I_2). When balancing half equations that produce these elements, two ions are needed to make one diatomic molecule.

For example, when H_2 is formed from H^+ ions, two H^+ ions are needed which both gain one electron and so two electrons are gained altogether:

$$2H^+ + 2e^- \rightarrow H_2$$

When O_2 is formed from O^{2-} ions, two O^{2-} ions are needed which both lose two electrons and so four electrons are lost altogether:

$$2O^{2-} - 4e^- \rightarrow O_2 \qquad \text{or} \qquad 2O^{2-} \rightarrow O_2 + 4e^-$$

In half equations, the total electric charge on the left-hand side must equal the total electric charge on the right-hand side of the equation. This can be used to check that the half equation is balanced.

In this example, both the left and right-hand sides of the equation add up to the same overall charge (which is 0 in this case):

$$Al^{3+} + 3e^- \rightarrow Al$$

total charge: $\underbrace{3+ \quad 3-}_{0}$ 0

left-hand side   right-hand side

Oxidation takes place when a substance loses electrons. Reduction takes place when a substance gains electrons. One way to remember this is the phrase OIL RIG (Figure 11.5).

Electrolysis

Half equations can be written for the reactions that take place at each electrode in electrolysis. Some examples are shown in Table 11.14.

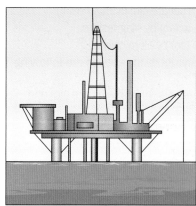

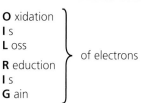

O xidation
I s
L oss
R eduction
I s
G ain
} of electrons

▲ Figure 11.5

Table 11.14

Substance (in molten state)	Negative electrode reaction		Positive electrode reaction	
	What happens	Half equation	What happens	Half equation
Copper sulfide	Cu^{2+} ions gain electrons to form Cu	$Cu^{2+} + 2e^- \rightarrow Cu$	S^{2-} ions lose electrons to form S	$S^{2-} - 2e^- \rightarrow S$ (or $S^{2-} \rightarrow S + 2e^-$)
Sodium chloride	Na^+ ions gain electrons to form Na	$Na^+ + e^- \rightarrow Na$	Cl^- ions lose electrons to form Cl_2	$2Cl^- - 2e^- \rightarrow Cl_2$ (or $2Cl^- \rightarrow Cl_2 + 2e^-$)
Aluminium oxide	Al^{3+} ions gain electrons to form Al	$Al^{3+} + 3e^- \rightarrow Al$	O^{2-} ions lose electrons to form O_2	$2O^{2-} - 4e^- \rightarrow O_2$ (or $2O^{2-} \rightarrow O_2 + 4e^-$)

Displacement reactions

Two half equations can be written for displacement reactions.

For example, in the reaction where zinc displaces copper from copper sulfate solution the overall ionic equation is:

$$Zn + Cu^{2+} \rightarrow Zn^{2+} + Cu$$

The two half equations for this are:

Zn atoms lose electrons to form Zn^{2+} ions:

$$Zn - 2e^- \rightarrow Zn^{2+} \text{ (or } Zn \rightarrow Zn^{2+} + 2e^-)$$

Cu^{2+} ions in $CuSO_4$ gain electrons to form Cu atoms:

$$Cu^{2+} + 2e^- \rightarrow Cu$$

For example, in the reaction when copper displaces silver from silver nitrate solution the overall ionic equation is:

$$Cu + 2Ag^+ \rightarrow Cu^{2+} + 2Ag$$

The two half equations for this are:

Cu atoms lose electrons to form Cu^{2+} ions:

$$Cu - 2e^- \rightarrow Cu^{2+} \text{ (or } Cu \rightarrow Cu^{2+} + 2e^-)$$

Ag^+ ions in $AgNO_3$ gain electrons to form Ag atoms:

$$Ag^+ + e^- \rightarrow Ag$$

Test yourself

17 Write a balanced half equation for each of the following conversions.
   a) $Mg^{2+} \rightarrow Mg$
   b) $Se^{2-} \rightarrow Se$
   c) $K^+ \rightarrow K$
   d) $Br^- \rightarrow Br_2$
   e) $O^{2-} \rightarrow O_2$
   f) $H^+ \rightarrow H_2$

18 Write two half equations to show what happens in the following displacement reactions.
   a) displacement of copper from copper(II) sulfate (aq) by magnesium
   b) displacement of silver from silver nitrate (aq) by magnesium
   c) displacement of zinc from zinc sulfate (aq) by aluminium

Glossary

Accurate Result close to the true value

Acid Solution with a pH less than 7; produces H^+ ions in water

Activation energy The minimum energy particles must have to react

Addition polymerisation Reaction where many small molecules are joined together to make a long chain molecule and nothing else is produced

Addition reaction Reaction in which atoms bond to the atoms in a C=C double bond to form a saturated molecule

Alcohols A homologous series of compounds containing the functional groups –OH

Alkali metals The elements in Group 1 of the periodic table (including lithium, sodium and potassium)

Alkali Solution with a pH more than 7; produces OH^- ions in water

Alkanes A homologous series of saturated hydrocarbons with the general formula C_nH_{2n+2}

Alkenes A homologous series of unsaturated hydrocarbons with the general formula C_nH_{2n}

Alloy A mixture of a metal with small amounts of other elements, usually other metals

Amino acids Molecules containing both a carboxylic acid and an amine functional group

Anion Negative ion

Anode An electrode where oxidation takes place (oxidation is the loss of electrons) – in electrolysis, it is the positive electrode

Aqueous Dissolved in water

Atom The smallest part of an element that can exist. A particle with no electric charge made up of a nucleus containing protons and neutrons surrounded by electrons in energy levels

Atom economy (atom utilisation) A way of measuring what percentage of the mass of all the atoms in the reactants ends up in the desired product

Atomic number Number of protons in an atom

Avogadro constant The number of atoms, molecules or ions in one mole of a given substance (the value of the Avogadro constant is 6.02×10^{23})

Battery Two or more chemical cells connected together

Bioleaching The use of bacteria to produce soluble metal compounds from insoluble metal compounds

Biomass A resource made from living or recently living creatures

Burette A glass tube with a tap and scale for measuring liquids to the nearest $0.1\,cm^3$

Carbohydrates Biological molecules containing carbon, hydrogen and oxygen

Carbon footprint The amount of carbon dioxide and other greenhouse gases given out over the full life cycle of a product, service or event

Carbon neutral Fuels and processes whose use results in zero net release of greenhouse gases to the atmosphere

Carboxylic acids A homologous series of compounds containing the functional group –COOH

Catalyst Substance that speeds up a chemical reaction but is not used up

Cathode An electrode where reduction takes place (reduction is the gain of electrons) – in electrolysis it is the negative electrode

Cation Positive ion

Cell Two electrodes in an electrolyte used to generate electricity

Closed system A system where no substances can get in or out

Complete combustion When a substance burns with a good supply of oxygen

Compound Substance made from different elements chemically bonded together

Concentrated A solution in which there is a lot of solute dissolved

Concordant Results that are very close together

Condensation polymerisation Reaction where many small molecules are joined together to make a long chain molecule plus a small molecule (e.g. water)

Conservation of mass In a reaction, the total mass of the reactants must equal the total mass of the products

Corrosion The destruction of materials by chemical reactions of substances in the environment

Covalent bond Two shared electrons joining atoms together

Cracking The thermal decomposition of long alkanes into shorter alkanes and alkenes

Delocalised Free to move around

Desalination Process to remove dissolved substances from sea water

Diatomic molecule A molecule containing two atoms

Dilute A solution in which there is a small amount of solute dissolved

Discharge Gain or lose electrons to become electrically neutral

Displacement reaction Reaction where a more reactive element takes the place of a less reactive element in a compound

Displayed formula Drawing of a molecule showing all atoms and bonds

Dynamic equilibrium System where both the forward and reverse reactions are taking place simultaneously and at the same rate

Electrolysis Decomposition of ionic compounds using electricity

Electrolyte A liquid that conducts electricity

Electron Negatively charged particle found in energy levels (shells) surrounding the nucleus inside atoms

Element A substance containing only one type of atom; a substance that cannot be broken down into simpler substances by chemical methods

End point The moment when the indicator changes colour in a titration showing that the moles of acid and alkali are equal

Endothermic reactions Reaction where thermal energy is transferred from the surroundings to the chemicals and so the temperature decreases

Energy level (shell) The region an electron occupies surrounding the nucleus inside an atom

Enzymes Molecules that act as catalysts in biological systems

Esters A homologous series of compounds containing the functional group –COO–

Excess When the amount of a reactant is greater than the amount that can react

Exothermic reactions Reaction where thermal energy is transferred from the chemicals to the surroundings and so the temperature increases

Filtrate Liquid that comes through the filter paper during filtration

Finite resource A resource that cannot be replaced once it has been used

Flammability How easily a substance catches fire; the more flammable, the more easily it catches fire

Formulation A mixture that has been designed as a useful product

Fraction A mixture of molecules with similar boiling points

Fractional distillation A method used to separate miscible liquids with different boiling points

Fuel cell A chemical cell with a continuous supply of chemicals to fuel the cell

Fullerenes Family of carbon molecules each with carbon atoms linked in rings to form a hollow sphere or tube

Functional group Atom or group of atoms responsible for most of the chemical reactions of a compound

Giant lattice A regular structure containing a massive number of particles that continues in all directions throughout the structure

Global warming An increase in the temperature at the Earth's surface

Greenhouse gas A gas that absorbs long wavelength infrared radiation given off by the Earth but does not absorb the Sun's radiation

Halides Compounds made from Group 7 elements

Halogens The elements in Group 7 of the periodic table (including fluorine, chlorine, bromine and iodine)

Hazard Something that could cause harm

Homologous series A family of compounds with the same general formula, the same functional group and similar chemical properties

Hydration Addition of water

Hydrocarbon A compound containing hydrogen and carbon only

Hydrogenation Addition of hydrogen

Hypothesis A proposal intended to explain certain facts or observations.

Immiscible Liquids that do not mix together and separate into layers

Incomplete combustion When a substance burns with a poor supply of oxygen

Inert electrodes Electrodes that allow electrolysis to take place but do not react themselves

Intermolecular forces Weak forces between molecules

Ion An electrically charged particle containing different numbers of protons and electrons

Ionic bonding The electrostatic attraction between positive and negative ions

Ionic equation Balanced equation for reaction that omits any spectator ions

Isotopes Atoms with the same number of protons, but a different number of neutrons

Leachate A solution produced by leaching or bioleaching

Leaching The use of dilute acid to produce soluble metal compounds from insoluble metal compounds

Life cycle assessment An examination of the impact of a product on the environment throughout its life

Limiting reactant The reactant in a reaction that determines the amount of products formed. Any other reagents are in excess and will not all react

Malleable Can be hammered into shape

Mass number Number of protons plus the number of neutrons in an atom

Meniscus The curve at the surface of a liquid in a container

Metallic bonding The attraction between the nucleus of metal atoms and delocalised electrons

Miscible Liquids that mix together

Mixture More than one substance that are not chemically joined together

Mole Measurement of the amount of a substance

Molecule Particle made from atoms joined together by covalent bonds

Monomer Small molecules that are joined together to make a polymer

Nanoparticles Structures that are between 1 and 100 nm in size, typically containing a few hundred atoms

Nanoscience The study of nanoparticles

Neutralisation A reaction that uses up some or all of the H^+ ions from an acid

Neutron Neutral particle found inside the nucleus of atoms

Noble gases The elements in Group 0 of the periodic table (including helium, neon and argon)

Nucleotide One of four different molecules that bond together to make up DNA

Nucleus Central part of an atom containing protons and neutrons

Ore A rock from which a metal can be extracted for profit

Oxidation A reaction where a substance gains oxygen and/or loses electrons

Peer-reviewed When scientific research is studied and commented on by experts in the same area of science to check that it is scientifically valid

Percentage yield The amount of product formed in a reaction compared to the maximum theoretical mass that could be produced as a percentage

Phytomining The use of plants to absorb metal compounds from soil as part of metal extraction

Pipette A glass tube used to measure volumes of liquids with a very small margin of error

Polymer Long chain molecule made from joining lots of small molecules together

Potable water Water that is safe to drink

Precipitate A solid formed when solutions are mixed together

Precise Measurements which have little spread around the mean value

Proteins Polymer molecules made from lots of different amino acids joined together

Proton Positively charged particle found inside the nucleus of atoms

Pure substance A single element or compound that is not mixed with any other substance

Random error Errors that cause a measurement to differ from the true value by different amounts each time

Redox reaction A reaction where both reduction and oxidation take place

Reduction A reaction where a substance loses oxygen and/or gains electrons

Relative atomic mass The average mass of atoms of an element taking into account the mass and amount of each isotope it contains on a scale where the mass of ^{12}C (on a scale where the mass of a ^{12}C atom is 12)

Relative formula mass The sum of the relative atomic masses of all the atoms shown in the formula (often referred to as *formula mass*)

Renewable resource A resource that we can replace once we have used it

Repeatable Similar results are achieved when an experiment is repeated by the same person

Reproducible Similar results are achieved when an experiment is repeated by another person or by a different method is used

Residue Solid left on the filter paper during filtration

Resolution The smallest change a piece of apparatus can measure

Risk An action involving a hazard that might result in danger

Rusting The corrosion of iron or steel

Sacrificial protection Preventing a metal from corrosion by attaching it to a more reactive metal that corrodes in its place

Saturated (in the context of solutions) A solution in which no more solute can dissolve at that temperature

Saturated (in the context of organic chemistry) A molecule that only contains single covalent bonds

Separating funnel Glass container with a tap used to separate immiscible liquids

Spectator ions Ions that do not take part in a reaction and do not appear in the ionic equation for the reaction

States of matter These are solid, liquid and gas

Strong acid Acid in which all the molecules break into ions in water

Sustainable development Using resources to meet the needs of people today without preventing people in the future from meeting theirs

Systematic error Errors that cause a measurement to differ from the true value by a similar amount each time

Thermal decomposition Reaction where heat causes a substance to break down into simpler substances

Thermosetting polymer A polymer with covalent bonds between polymer chains that does not soften or melt when heated

Thermosoftening polymer A polymer with no bonds between polymer chains that softens and melts when heated

Turbidity The cloudiness of a solution

Uncertainty The range of measurements within which the true value can be expected to lie.

Unsaturated (in the context of organic chemistry) A molecule that contains one or more double covalent bonds

Valid results Results from a fair test in which only one variable is changed

Viscosity How easily a liquid flows; the higher the viscosity, the less easily it flows

Water stress A shortage of fresh water

Weak acid Acid in which only a small fraction of the molecules break into ions in water

Yield The amount of product formed in a reaction

Index

Acknowledgements

The Publisher would like to thank the following for permission to reproduce copyright material.

AQA material is reproduced by permission of AQA.

p. 1 and 30 Richard Grime; p. 3 T © johny007pan - Fotolia; p. 3 B © Dionisvera - Fotolia; p. 10 © SCIENCE PHOTO LIBRARY; p. 14 © sciencephotos / Alamy; p. 15 L © sciencephotos / Alamy; p. 15 M and 100 T © TREVOR CLIFFORD PHOTOGRAPHY/SCIENCE PHOTO LIBRARY; p. 15 R © The Open University; p. 16 © MARTYN F. CHILLMAID/SCIENCE PHOTO LIBRARY; p. 17 © 2005 Richard Megna - Fundamental Photographs; p. 19 T © ASampedro - Fotolia; p. 19 B © MARK SYKES/SCIENCE PHOTO LIBRARY; p. 20 © SSPL/Science Museum/Getty Images; p. 22 L © SCIENCE PHOTO LIBRARY; p. 22 M © ANDREW LAMBERT PHOTOGRAPHY/SCIENCE PHOTO LIBRARY; p. 22 R © hriana - Fotolia; p. 23 © ANDREW LAMBERT PHOTOGRAPHY/SCIENCE PHOTO LIBRARY; p. 24 © ANDREW LAMBERT PHOTOGRAPHY/SCIENCE PHOTO LIBRARY; p. 25 L Richard Grime; p. 25 M Richard Grime; p. 25 R Richard Grime; p. 32 © Rich Legg - iStock via Thinkstock/Getty Images; p. 33 © Vera Kuttelvaserova - Fotolia; p. 37 © CHARLES D. WINTERS/SCIENCE PHOTO LIBRARY; p. 43 © Dzarek - iStock via Thinkstock/Getty Images; p. 44 © Martinan - Fotolia; p. 45 and 106 R © Alexandru Dobrea - Hemara via Thinkstock/Getty Images; p. 45 © ulkan - iStock via Thinkstock/Getty Images; p. 52 B © Purestock via Thinkstock/Getty Images; p. 52 T © Spiegl/ullstein bild via Getty Images; p. 53 © CHRIS KNAPTON/SCIENCE PHOTO LIBRARY; p. 54 © Igor_M - iStock via Thinkstock/Getty Images; p. 55 and 59 L © koosen - iStock via Thinkstock/Getty Images; p. 56 © karaboux - Fotolia; p. 59 R © CHARLES D. WINTERS/SCIENCE PHOTO LIBRARY; p. 61 © Chad Baker - DigitalVision via Thinkstock/Getty Images; p. 62 Richard Grime; p. 63 © psphotograph - iStock via Thinkstock/Getty Images; p. 64 © Blend Images / Alamy Stock Photo; p. 74 © studiomode / Alamy Stock Photo; p. 76 © W. Oelen via Wikipedia Commons (https://creativecommons.org/licenses/by-sa/3.0/deed.en); p. 78 © Mark_KA - iStock via Thinkstock/Getty Images; p. 85 Richard Grime; p. 88 © MARTYN F. CHILLMAID/SCIENCE PHOTO LIBRARY; p. 99 © ANDREW LAMBERT PHOTOGRAPHY/SCIENCE PHOTO LIBRARY; p. 100 B © MARTYN F. CHILLMAID/SCIENCE PHOTO LIBRARY; p. 101 T © Alvey & Towers Picture Library / Alamy; p. 101 B © CHARLES D. WINTERS/SCIENCE PHOTO LIBRARY; p. 106 L © Isidre blanc via Wikipedia (http://creativecommons.org/licenses/by-sa/3.0/); p. 108 B © Martin Shields / Alamy Stock Photo; p. 109 © Getty Images/iStockphoto/Thinkstock; p. 114 © GEOFF KIDD/SCIENCE PHOTO LIBRARY; p. 116 © Hodder Education; p. 126 T © Prill Mediendesign & Fotografie - iStock via Thinkstock/Getty Images; p. 126 BR © sciencephotos / Alamy Stock Photo; p. 126 BL © PRHaney via Wikipedia (http://creativecommons.org/licenses/by-sa/3.0/); p. 126 B Richard Grime; p. 128 © David J. Green / Alamy Stock Photo; p. 129 T Courtesy of Nephron via Wikipedia Commons (https://creativecommons.org/licenses/by-sa/3.0/deed.en); p. 129 B © studiomode / Alamy Stock Photo; p. 130 T © Solent News & Photo Agency/REX Shutterstock; p. 130 M © ANDREW LAMBERT PHOTOGRAPHY/SCIENCE PHOTO LIBRARY; p. 130 B © MBI / Alamy Stock Photo; p. 131 © Creatas Images - Creatas via Thinkstock/Getty Images; p. 137 TL © Admiral_Aladeen via Reddit (http://i.imgur.com/CdKB6UW.jpg); p. 137 TC Duracell; p. 137 TR © Ensup - iStock via Thinkstock/Getty Images; p. 137 CL © Helen Sessions / Alamy Stock Photo; p. 137 BL © Yiap See fat - Hemara via Thinkstock/Getty Images; p. 138 T © phittavas phupakdee - 123RF; p. 138 B © Alex Segre / Alamy Stock Photo; p. 144 © ANDREW LAMBERT PHOTOGRAPHY/SCIENCE PHOTO LIBRARY; p. 146 © Linda Macpherson - Hemara via Thinkstock/Getty Images; p. 151 © bulentozber - iStock via Thinkstock/Getty Images; p. 154 © Mediablitzimages / Alamy Stock Photo; p. 159 T © MARTYN F. CHILLMAID/SCIENCE PHOTO LIBRARY; p. 159 M © GIPhotoStock/SCIENCE PHOTO LIBRARY; p. 159 B © MARTYN F. CHILLMAID/SCIENCE PHOTO LIBRARY; p. 170 © TomasSereda - iStock via Thinkstock/Getty Images; p. 179 L © ANDREW LAMBERT PHOTOGRAPHY/SCIENCE PHOTO LIBRARY; p. 179 R © ANDREW LAMBERT PHOTOGRAPHY/SCIENCE PHOTO LIBRARY; p. 182 T © Mediablitzimages / Alamy Stock Photo; p. 182 B © CHARLES D. WINTERS/SCIENCE PHOTO LIBRARY; p. 184 © Dmytro Titov - 123RF; p. 185 © Eric Gevaert via 123RF; p. 201 © Keith Brofsky - Photodisc via Thinkstock/Getty Images; p. 203 T © Alistair Heap / Alamy Stock Photo; p. 203 B © Steve Mann - Hemara via Thinkstock/Getty Images; p. 204 TL © Kevin Britland / Alamy Stock Photo; p. 204 TR © Nigel Cattlin / Alamy Stock Photo; p. 204 BL © FOOD DRINK AND DIET/MARK SYKES / Alamy Stock Photo; p. 204 BR © René van den Berg / Alamy Stock Photo; p. 207 © MARTYN F. CHILLMAID/SCIENCE PHOTO LIBRARY; p. 207 © MARTYN F. CHILLMAID/SCIENCE PHOTO LIBRARY; p. 207 © SCIENCE PHOTO LIBRARY; p. 207 © ANDREW LAMBERT PHOTOGRAPHY/SCIENCE PHOTO LIBRARY; p. 209, 215 © DAVID TAYLOR/SCIENCE PHOTO LIBRARY; p. 209, 215 © MARTYN F. CHILLMAID/SCIENCE PHOTO LIBRARY; p. 209, 215 © ANDREW LAMBERT PHOTOGRAPHY/SCIENCE PHOTO LIBRARY; p. 209 © DAVID TAYLOR/SCIENCE PHOTO LIBRARY; p. 209 Richard Grime; p. 209 BC Richard Grime; p. 209 BR Richard Grime; p. 210 TL Richard Grime; p. 210 TC Richard Grime; p. 210 TR Richard Grime; p. 210 M Richard Grime; p. 210 B Richard Grime; p. 211 L Richard Grime; p. 211 M Richard Grime; p. 211 R Richard Grime; p. 214 © Jupiterimages via Thinkstock/Getty Images; p. 221 © ANDREW LAMBERT PHOTOGRAPHY/SCIENCE PHOTO LIBRARY; p. 221 © MARTYN F. CHILLMAID/SCIENCE PHOTO LIBRARY; p. 221 © TREVOR CLIFFORD PHOTOGRAPHY/SCIENCE PHOTO LIBRARY; p. 221 © GIPhotoStock/SCIENCE PHOTO LIBRARY; p. 221 Richard Grime; p. 223 © dell640 - iStock via Thinkstock/Getty Images; p. 225 © Byelikova_Oksana - iStock via Thinkstock/Getty Images; p. 226 © Nature Picture Library / Alamy Stock Photo; p. 229 L © jenoche - iStock via Thinkstock/Getty Images; p. 229 R © tfoxfoto - iStock via Thinkstock/Getty Images; p. 230 T © Ingram Publishing via Thinkstock/Getty Images; p. 230 B The cover of volume 81 of Environment International was published in AQA GCSE Chemistry, Copyright Elsevier; p. 231 L © Science & Society Picture Library via Getty Images; p. 231 R Richard Grime; p. 232 L © Louis Levy, 1906; p. 232 M © Steve Speller / Alamy Stock Photo; p. 232 R © Joseph Nickischer - iStock via Thinkstock/Getty Images; p. 232 B © Mr.Lukchai Chaimongkon - iStock via Thinkstock/Getty Images; p. 233 © erectus - Fotolia; p. 234 © GIPhotoStock X / Alamy Stock Photo; p. 238 L © INTERFOTO / Alamy Stock Photo; p. 238 R © Peter Barritt / Alamy Stock Photo; p. 244 © NASA; p. 246 © Harvepino - iStock via Thinkstock/Getty Images; p. 248 © jimiknightley -iStock via Thinkstock/Getty Images; p. 249 © Ros Drinkwater / Alamy Stock Photo; p. 250 © Chalabala - iStock via Thinkstock/Getty Images; p. 252 © Michael Blann - DigitalVision via Thinkstock/Getty Images; p. 257 T © kafl - iStock via Thinkstock/Getty Images; p. 257 B © gl0ck - iStock via Thinkstock/Getty Images; p. 258 T © Dmitrii Bachtub / Alamy Stock Photo; p. 258 B © Mr_Marc - iStock via Thinkstock/Getty Images; p. 259 T © Pekka Jaakkola - Hemera via Thinkstock/Getty Images; p. 259 B © Agencja Fotograficzna Caro / Alamy Stock Photo; p. 260 © Artfoliophoto - iStock via Thinkstock/Getty Images; p. 261 L Richard Grime; p. 261 R © sciencephotos / Alamy Stock Photo; p. 263 L © Nomadsoul1 - iStock via Thinkstock/Getty Images; p. 263 R © Geri Lavrov/Ocean/Corbis; p. 264 T © Cephas Picture Library / Alamy Stock Photo; p. 264 B © Acker/Bloomberg via Getty Images; p. 265 © TongRo Images via Thinkstock/Getty Images; p. 266 Public Domain; p. 274 © pilip76 - iStock via Thinkstock/Getty Images; p. 276 © Zoonar RF – Thinkstock

T, top, B, bottom, L, left; R, right; M, middle

Every effort has been made to trace all copyright holders, but if any have been inadvertently overlooked, the Publisher will be pleased to make the necessary arrangements at the earliest opportunity.

Colette Samson

Alex et Zoé 1

et compagnie

Cahier d'activités

CLE
INTERNATIONAL

Bonjour ! Comment tu t'appelles ?

1A

Ecris une lettre et adresse-la à ton ou ta camarade !

Livre de l'élève p. 2
GP p. 6

Bonjour ...., ça va ?

Au revoir !

....

1B

Complète les bulles !

Livre de l'élève p. 2
GP p. 6

~~Salut !~~ Bonjour ! Ça va ? Bonjour monsieur ! Bonjour madame ! Au revoir !

Unité **1** LEÇON 2

1
Onubroj !
. . . . !

2
TAULS !
. . . . !

3
ua Vreiro !
. . . . !

2A

Décode les messages
secrets et écris-les !

Livre de l'élève p. 3
GP p. 8

4

Bonjourcommentçavaçavabienmerciettoi

. . . . ?

. . . . !

5

b=🌲 e=🐘 i=☂ j=🐼 m=🐰 n=🌸 o=🚀 u=🎈 s=🌙 r=🎁

🌲🚀🌸🐼🎈🎁 🚀🌸🌙☂🐘🐰🎈🎁 !

. . . . !

Tu as besoin de :

Alice

Salut !
Je m'appelle
Alice

2B

Fabrique ton badge !

Livre de l'élève p. 3
GP p. 8

3

3A

Relie les nombres
aux mots !

Livre de l'élève p. 4
GP p. 10

3B

Compte et écris
le nombre de doigts !

Livre de l'élève p. 4
GP p. 10

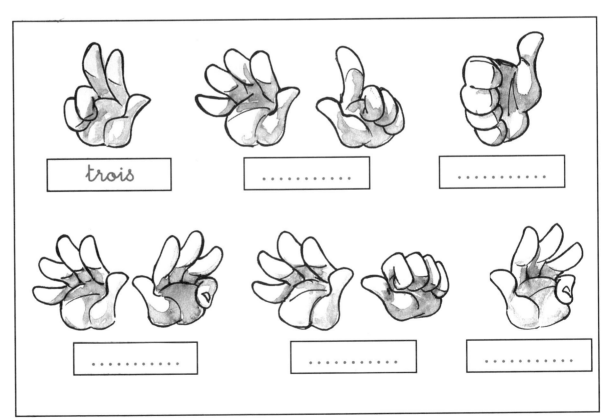

Unité 1 LEÇON 4

Cahier de vie

Bonjour, ça va ?   *Comment tu t'appelles ?*

. . . .

4A

Tu sais répondre
à ces questions ?

Livre de l'élève p. 5
GP p. 12

Test

1	*oui / non*	4	. . . .	9	. . . .
2	. . . .	5	. . . .	10	. . . .
3	. . . .	6	. . . .		
		7	. . . .		
		8	. . . .		

MON SCORE : ... /10

4B

Tu sais dire ces
nombres en français ?
oui / non

Auto-évaluation, Unité 1

Super !

Pas mal !

À revoir !

4C

Evalue ton travail !

Dico-mémento

Tu as besoin de :

Ecris sur chaque page les lettres de l'alphabet,
puis découpe les mots page 63 et colle-les dans ton dico-mémento !

4D

Fabrique ton
dico-mémento et
contrôle ce que tu
sais avec ton voisin
ou ta voisine !

A a
B b
C c
D d
E e
F f
G g
H h
I i
J j
K k
L l
M m
N n
O o
P p
Q q
R r
S s
T t
U u
V v
W w
X x
Y y
Z z

Unité 2 LEÇON 1

Tu as quel âge ?

Va interviewer tes camarades !

Livre de l'élève p. 6
GP p. 14

Comment tu t'appelles ? Tu as quel âge ?	
Prénom	Âge
CHARA	5
CIARAN	F
ERIKa	6
SAMMY	5 1/2

Regarde et écris sous les dessins !

Livre de l'élève p. 6
GP p. 14

1 huit ans

2 ...6...

3 ....P....

4 .....F.....

j'ai	un frère	une sœur	trois frères	deux sœurs
	je n'ai pas de frère		je n'ai pas de sœur	

Bonjour .... !

Je m'appelle .... J'ai .... ans.

J'ai un .... et ....

Je n'ai pas de ....

Et toi ?

Salut !

....

2A

Ecris une lettre et adresse-la à ton ou ta camarade !

Livre de l'élève p. 7
GP p. 16

deux sœurs

....

2B

Ecris le nombre de frères et sœurs !

Livre de l'élève p. 7
GP p. 16

....

....

3A

Regarde et écris !

Livre de l'élève p. 8
GP p. 18

un chat - (un) chien - (un) hamster - (une) perruche - un poisson rouge - une tortue

J'ai ....

Je n'ai pas de ....

3B

Compte et écris le
nombre d'animaux !

Livre de l'élève p. 8
GP p. 18

trois chats, ...

....

8

Cahier de vie

Comment tu t'appelles ?

Tu as quel âge ?

....

....

Tu as un frère ? une sœur ?

Tu as un chat ? un chien ?

....

....

Tu sais répondre
à ces questions ?

Livre de l'élève p. 9
GP p. 20

Test

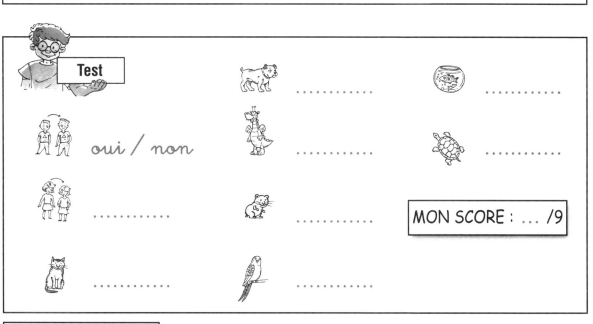

oui / non

MON SCORE : ... /9

Tu sais dire ces mots
en français ? oui / non

Auto-évaluation, Unité 2

 Super !

 Pas mal !

 À revoir !

Evalue ton travail !

Dico-mémento

Découpe les mots
et colle-les
puis contrôle
ce que tu sais !

Unité 3 LEÇON 1

Qu'est-ce que c'est ?

1A

Dessine ta trousse, ta gomme, etc. et relie les mots !

Livre de l'élève p. 10
GP p. 22

ma trousse

mon crayon

ma gomme

mon stylo

ma règle

1B

Regarde et écris !

Livre de l'élève p. 10
GP p. 22

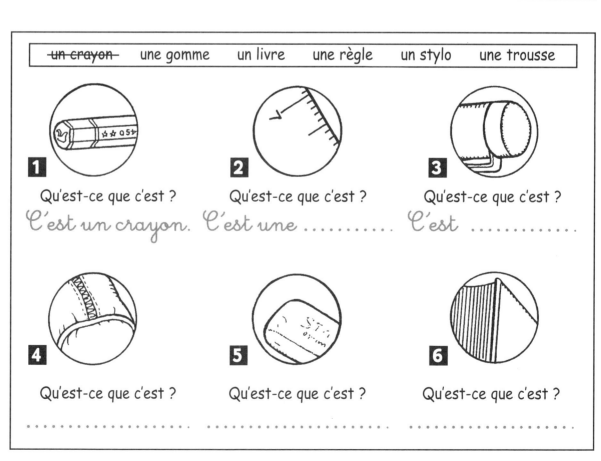

un crayon   une gomme   un livre   une règle   un stylo   une trousse

1 Qu'est-ce que c'est ?
C'est un crayon.

2 Qu'est-ce que c'est ?
C'est une ....

3 Qu'est-ce que c'est ?
C'est ....

4 Qu'est-ce que c'est ?

5 Qu'est-ce que c'est ?

6 Qu'est-ce que c'est ?

....

Retrouve le nom
des couleurs,
colorie les cailloux
et écris les mots !

Livre de l'élève p. 11
GP p. 24

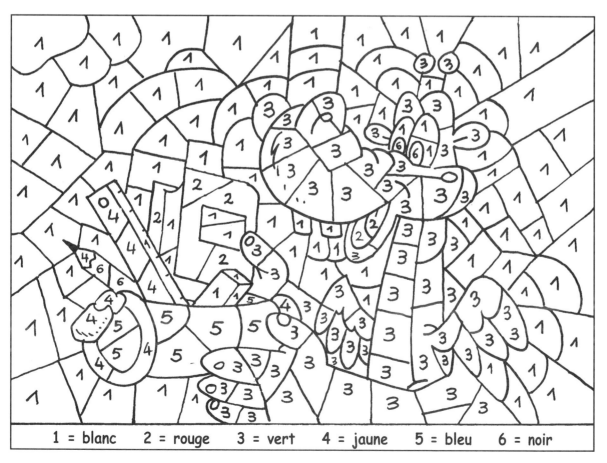

Colorie les cases !
Qui est-ce ?

Livre de l'élève p. 11
GP p. 24

1 = blanc   2 = rouge   3 = vert   4 = jaune   5 = bleu   6 = noir

Unité 3 LEÇON 3

3A

Lis et écris les numéros !

Livre de l'élève p. 12
GP p. 26

☐ Prends ton livre !
☑ Prête-moi ton livre !

☐ Pose ton livre !
☐ Tiens, voilà mon livre !

3B

Colorie et écris !

Livre de l'élève p. 12
GP p. 26

blanc (blanche) - bleu(e) - jaune - noir(e) - rouge - vert(e)

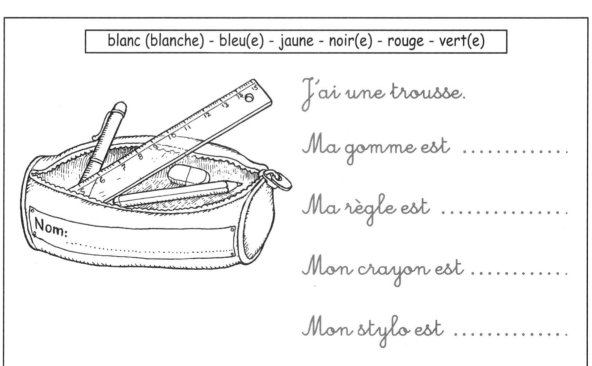

J'ai une trousse.

Ma gomme est ....

Ma règle est ....

Mon crayon est ....

Mon stylo est ....

Unité 3 — LEÇON 4

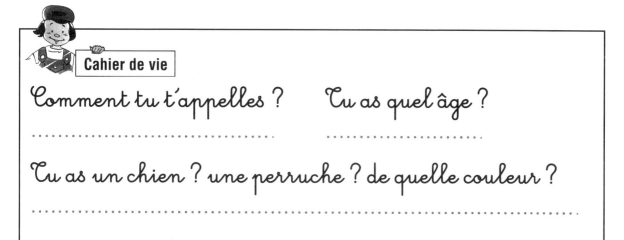

Cahier de vie

Comment tu t'appelles ?   Tu as quel âge ?

....

Tu as un chien ? une perruche ? de quelle couleur ?

....

Tu sais répondre
à ces questions ?

Livre de l'élève p. 13
GP p. 28

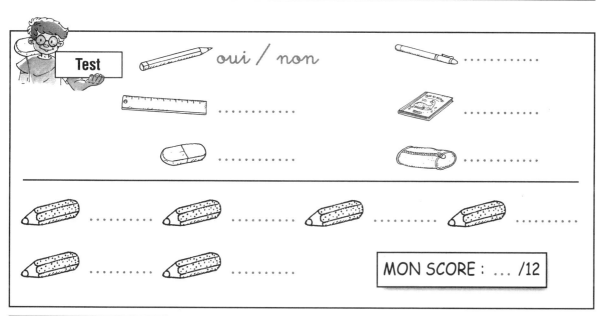

Test   oui / non

MON SCORE : ... /12

Tu sais dire
ces mots en français ?
oui / non

Combien de couleurs
sais-tu dire
en français ?
Colorie les crayons !

Auto-évaluation, Unité 3

Super !   Pas mal !   À revoir !

Evalue ton travail !

Dico-mémento

Découpe les mots
et colle-les,
puis contrôle
ce que tu sais !

Qu'est-ce que tu fais ?

dauphin - dragon - ~~hamster~~ - éléphant - ours - papillon - perruche - tigre - tortue

1 hamster
2 tortue
3 dauphin
4 dragon
5 éléphant
6 ours
7 tigre
8 papillon
9 perruche

Tu as besoin de :

Je suis un dauphin   Je suis un tigre

un chat - un chien - un dauphin - un dragon - un éléphant - un hamster - un ours - un papillon - une perruche - un poisson rouge - un tigre - une tortue

je suis un(e)	je danse	je marche	je nage	je saute	je vole	comme un(e)

Bonjour ....!

Ça va ? Aujourd'hui, ...

....

Salut !

....

2A

Ecris une lettre et adresse-la à ton ou ta camarade !

Livre de l'élève p. 17
GP p. 32

un chat - un chien - un dauphin - un dragon - un éléphant - un hamster - un ours - un papillon - une perruche - un poisson rouge - un tigre - une tortue

Qui es-tu ? Qu'est-ce que tu fais ?

Prénoms							
....							
....							
....							
....							
....							
....							

2B

Va interviewer tes camarades !

Livre de l'élève p. 17
GP p. 32

Unité 4 — LEÇON 4

Cahier de vie

Qu'est-ce que tu fais ? Tu chantes ? Tu joues? Tu comptes ?

Je ....

Tu nages comme un dauphin? Tu marches comme un ours ?

Je .... comme .... !

 4B

Tu sais dire ces
mots en français ?
oui / non

Test

oui / non

MON SCORE : ... /15

Auto-évaluation, Unité 4

 Super !

 Pas mal !

 À revoir !

Dico-mémento

 4D

Découpe les mots
et colle-les,
puis contrôle
ce que tu sais !

Qu'est-ce que tu veux ?

1A

Regarde et écris !

Livre de l'élève p. 20
GP p. 38

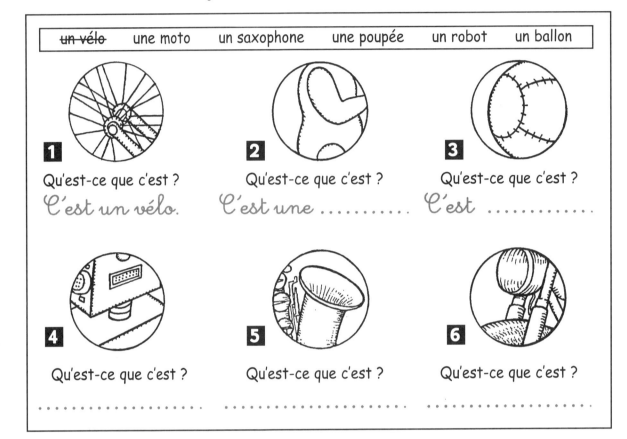

| un vélo | une moto | un saxophone | une poupée | un robot | un ballon |

1 Qu'est-ce que c'est ?
C'est un vélo.

2 Qu'est-ce que c'est ?
C'est une ....

3 Qu'est-ce que c'est ?
C'est ....

4 Qu'est-ce que c'est ?
....

5 Qu'est-ce que c'est ?
....

6 Qu'est-ce que c'est ?
....

1B

Ecris une lettre et
adresse-la à ton
ou ta camarade !

Livre de l'élève p. 20
GP p. 38

un vélo - une moto - un saxophone - une guitare - une poupée - un robot - un ballon
un frère - une sœur - un chat - un chien - un hamster - une perruche
un poisson rouge - une tortue - une gomme - un stylo - une trousse - une règle

Bonjour ....!

J'ai ....

....

Je n'ai pas de ....

....

Au revoir !

....

Livre de l'élève p. 21
GP p. 40

2A

Retrouve les mots et entoure-les dans la grille !

→   ↓   ↗

trousse	éléphant	chien
règle	robot	ours
poisson	livre	dauphin
hamster	saxophone	vélo
papillon	tigre	
dragon		
stylo		↘
chat		
ballon		~~crayon~~

2B

Lis et numérote les phrases !

Livre de l'élève p. 21
GP p. 40

JOYEUX ANNIVERSAIRE !

5 Pour mon anniversaire, je veux danser.   ☐ Pour mon anniversaire, je veux boire.

☐ Pour mon anniversaire, je veux manger.   ☐ Pour mon anniversaire, je veux jouer.

☐ Pour mon anniversaire, je veux dormir.   ☐ Pour mon anniversaire, je veux chanter.

3A

A quoi joue une fille ?
A quoi joue un garçon
selon toi ?
Mets une croix dans
les cases !

Livre de l'élève p. 22
GP p. 42

Lis et colorie
les jouets !

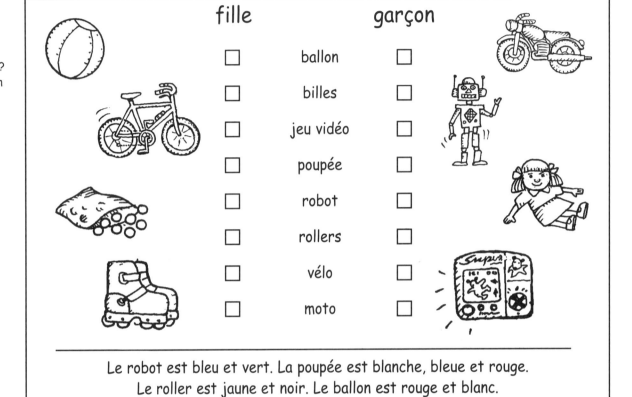

	fille		garçon
	☐	ballon	☐
	☐	billes	☐
	☐	jeu vidéo	☐
	☐	poupée	☐
	☐	robot	☐
	☐	rollers	☐
	☐	vélo	☐
	☐	moto	☐

Le robot est bleu et vert. La poupée est blanche, bleue et rouge.
Le roller est jaune et noir. Le ballon est rouge et blanc.
Le jeu vidéo est jaune. Le vélo est vert. Une bille est bleue. La moto est rouge.

3B

Reconstitue
les phrases !

Livre de l'élève p. 22
GP p. 42

faire ~~Je veux~~ ~~aux billes.~~
Je veux Je veux
~~jouer~~ du roller.
du saxophone.
jouer au ballon. Je veux jouer
Je veux de la guitare.
Je veux jouer
faire faire Je veux
jouer Je veux du vélo.
à la poupée. de la moto.

1. *Je veux jouer aux billes.*   5. . . . .
2. . . . .   6. . . . .
3. . . . .   7. . . . .
4. . . . .   8. . . . .

Cahier de vie

Qu'est-ce que tu fais ? Tu joues au ballon ? à la poupée ?

. . . .

Qu'est-ce que tu veux pour ton anniversaire ?

. . . .

Qu'est-ce que tu veux faire pour ton anniversaire ?

. . . .

4A

Tu sais répondre
à ces questions ?

Livre de l'élève p. 23
GP p. 44

Test

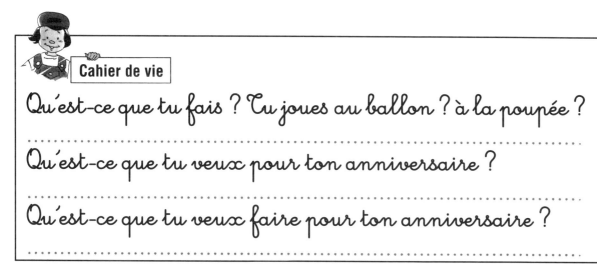

oui / non

MON SCORE : ... /14

4B

Tu sais dire ces
mots en français ?
oui / non

Auto-évaluation, Unité 5

 **Super !**

 **Pas mal !**

 **À revoir !**

4C

Evalue ton travail !

 **Dico-mémento**

4D

Découpe les mots
et colle-les,
puis contrôle
ce que tu sais !

Noël ? Qu'est-ce qu'il y a à Noël ?

1A

Ecris les mots !

Livre de l'élève p. 24
GP p. 46

le Père Noël   un vélo   des rollers   un ballon   une poupée   une guitare
des billes   un jeu vidéo   un robot   des chocolats   une bûche de Noël

le Père Noël

1B

Relie les nombres aux mots !

Livre de l'élève p. 24
GP p. 46

un   deux   trois   quatre   cinq   six   sept   huit   neuf   dix   onze   douze

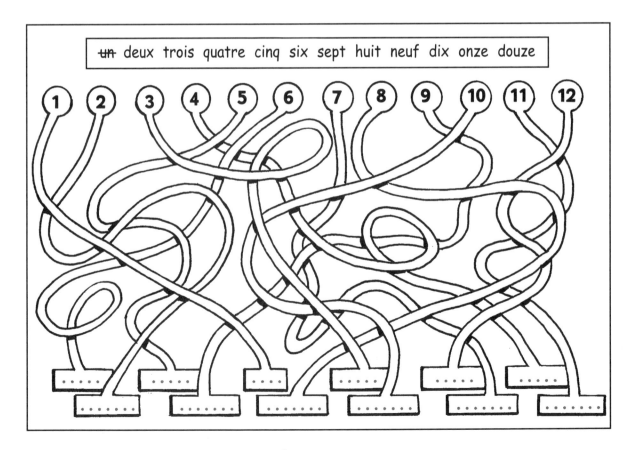

1   2   3   4   5   6   7   8   9   10   11   12

des rollers   un télescope   un ours   un appareil photo

1

Je voudrais ....

2

....

3

....

4

....

2B

Ecris ta lettre
au Père Noël
et décore-la !

Livre de l'élève p. 25
GP p. 48

un livre   un stylo   un saxophone   une guitare   un chat   un chien   un ours
un dauphin   un ballon   des billes   un jeu vidéo   une poupée   un robot
des rollers   un vélo   des chocolats   un télescope   un appareil photo

Cher Père Noël !

Je voudrais ....

Je voudrais aussi ....

....

Merci !

....

3A

Décris ce qu'il y a !

Livre de l'élève p. 26
GP p. 50

1. Il y a un ballon. **2.** Il y a une ....

3. Il y a .... **4.** .... **5.** ....

6. .... **7.** ....

3B

Regarde l'image dans
le livre page 26.
Ecris les mots !

Livre de l'élève p. 26
GP p. 50

b	billes
	— — — — —
	— — — — — (de Noël)
	— — — — —
	— — — — — (de Noël)
c	— — — —
	— — — — —
	— — — — — — — —
	— — — — — —
d	— — — — — —
	— — — — —
e	— — — — — — — —
	— — — — — —

g	— — — — — — — — —
	— — — — — — —
p	— — — — — — —
	— — — — — — —
	— — — — — — —
	— — — — — —
t	— — — — — — — —
	— — — — — —
	— — — — — — —
v	— — — —

Unité **6** LEÇON 4

Cahier de vie

Qu'est-ce que tu voudrais pour Noël ?

Pour Noël, je voudrais ....

....

....

4A

Tu sais répondre
à cette question ?

Livre de l'élève p. 27
GP p. 52

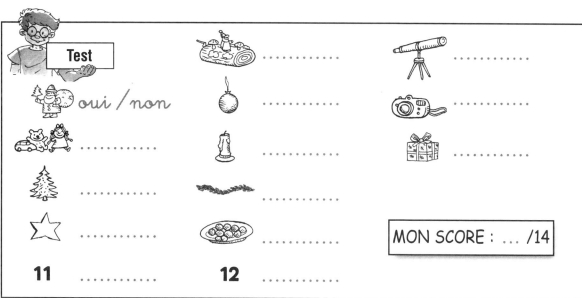

Test

oui / non

11 ....

12 ....

MON SCORE : ... /14

4B

Tu sais dire ces mots
en français ?
oui / non

Auto-évaluation, Unité 6

Super !

Pas mal !

À revoir !

4C

Evalue ton travail !

Dico-mémento

4D

Découpe les mots
et colle-les,
puis contrôle
ce que tu sais !

25

Qu'est-ce que tu aimes ?

1A

Lis et écris les noms !

Livre de l'élève p. 30
GP p. 54

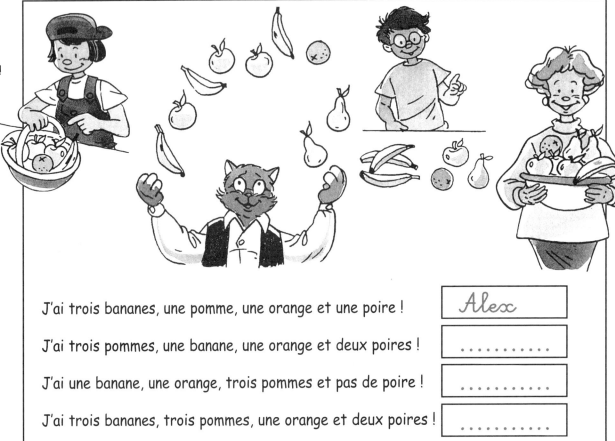

J'ai trois bananes, une pomme, une orange et une poire ! | Alex |

J'ai trois pommes, une banane, une orange et deux poires ! | . . . . |

J'ai une banane, une orange, trois pommes et pas de poire ! | . . . . |

J'ai trois bananes, trois pommes, une orange et deux poires ! | . . . . |

1B

Retrouve le nom des
fruits et écris les
mots !

Livre de l'élève p. 30
GP p. 54

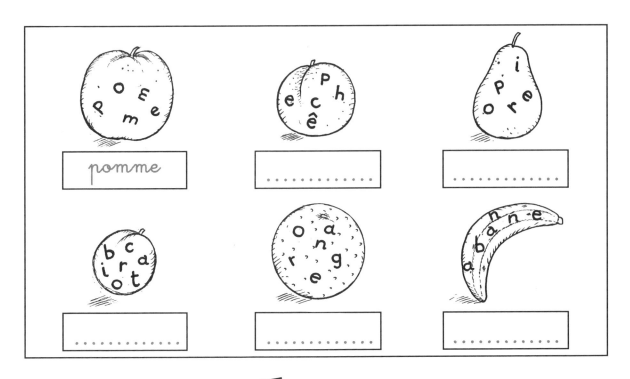

| pomme |

| . . . . |

| . . . . |

| . . . . |

| . . . . |

| . . . . |

2A

Dessine et écris !
Puis adresse
ta lettre à ton ou
ta camarade !

Livre de l'élève p. 31
GP p. 56

J'aime	Je n'aime pas

J'aime les frites. J .... le poisson. J ....la salade. J .... le fromage. J .... le poulet. J .... les pommes. J .... les bananes. J .... les oranges. J .... les gâteaux.

2B

Ecris les mots !

Livre de l'élève p. 31
GP p. 56

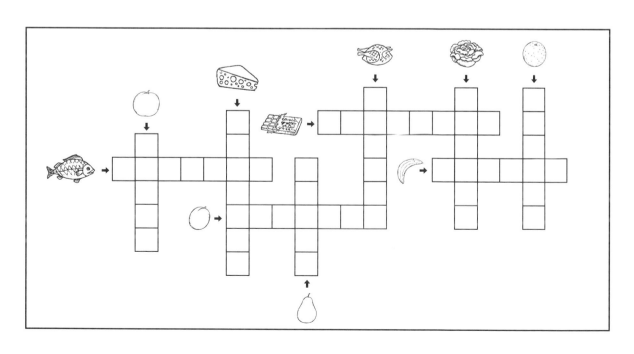

3A

Ecris !

Livre de l'élève p. 32
GP p. 58

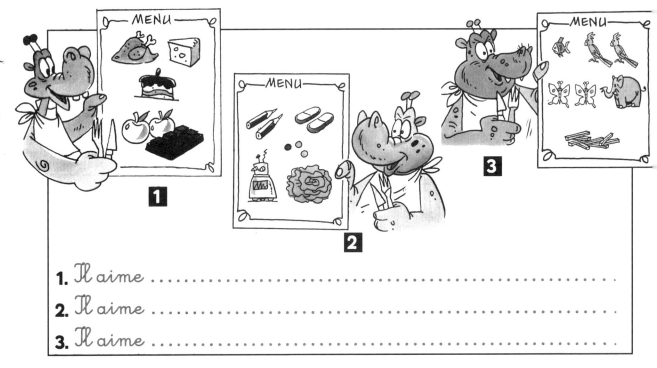

1

2

3

1. Il aime ....

2. Il aime ....

3. Il aime ....

3B

Va interviewer tes
camarades et écris !

Livre de l'élève p. 32
GP p. 58

| | la salade | le poisson | le poulet | les frites |
| le fromage | les pommes | les bananes | les gâteaux |

Tu aimes? / Prénoms								
....								
....								
....								
....								
....								

.... aime le poulet, les frites et les bananes.

.... aime ....

.... aime ....

.... aime ....

.... aime ....

Cahier de vie

Qu'est-ce que tu aimes ?

J ....

Qu'est-ce que tu n'aimes pas ?

J ....

4A

Tu sais répondre
à ces questions ?

Livre de l'élève p. 33
GP p. 60

Test

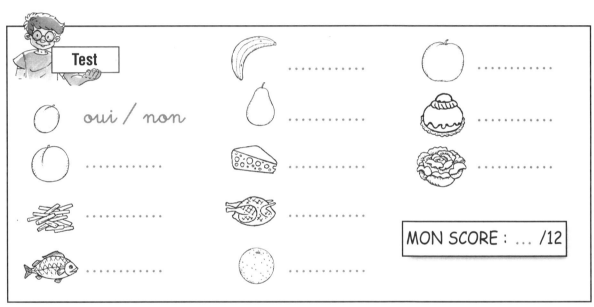

oui / non

MON SCORE : ... /12

4B

Tu sais dire ces mots
en français ? oui / non

Auto-évaluation, Unité 7

 Super !  Pas mal !  À revoir !

4C

Evalue ton travail !

 **Dico-mémento**

4D

Découpe les mots
et colle-les,
puis contrôle
ce que tu sais !

Unité 8 LEÇON 1

Qu'est-ce que tu sais faire ?

1A

Lis et colorie
les nombres !

Livre de l'élève p. 34
GP p. 62

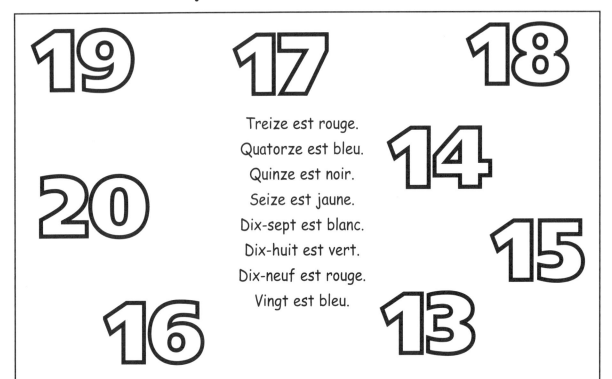

19 17 18

Treize est rouge.
Quatorze est bleu.
Quinze est noir.
Seize est jaune.
Dix-sept est blanc.
Dix-huit est vert.
Dix-neuf est rouge.
Vingt est bleu.

14 20 15

16 13

1B

Réponds et écris !

Livre de l'élève p. 34
GP p. 62

| lire | écouter de la musique | compter | dessiner | jouer au ballon | nager |
| jouer à la poupée | jouer aux billes | faire du vélo | dormir | danser | chanter |

Ce que j'aime faire

🙂 beaucoup										
😐 un peu										
🙁 pas du tout										

J'aime ....

Ecris une lettre et adresse-la à ton ou ta camarade !

Livre de l'élève p. 35
GP p. 64

Bonjour .... !

J'aime ....

Je n'aime pas ....

Et toi ? Qu'est-ce que tu aimes faire ?

Au revoir !

....

Ecris !

Livre de l'élève p. 35
GP p. 64

faire du cheval   jongler   faire la cuisine   jouer de la flûte   faire du ski   faire du judo

Mamie sait ....
Alex ne sait pas ....
Loulou ....
Zoé ....
Croquetout ....
Basile ....

Unité **8** LEÇON 3

3A

Va interviewer tes camarades et écris !

Livre de l'élève p. 36
GP p. 66

Est-ce que tu sais...?

Prénoms	faire du cheval	faire la cuisine	jongler	sauter à la corde	jouer aux billes	nager	danser	faire du vélo	jouer de la flûte
. . . .									
. . . .									
. . . .									
. . . .									
. . . .									

sait = *oui*

ne sait pas = *non*

. . . . . *sait* . . . . *Il / Elle ne sait pas* . . . .
. . . . . *sait* . . . . *Il / Elle ne sait pas* . . . .
. . . . . *sait* . . . . *Il / Elle ne sait pas* . . . .
. . . . . *sait* . . . . *Il / Elle ne sait pas* . . . .
. . . . . *sait* . . . . *Il / Elle ne sait pas* . . . .

3B

Regarde l'image dans le livre page 36. Ecris les mots !

Livre de l'élève p. 36
GP p. 66

f	(jouer de la) _f l û t e_ (jouer au) _ _ _ _ _ _ _ _
c	_ _ _ _ _ _ _ _ _ _ _ _ _ _ (faire du) _ _ _ _ _ _ (faire la) _ _ _ _ _ _ _
g	(jouer de la) _ _ _ _ _ _ _ (manger du) _ _ _ _ _ _
j	_ _ _ _ _ _ _ (faire du) _ _ _ _

n	_ _ _ _ _
s	(jouer du) _ _ _ _ _ _ _ (faire du) _ _ _ _ _ _ _ _ _ (à la corde)
r	(faire du) _ _ _ _ _ _
t	(jouer au) _ _ _ _ _ _

Unité **8** LEÇON 4

Cahier de vie

Qu'est-ce que tu sais faire ?

. . . .

Qu'est-ce que tu ne sais pas faire ?

. . . .

4A

Tu sais répondre
à ces questions ?

Livre de l'élève p. 37
GP p. 68

Test

oui / non

13 . . . .   14 . . . .   15 . . . .

16 . . . .   17 . . . .   18 . . . .

19 . . . .   20 . . . .   MON SCORE : ... /18

4B

Tu sais dire
ces mots en français ?
oui / non

Tu sais dire ces
nombres en français ?
oui / non

Auto-évaluation, Unité 8

 Super !

 Pas mal !

 À revoir !

4C

Evalue ton travail !

Dico-mémento

4D

Découpe les mots
et colle-les,
puis contrôle
ce que tu sais !

Qu'est-ce que tu mets aujourd'hui ?

1A

Complète les bulles !

Livre de l'élève p. 38
GP p. 70

Je mets   une robe   un pantalon   un pull   une jupe   un jean   un tee-shirt

1B

Relie les mots aux vêtements !

Livre de l'élève p. 38
GP p. 70

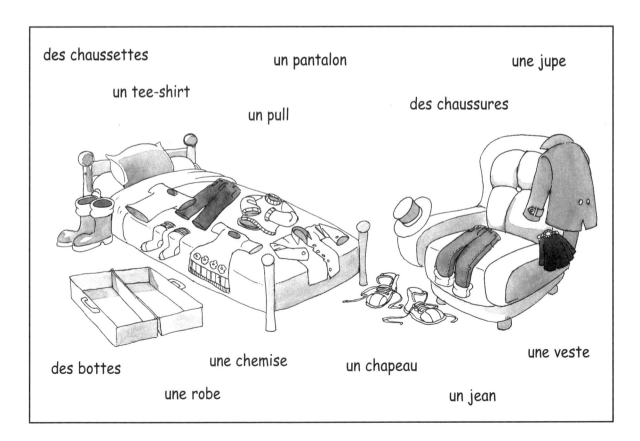

des chaussettes   un pantalon   une jupe   un tee-shirt   des chaussures   un pull   des bottes   une chemise   un chapeau   une veste   une robe   un jean

~~beige~~   orange   gris   rose   marron   violet

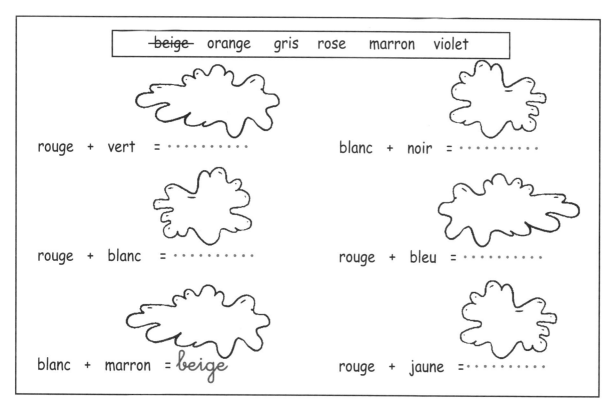

rouge + vert = · · · · · · · · · ·

blanc + noir = · · · · · · · · · ·

rouge + blanc = · · · · · · · · · ·

rouge + bleu = · · · · · · · · · ·

blanc + marron = *beige*

rouge + jaune = · · · · · · · · · ·

Mamie a une robe verte,
une veste bleue,
un chapeau rose et
des chaussures noires.

Alex a un pantalon gris,
un pull rouge,
des bottes jaunes
et un bonnet orange.

Zoé a un jean bleu,
un tee-shirt violet,
des chaussettes orange et
des baskets beiges.

3A

Ecris une lettre et adresse-la à ton ou ta camarade !

Livre de l'élève p. 40
GP p. 74

| chaussures | chemise | gilet | jean | jupe | pantalon | pull | robe | tee-shirt | veste |

| blanc (blanche) | beige | bleu(e) | gris(e) | jaune | noir(e) |
| rose | rouge | vert(e) | orange | marron | violet(te) |

Bonjour .... !

Aujourd'hui j'ai un tee-shirt ....

....

....

Et toi, qu'est-ce que tu as ?

Salut ! À bientôt !

....

3B

Colorie les personnages et complète le texte !

Livre de l'élève p. 40
GP p. 74

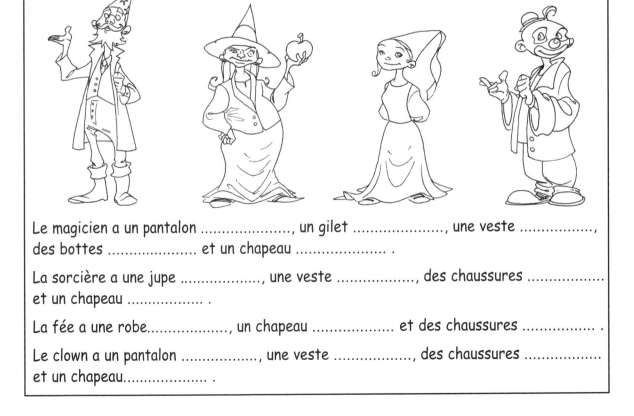

Le magicien a un pantalon ...., un gilet ...., une veste ...., des bottes .... et un chapeau ....

La sorcière a une jupe ...., une veste ...., des chaussures .... et un chapeau ....

La fée a une robe................., un chapeau .... et des chaussures ....

Le clown a un pantalon ...., une veste ...., des chaussures .... et un chapeau................... .

4A

Tu sais répondre
à cette question ?

Livre de l'élève p. 41
GP p. 76

Cahier de vie

Qu'est-ce que tu mets pour le Carnaval ?

Je mets ....

Je suis ....

4B

Tu sais dire ces mots
en français ? oui / non

Test

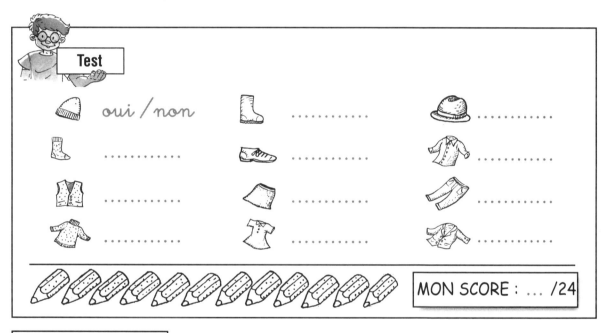

oui / non

Combien de couleurs
sais-tu dire en français ?
Colorie les crayons !

MON SCORE : ... /24

Auto-évaluation, Unité 9

 Super !

 Pas mal !

 À revoir !

4C

Evalue ton travail !

Dico-mémento

4D

Découpe les mots
et colle-les,
puis contrôle
ce que tu sais !

Qu'est-ce que tu prends au petit déjeuner ?

Ecris les mots !

Livre de l'élève p. 44
GP p. 78

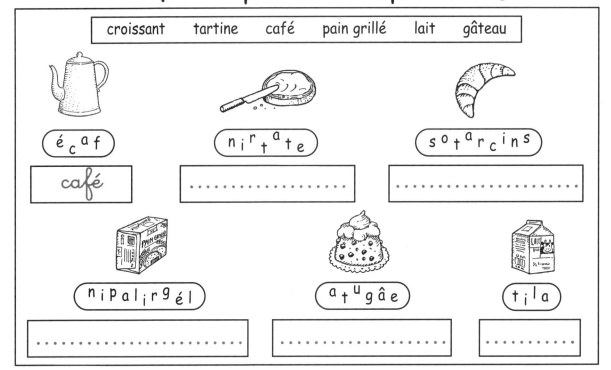

croissant   tartine   café   pain grillé   lait   gâteau

é c ᵃ f

café

n i ᵣ ᵗ ᵃ t e

....

s ᵒ ᵗ ᵃ ᵣ c i n ˢ

....

n i ᵖ a l ᵢ ᵣ ᵍ é l

....

ᵃ ᵗ ᵘ g â e

....

t ᵢ l a

....

Complète les phrases !

Livre de l'élève p. 44
GP p. 78

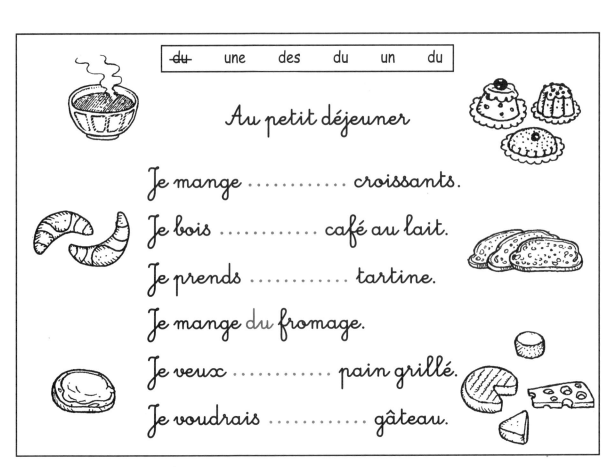

~~du~~   une   des   du   un   du

Au petit déjeuner

Je mange .... croissants.

Je bois .... café au lait.

Je prends .... tartine.

Je mange du fromage.

Je veux .... pain grillé.

Je voudrais .... gâteau.

une tartine
du lait
du chocolat
du pain grillé
des céréales
de la confiture
du thé
un croissant
un œuf
du jus d'orange

2A

Relie les mots aux dessins !

Livre de l'élève p. 45
GP p. 80

a	a n n i v e r s a i r e
	_ _ _ _ _ _ _ _ ◯
e	_ _ _ _ _ _ _
	_ _ _ _ _ _ _
	_ _ _ _ _ _
i	_ _ _ _ _
o	_ _ _ _ _
	_ _ _ _ _
	_ _ _ _ _

u	_ _ **1**
h	_ _ _ _ _ _
	_ _ _ _ _ **8**
j	_ _ _ _
	_ _ _ _ (-vidéo)
	_ _ _ _
	_ _ _ _
	_ _ _ (d'orange)

2B

Ecris les mots !

Livre de l'élève p. 45
GP p. 80

3A

Décris ton petit déjeuner, dessine-le et envoie ta lettre à ton ou ta camarade !

Livre de l'élève p. 46
GP p. 82

Mon petit déjeuner

Au petit déjeuner, je prends ....

Je prends aussi ....

Et toi ?

Salut !

....

3B

Va interviewer tes camarades !

Livre de l'élève p. 46
GP p. 82

Qu'est-ce que tu prends au petit déjeuner ? Tu prends du café ? du thé ?

Prénoms	du café	du thé	du chocolat	du jus d'orange	du lait	du pain	de la confiture	des céréales	un œuf	
....										
....										
....										
....										
....										

...........prend...............................

...........prend...............................

...........prend...............................

...........prend...............................

...........prend...............................

4A

Tu sais répondre
à ces questions ?

Livre de l'élève p. 47
GP p. 84

Cahier de vie

Qu'est-ce que tu bois au petit déjeuner ?

J...

Qu'est-ce que tu manges au petit déjeuner ?

J...

4B

Tu sais dire ces mots
en français ? oui / non

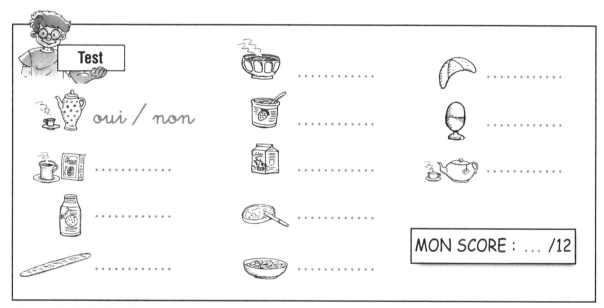

Test

oui / non

MON SCORE : ... /12

Auto-évaluation, Unité 10

4C

Evalue ton travail !

 **Super !**

 **Pas mal !**

 **À revoir !**

4D

Découpe les mots
et colle-les,
puis contrôle
ce que tu sais !

 **Dico-mémento**

Quelle heure est-il ?

1A

Lis les phrases et écris les jours de la semaine !

Livre de l'élève p. 48
GP p. 86

lundi

☒ Lundi, je fais du vélo.

☐ Mardi, je dessine.

☐ Mercredi, je regarde la télévision.

☐ Jeudi, j'écoute de la musique.

☐ Vendredi, je joue de la flûte.

☐ Samedi, je vais nager.

☐ Dimanche, je vais au cinéma.

1B

Retrouve les mots et entoure-les dans la grille !

Livre de l'élève p. 48
GP p. 86

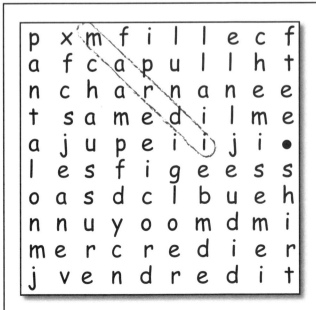

samedi	chaussure	gilet
pull	jeudi	jeudi
jupe	tee-shirt	robe
vendredi	pantalon	école
fille	chemise	
mercredi	jean	
	lundi	

~~mardi~~

Le numéro vingt et un est un appareil photo !
Le numéro vingt-deux est un ballon !
Le numéro vingt-trois est une guitare !
Le numéro vingt-quatre est un bonnet !
Le numéro vingt-cinq est un clown !
Le numéro vingt-six est un jeu vidéo !
Le numéro vingt-sept est un chat !
Le numéro vingt-huit est un chapeau !
Le numéro vingt-neuf est un livre !
Le numéro trente est une flûte !
Le numéro trente et un est un ours !
Le numéro trente-deux est un robot !

2A

Lis et écris
les numéros !

Livre de l'élève p. 49
GP p. 88

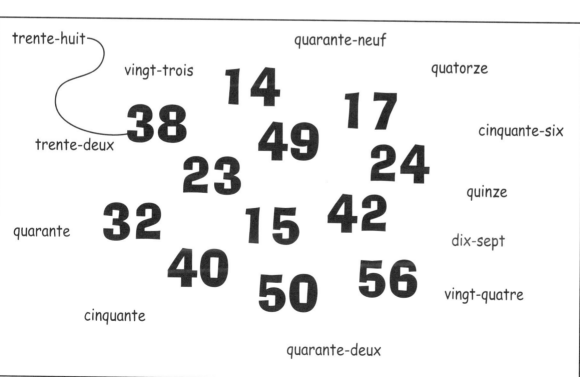

trente-huit

quarante-neuf

vingt-trois

quatorze

14

38

17

49

cinquante-six

trente-deux

23

24

quarante

32

15

42

quinze

dix-sept

40

56

50

vingt-quatre

cinquante

quarante-deux

2B

Relie les nombres
aux mots !

Livre de l'élève p. 49
GP p. 88

3A

Ecris l'heure qu'il est !

Livre de l'élève p. 50
GP p. 90

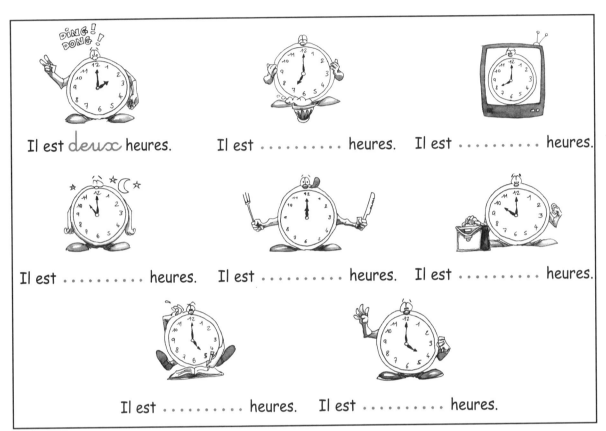

Il est *deux* heures.   Il est . . . . heures.   Il est . . . . heures.

Il est . . . . heures.   Il est . . . . heures.   Il est . . . . heures.

Il est . . . . heures.   Il est . . . . heures.

3B

Dessine l'heure qu'il est !

Livre de l'élève p. 50
GP p. 90

Il est une heure.   Il est trois heures.   Il est six heures.

Il est neuf heures.   Il est onze heures.   Il est quatre heures.

Cahier de vie

Qu'est-ce que tu fais lundi ?

....

Qu'est-ce que tu fais mercredi ?

....

Et dimanche ?

....

4A

Tu sais répondre
à ces questions ?

Livre de l'élève p. 51
GP p. 92

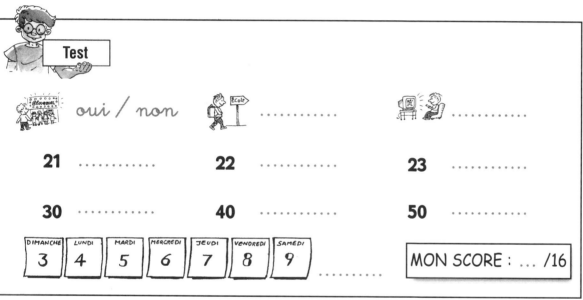

Test

oui / non ....

21 .... **22** .... **23** ....

30 .... **40** .... **50** ....

DIMANCHE	LUNDI	MARDI	MERCREDI	JEUDI	VENDREDI	SAMEDI
3	4	5	6	7	8	9

....

MON SCORE : ... /16

4B

Tu sais dire ces mots
en français ? oui / non

Tu sais dire ces
nombres en français ?
oui / non

Tu sais dire les jours
de la semaine
en français ?
oui / non

Auto-évaluation, Unité 11

 **Super !**

 **Pas mal !**

 **À revoir !**

4C

Evalue ton travail !

 **Dico-mémento**

4D

Découpe les mots
et colle-les,
puis contrôle
ce que tu sais !

Tu as les yeux de quelle couleur ?

Lis les consignes et écris le numéro de l'image !

Livre de l'élève p. 52
GP p. 94

☑ Sautez sur un pied !   ☐ Secouez les mains !   ☐ Tournez la tête !

☐ Levez les bras !   ☐ Secouez les bras !   ☐ Pliez les jambes !

Trouve les mots et écris-les !

Livre de l'élève p. 52
GP p. 94

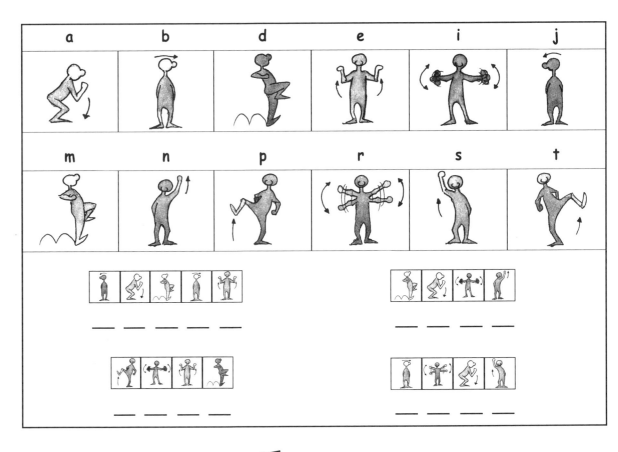

2A

Dessine ton monstre, décris-le et donne-lui un nom !

Livre de l'élève p. 53
GP p. 96

tête

bras

main

jambe

pied

Je m'appelle ....!

Voilà mon monstre :

Il a ....tête , ...

.............bras et .....

....

..........................!

Blirp

2B

Lis, écris le nom des monstres et colorie-les !

Livre de l'élève p. 53
GP p. 96

1. Voilà Blarp : il a deux bras bleus, six jambes orange et six pieds noirs.
2. Blourp est vert et jaune. Il a deux têtes, une jambe et six mains.
3. Blirp est rose. Il a deux bras, deux mains et quatre jambes.
4. Voilà Bleurp : il a une tête grise, un corps vert, un bras et une main rouges, deux jambes et deux pieds gris.
5. Blorp a une tête blanche, trois bras roses, cinq jambes et cinq pieds violets.

3A

Dessine ton portrait et envoie-le à ton ou ta camarade !

Livre de l'élève p. 54
GP p. 98

Mon portrait

J'ai les cheveux .... et les yeux...............
Et toi ?

À plus tard ! ....

3B

Ecris les mots !

Livre de l'élève p. 54
GP p. 98

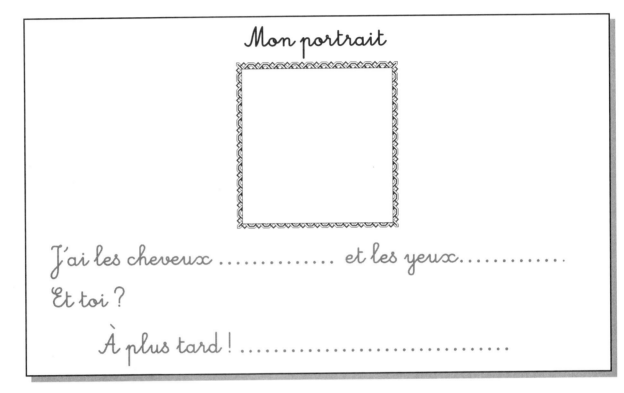

la veste	la tête	la chemise	le lait	les cheveux	l'œuf	la robe	le fromage
le nez	le chapeau	l'oreille	le poulet	la jupe	le bras	la confiture	la main
le gilet	le gâteau	le pull	la jambe	le poisson	le pantalon	la salade	le pied

vêtements **aliments** **corps**

la veste

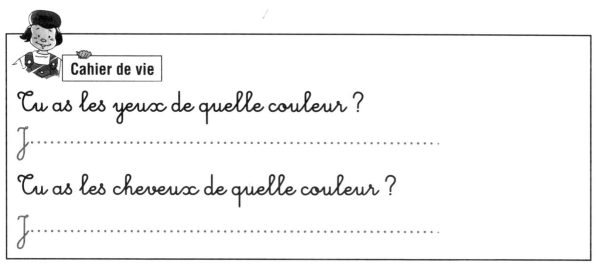

Cahier de vie

Tu as les yeux de quelle couleur ?

J...

Tu as les cheveux de quelle couleur ?

J...

4A

Tu sais répondre
à ces questions ?

Livre de l'élève p. 55
GP p. 100

Test

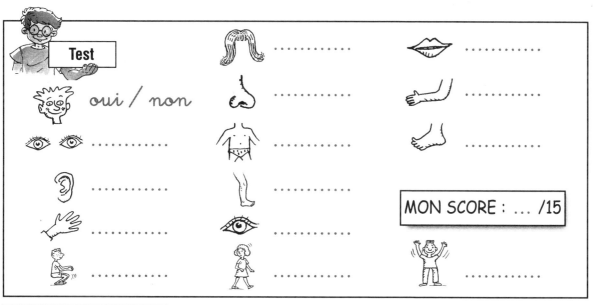

oui / non

MON SCORE : ... /15

4B

Tu sais dire ces mots
en français ? oui / non

Tu sais donner ces
consignes ? oui / non

Auto-évaluation, Unité 12

Super !

Pas mal !

À revoir !

4C

Evalue ton travail !

Dico-mémento

4D

Découpe les mots
et colle-les,
puis contrôle
ce que tu sais !

Unité **13** LEÇON 1

Où es-tu ?

1A

Relie les animaux à leur cri et écris leur nom !

Livre de l'élève p. 58
GP p. 102

| l'âne | le canard | le coq | le mouton | la poule | la vache |

Bêêê !

Cot-cot codé !

Coin-coin !

Cocorico !

I-an !

Meuh !

l' ....

le ....

le ....

la ....

le ....

la ....

1B

Complète le texte et colorie les animaux !

Livre de l'élève p. 58
GP p. 102

| ~~chante~~ | coq | poule | canard | vache | âne | |
| | mange | donne du lait | nage | donne des œufs | | des pommes |

À la ferme, il y a un .... rouge, noir et orange. Il *chante*.

Il y a une .... blanche. Elle ....

Il y a un .... vert, jaune et marron. Il ....

Il y a une .... marron. Elle ....

Il y a un .... gris. Il .... du pain et .... !

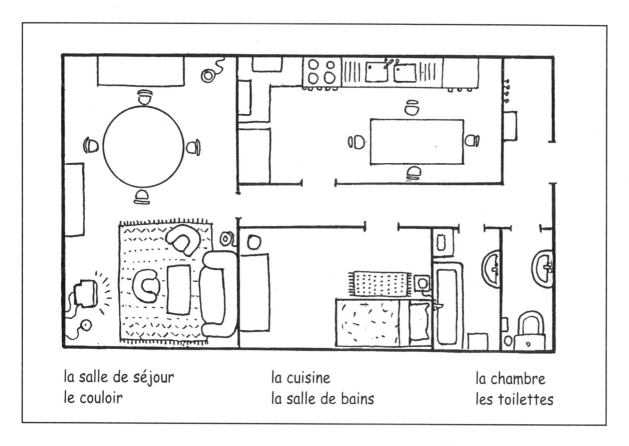

la salle de séjour
le couloir

la cuisine
la salle de bains

la chambre
les toilettes

2A

Relie les mots aux dessins !

Livre de l'élève p. 59
GP p. 104

2B

Ecris les mots !

Livre de l'élève p. 59
GP p. 104

3A

Complète
les phrases !

Livre de l'élève p. 60
GP p. 106

salle de bains - télévision - cuisine - billes - salle de séjour - cuisine - chambre - musique

Où es-tu ?

☐ Je suis dans la . . . . : je regarde la . . . . !
☐ Je suis dans la . . . . : je fais la . . . . !
☐ Je suis dans la . . . . : j'écoute de la . . . . !
☐ Je suis dans la . . . . : je joue aux . . . . !

3B

Lis et dessine
les objets !

Livre de l'élève p. 60
GP p. 106

Il y a un gâteau et du chocolat dans la chambre ! Il y a une poupée et un stylo dans la cuisine ! Il y a une pomme et une banane dans la salle de bains ! Il y a un ballon dans les toilettes ! Il y a des chaussettes dans la salle de séjour !

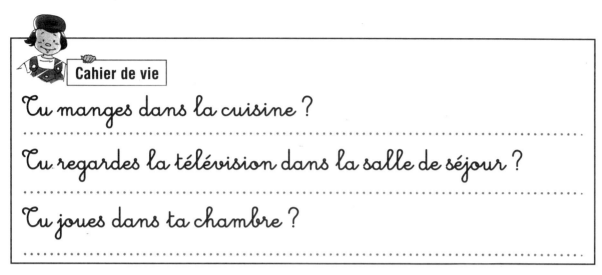

Cahier de vie

Tu manges dans la cuisine ?

....

Tu regardes la télévision dans la salle de séjour ?

....

Tu joues dans ta chambre ?

....

Tu sais répondre
à ces questions ?

Livre de l'élève p. 61
GP p. 108

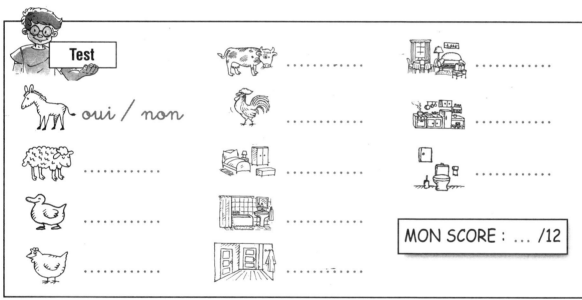

Test

oui / non

....

....

....

....

MON SCORE : ... /12

Tu sais dire ces
mots en français ?
oui / non

Auto-évaluation, Unité 13

 **Super !**

 **Pas mal !**

 **À revoir !**

Evalue ton travail !

 **Dico-mémento**

Découpe les mots
et colle-les,
puis contrôle
ce que tu sais !

Où vas-tu ?

1A

Relie les mots aux dessins !

Livre de l'élève p. 62
GP p. 110

1B

Complète les bulles !

Livre de l'élève p. 62
GP p. 110

2A

Lis et numérote
les phrases !

Livre de l'élève p. 63
GP p. 112

3 Aujourd'hui, je vais à l'école en taxi !

☐ Et moi, je vais à la ferme à vélo !

☐ Aujourd'hui, je vais au cinéma en bus !

☐ Moi, je vais au bal à cheval !

☐ Je vais à la piscine à moto !

☐ Je vais au zoo en train !

2B

Va interviewer
tes camarades !

Livre de l'élève p. 63
GP p. 112

Tu aimes ? / Prénoms	prendre le bus	prendre le métro	prendre la voiture	prendre le train	prendre le bateau	prendre le taxi	faire du vélo	aller à pied
. . . .								
. . . .								
. . . .								
. . . .								
. . . .								

. . . . aime . . . .

. . . . aime . . . .

. . . . aime . . . .

. . . . aime . . . .

. . . . aime . . . .

3A

Ecris une lettre et
envoie-la à ton ou ta
camarade !

Livre de l'élève p. 64
GP p. 114

| en bateau | en bus | en métro | à pied | en taxi | en train | à vélo | en voiture |

Bonjour ....!

Je vais à l'école ....

Je voudrais aussi aller à l'école ....!

Et toi ?

Au revoir !

....

3B

Ecris les mots !

Livre de l'élève p. 64
GP p. 114

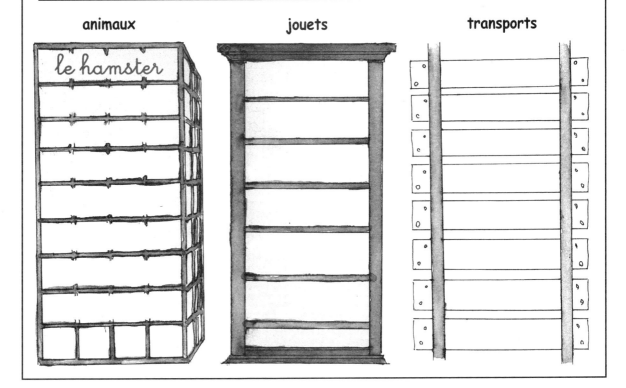

~~le hamster~~ le ballon   la voiture   le tigre   les billes   le métro   la tortue
le robot   le taxi   le télescope   le dauphin   le bateau   le papillon   la moto
le jeu vidéo   la perruche   le train   l'ours   la poupée   le vélo   l'éléphant   le bus

animaux   jouets   transports

le hamster

4A

Tu sais répondre
à ces questions ?

Livre de l'élève p. 65
GP p. 116

Cahier de vie

Comment vas-tu à l'école ? En bus ? À pied ? À vélo ?

. . . .

Tu préfères le train ou la voiture ?

. . . .

4B

Tu sais dire ces
mots en français ?
oui / non

Test

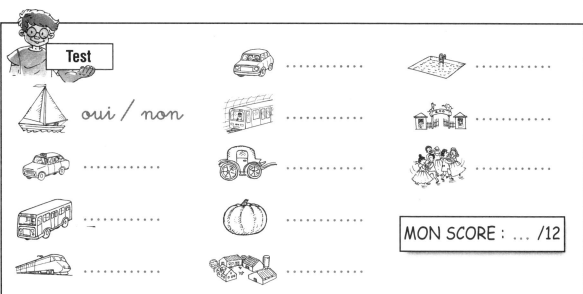

oui / non

MON SCORE : ... /12

Auto-évaluation, Unité 14

4C

Evalue ton travail !

 **Super !**

 **Pas mal !**

 **À revoir !**

4D

Découpe les mots
et colle-les,
puis contrôle
ce que tu sais !

Dico-mémento

Unité 15 LEÇON 1

On va à Paris ?

Ecris les mots
sous les photos !

Livre de l'élève p. 66
GP p. 118

l'Arc de Triomphe   le Louvre   Notre-Dame   la tour Eiffel

_____

_____

_____

_____

Lis et numérote
les phrases !

Livre de l'élève p. 66
GP p. 118

Je vais / nous allons   je prends / nous prenons   je suis / nous sommes
je veux / nous voulons

1 Je vais à Paris !

☐ Je suis une touriste !

☐ Je prends un taxi !

☐ Je veux voir la tour Eiffel !

☐ Nous sommes des touristes !

☐ Nous voulons voir la tour Eiffel !

☐ Nous allons à Paris !

☐ Nous prenons un taxi !

Aladin   le Roi Lion   le Roi des Singes   Troll   Alice   Baba Yaga
Afrique   Angleterre   Chine   Egypte   Norvège   Russie

2A

Relie les dessins
et écris les mots
en dessous !

Livre de l'élève p. 67
GP p. 120

Je viens de France !
Et toi, tu viens d'où ?

Je viens d .... !

2B

Dessine et écris !

Livre de l'élève p. 67
GP p. 120

3A

Ecris une lettre à ton correspondant ou à ta correspondante !

Livre de l'élève p. 68
GP p. 122

........................, le ....

Bonjour ...., ça va ?

Je m'appelle .... J'ai ...... ans.

Je viens de ....

J'ai les yeux .... et les cheveux ....

J'ai (Je n'ai pas de)   sœur   frère   .

J'ai (Je n'ai pas de) chien..................................

J'aime ....

J'aime aussi ....

Je n'aime pas ....

Je sais .... et aussi ....

Tu vas à l'école à quelle heure ? Moi, je vais à l'école à ....

Voilà mon école :

Tu connais Paris, la tour Eiffel, Notre-Dame ?
Ecris-moi vite !

Au revoir !

....

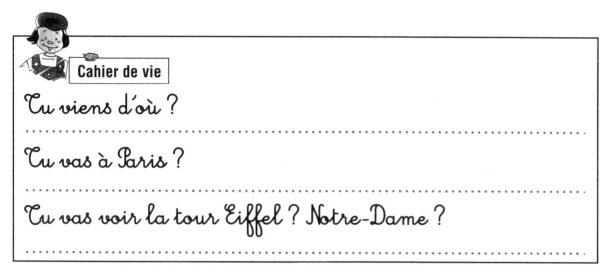

Cahier de vie

Tu viens d'où ?

....

Tu vas à Paris ?

....

Tu vas voir la tour Eiffel ? Notre-Dame ?

....

4A

Tu sais répondre
à ces questions ?

Livre de l'élève p. 69
GP p. 124

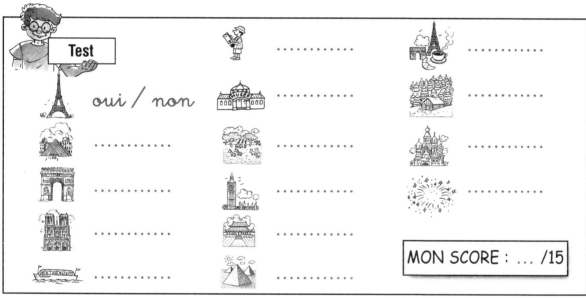

Test

oui / non

MON SCORE : ... /15

4B

Tu sais dire ces
mots en français ?
oui / non

Auto-évaluation, Unité 15

Super !

Pas mal !

À revoir !

4C

Evalue ton travail !

Dico-mémento

4D

Découpe les mots
et colle-les,
puis contrôle
ce que tu sais !

Unité 1

Bonjour monsieur !	Bonjour madame !	Salut !
Au revoir !	s'appeler (je m'appelle)	**1** un
2 deux	**3** trois	**4** quatre
5 cinq	**6** six	**7** sept
8 huit	**9** neuf	**10** dix

Unité 2

le frère	la sœur	le chat
le chien	le dragon	le hamster
la perruche	le poisson rouge	la tortue

Unité 3

le crayon	la gomme	le livre
la règle	le stylo	la trousse
blanc	bleu	jaune
noir	rouge	vert
prendre (je prends)	poser (je pose)	

Unité 4

le dauphin	l'éléphant	l'ours
le papillon	le tigre	danser (je danse)
marcher (je marche)	nager (je nage)	sauter (je saute)

voler (je vole)	jouer (je joue)	compter (je compte)
chanter (je chante)	faire de la moto (je fais)	la guitare
jouer de la guitare	le saxophone	jouer du saxophone

Unité 5

le ballon	la poupée	le robot
le vélo	la moto	la fille
le garçon	les rollers	les billes
le jeu vidéo	manger (je mange)	boire (je bois)
dormir (je dors)	jouer aux billes	jouer à la poupée
jouer au ballon	faire du vélo	faire du roller
la sorcière	le gâteau	vouloir (je veux) (je voudrais)

Unité 6

11 onze	**12** douze	le Père Noël
la bûche de Noël	les chocolats	les jouets
l'appareil photo	le téléscope	le sapin
la bougie	la boule de Noël	l'étoile
la guirlande	le cadeau	Noël

Unité 7

l'abricot	la banane	l'orange
la pêche	la poire	la pomme

	les frites		le fromage		le poisson
	le poulet		la salade		la galette
	le panier		aimer (j'aime)		

Unité8

13	treize	**14**	quatorze	**15**	quinze
16	seize	**17**	dix-sept	**18**	dix-huit
19	dix-neuf	**20**	vingt		faire du cheval
	faire la cuisine		dessiner (je dessine)		écouter de la musique
	jouer de la flûte		jouer au football		jongler (je jongle)
	faire du judo		lire (je lis)		sauter à la corde
	faire du ski		jouer au tennis		savoir (je sais)

Unité 9

	le bonnet		la botte		le chapeau
	la chaussette		la chaussure		la chemise
	le gilet		le jean		la jupe
	le pantalon		le pull		la robe
	le tee-shirt		la veste		beige
	gris		marron		orange
	rose		violet		le clown
	la fée		le magicien		le pirate

	le Pierrot		mettre (je mets)		le Carnaval

Unité 10

	le café		le café au lait		les céréales
	le chocolat		la confiture		le croissant
	le jus d'orange		le lait		l'œuf
	le pain		la tartine		le thé

Unité 11

	lundi		mardi		mercredi
	jeudi		vendredi		samedi
	dimanche		regarder la télévision (je regarde la télévision)		aller (je vais)
	l'école		le cinéma		le jour
	la semaine		le mois		l'année
21	vingt et un	**22**	vingt-deux	**23**	vingt-trois
30	trente	**40**	quarante	**50**	cinquante

Unité 12

	lever (je lève)		plier (je plie)		tourner (je tourne)
	secouer (je secoue)		le corps		le bras
	la main		la jambe		le pied
	la tête		les cheveux		l'œil
	les yeux		le nez		la bouche

| | l'oreille | | le monstre | | |

Unité 13

	la ferme		l'âne		le canard
	le coq		le mouton		la poule
	la vache		la chambre		le couloir
	la cuisine		la salle de bains		la salle de séjour
	les toilettes		cacher		Pâques

Unité 14

	le bateau		le bus		le métro
	le taxi		le train		la voiture
	le carrosse		la citrouille		le bal
	la piscine		le zoo		préférer

Unité 15

	la tour Eiffel	l'Arc de Triomphe		Notre-Dame
	le Louvre	le bateau-mouche		le touriste
	l'Afrique	l'Angleterre		la Chine
	l'Égypte	la France		la Norvège
	la Russie	le musée		le tapis volant
	le feu d'artifice			

Édition : Martine Ollivier
Couverture : Daniel Musch
Illustration de couverture : Jean-Claude Bauer

Maquette intérieure : Planète Publicité
Illustrations : Jean-Claude Bauer
           Nathanaël Bronn
           Volker Theinhardt

Recherche iconographique : Nadine Gudimard
Crédits photos : p58g : Hoa Qui/ S. Grandadam - p.58mh : Hoa Qui/M. Renaudeau
- p.58mb : Hoa Qui/M. Renaudeau - p.58d : Hoa Qui/J.F. Lanzarone

Imprimé par IFC. Saint-Germain-du-Puy 18390. N° d'imprimeur : 02/851
N° éditeur : 10098545 - (V) - (64) septembre 2002
Dépôt légal : septembre 2000. Imprimé en France